W9-BVX-020

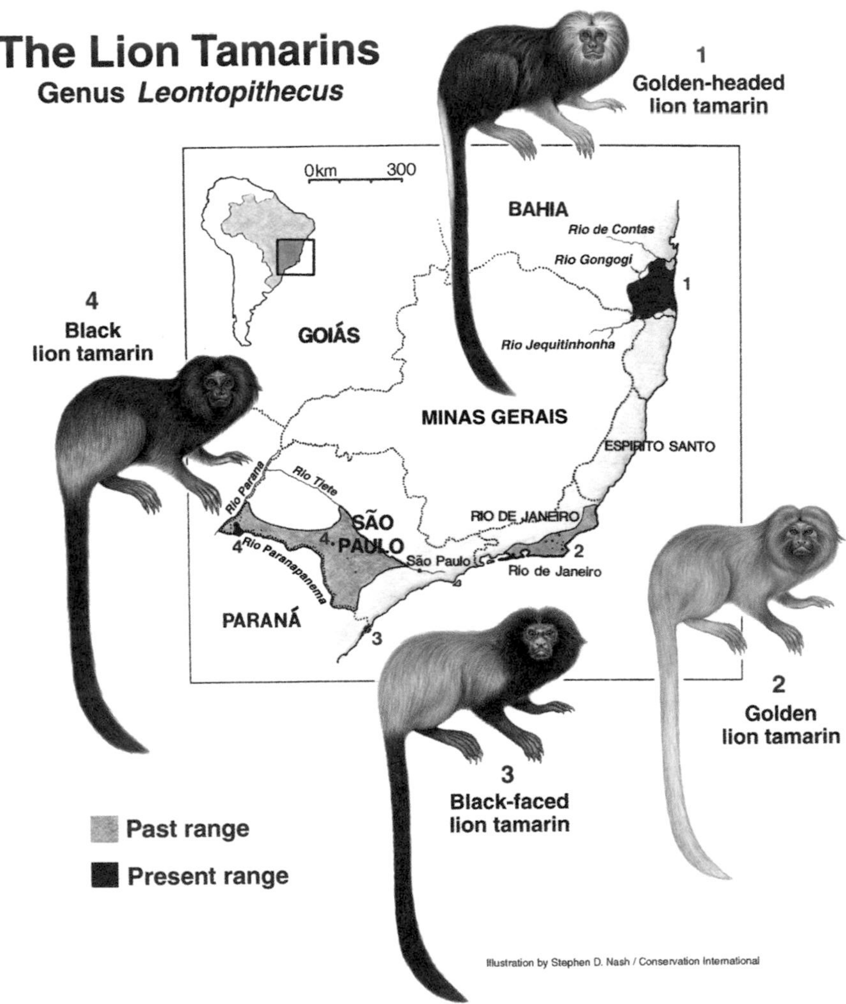
The Lion Tamarins
Genus *Leontopithecus*
1
Golden-headed
lion tamarin
0km 300
BAHIA
Rio de Contas
Rio Gongogi
1
GOIÁS
Rio Jequitinhonha
4
Black
lion tamarin
MINAS GERAIS
ESPIRITO SANTO
Rio Parana
Rio Tiete
SÃO
PAULO
4
4
Rio Paranapanema
RIO DE JANEIRO
2
São Paulo
Rio de Janeiro
PARANÁ
3
2
Golden
lion tamarin
3
Black-faced
lion tamarin
Past range
Present range
Illustration by Stephen D. Nash / Conservation International

LION TAMARINS

ZOO AND AQUARIUM BIOLOGY AND CONSERVATION SERIES

Published in cooperation with the American Zoo and Aquarium Association

This series publishes innovative works in the field of zoo and aquarium biology, conservation, and philosophy. Books in the series cover a wide range of relevant topics, including, but not limited to, zoo- and aquarium-based field conservation, animal management science, public education, philosophy, and ethics. Volumes range from conceptual books such as *Ethics on the Ark: Zoos, Animal Welfare, and Wildlife Conservation* to taxon-specific titles such as *Komodo Dragons: Biology and Conservation.*

LION TAMARINS

Biology and Conservation

Edited by Devra G. Kleiman and Anthony B. Rylands

SMITHSONIAN INSTITUTION PRESS • Washington and London

Copy Editor: Betsy Hovey
Production Editor: E. Anne Bolen
Designer: Brian Barth

Library of Congress Cataloging-in-Publication Data

Lion tamarins / edited by Devra G. Kleiman and Anthony B. Rylands
p. cm. — (Zoo and aquarium biology and conservation series)
Includes bibliographical references (p.).
ISBN 1-58834-072-4 (alk. paper)
1. Leontopithecus. 2. Wildlife conservation. I. Kleiman, Devra G.
II. Rylands, Anthony B. III. Series

QL737.P92 L56 2002
599.8′4—dc21 2002021680

British Library Cataloguing-in-Publication Data is available

Manufactured in the United States of America
09 08 07 06 05 04 03 02 5 4 3 2 1

♾ The paper used in this publication meets the minimum requirements of the American National Standard for Information Sciences—Permanence of Paper for Printed Library Materials ANSI Z39.48-1984.

We dedicate this book to Adelmar F. Coimbra-Filho, a truly remarkable man who has always challenged us to do our best and to keep questioning, who has never swayed from his beliefs, and who has inspired so many to seek careers in primatology and conservation biology.

CONTENTS

PART THREE: CONSERVATION AND MANAGEMENT OF LION TAMARINS IN THE WILD

CONTRIBUTORS

Carlos Alberto da S. Almeida
Seção de Genética
Instituto Nacional de Câncer
Praça da Cruz Vermelha 23
20230-130 Rio de Janeiro
Rio de Janeiro, Brazil

Andrew J. Baker
Philadelphia Zoological Garden
3400 West Girard Avenue
Philadelphia, PA 19104, USA

Karen Bales
Department of Psychiatry
University of Illinois at Chicago
1601 W Taylor Street, Room 436
Chicago, IL 60612, USA

Jonathan D. Ballou
Department of Conservation Biology
National Zoological Park
Smithsonian Institution
Washington, DC 20008, USA

Maria Iolita Bampi
Departamento de Vida Silvestre
Diretoria de Ecossistemas
Instituto Brasileiro do Meio Ambiente e dos Recursos Naturais Renováveis (IBAMA)
SAIN Avenida L4 Norte
Edifício Sede
70800 Brasília, DF, Brazil

Benjamin B. Beck
National Zoological Park
Smithsonian Institution
Washington, DC 20008, USA

Ibsen de Gusmão Câmara
Avenida das Américas 2300-C40
Barra da Tijuca
22640 101 Rio de Janeiro
Rio de Janeiro, Brazil

Flávio C. Canavez
Seção de Genética
Instituto Nacional de Câncer

Praça da Cruz Vermelha 23
20230-130 Rio de Janeiro
Rio de Janeiro, Brazil

Maria Inês Castro
Conservation International
1919 M Street, NW
Washington, DC 20036, USA

Adelmar F. Coimbra-Filho
Rua Artur Araripe 60/901
Gávea
22450-020 Rio de Janeiro
Rio de Janeiro, Brazil

Laury Cullen, Jr.
Instituto de Pesquisas Ecológicas (IPÊ)
Caixa Postal 47
12960-000 Nazaré Paulista
São Paulo, Brazil

Kristel De Vleeschouwer
Center for Research and Conservation
Royal Zoological Society of Antwerp
Koningin Astridplein 26
B-2018 Antwerpen, Belgium

James M. Dietz
Department of Zoology
University of Maryland
College Park, MD 20742, USA

Lou Ann Dietz
World Wildlife Fund
1250 24th Street NW, Room 5047
Washington, DC 20037-1132, USA

Anthony Di Fiore
Department of Anthropology and New York Consortium in Evolutionary Primatology (NYCEP)
New York University
25 Waverly Place
New York, NY 10003, USA

Anna T. C. Feistner
Durrell Wildlife Conservation Trust
Les Augrès Manor
Trinity
Jersey JE3 5BP
Channel Islands, British Isles

Jeffrey A. French
Department of Psychology
University of Nebraska at Omaha
Omaha, NE 68182-0274, USA

Michael Heistermann
Department of Reproductive Biology
German Primate Center
Kellnerweg 4
D-37077 Göttingen, Germany

Maria Cecília M. Kierulff
Conservation International do Brasil
Rua Major Homem Del Rey 147
Cidade Nova
45650-000 Ilhéus
Bahia, Brazil

Devra G. Kleiman
Department of Conservation Biology
National Zoological Park
Smithsonian Institution
Washington, DC 20008, USA

Kristin Leus
Center for Research and Conservation
Royal Zoological Society of Antwerp
Koningin Astridplein 26
B-2018 Antwerpen, Belgium

Jeremy J. C. Mallinson
Durrell Wildlife Conservation Trust
Les Augrès Manor
Trinity
Jersey JE3 5BP
Channel Islands, British Isles

Andréia Martins
Associação Mico-Leão-Dourado
Caixa Postal 109.968
28.860-000 Casimiro de Abreu
Rio de Janeiro, Brazil

Cristiana Saddy Martins
Instituto de Pesquisas Ecológicas (IPÊ)
Caixa Postal 47
12960-000 Nazaré Paulista
São Paulo, Brazil

Kimran Miller
Department of Biology
University of Maryland
College Park, MD 20742, USA

Russell A. Mittermeier
Conservation International
1919 M Street, NW
Washington, DC 20036, USA

Richard J. Montali
National Zoological Park
Smithsonian Institution
Washington, DC 20008, USA

Miguel Ângelo M. Moreira
Seção de Genética
Instituto Nacional de Câncer
Praça da Cruz Vermelha 23
20230-130 Rio de Janeiro
Rio de Janeiro, Brazil

Antônio Christian de A. Moura
Departamento de Sistemática e Ecologia—CCEN
Universidade Federal de Paraíba
58.059-900 João Pessoa
Paraíba, Brazil

Suzana M. Padua
Instituto de Pesquisas Ecológicas (IPÊ)
Caixa Postal 47
12960-000 Nazaré Paulista
São Paulo, Brazil

Fernando C. Passos
Departamento Zoologia
Universidad Federal do Paraná
Caixa Postal 19020
81531-990 Curitiba
Paraná, Brazil

Luiz Paulo de Souza Pinto
Conservation International do Brasil
Av. Getúlio Vargas, 1300-7° andar
30112-021 Belo Horizonte
Minas Gerais, Brazil

Alcides Pissinatti
Centro de Primatologia do Rio de Janeiro (CPRJ)
Fundação Estadual de Engenharia do Meio Ambiente (FEEMA)
Rua Fonseca Teles 121/1624
São Cristóvão
20940-200 Rio de Janeiro
Rio de Janeiro, Brazil

Fabiana Prado
Instituto de Pesquisas Ecológicas (IPÊ)
Caixa Postal 47
12960-000 Nazaré Paulista
São Paulo, Brazil

Eluned C. Price
Durrell Wildlife Conservation Trust
Les Augrès Manor
Trinity
Jersey JE3 5BP
Channel Islands, British Isles

Paula Procópio de Oliveira
Programa de Translocação
Associação Mico-Leão-Dourado
Caixa Postal 109.995
28.860-000 Casimiro de Abreu
Rio de Janeiro, Brazil

Becky E. Raboy
Department of Biology
University of Maryland
College Park, MD 20742, USA

Denise Marçal Rambaldi
Associação Mico-Leão-Dourado
Caixa Postal 109.968
28.860-970 Casimiro de Abreu
Rio de Janeiro, Brazil

Lisa G. Rapaport
Department of Anthropology
University of New Mexico
Albuquerque, NM 87131, USA

Carlos R. Ruiz-Miranda
Lab. de Ciências Ambientais
Centro de Biociências
e Biotecnologia
Universidade Estadual Norte
Fluminense
Avenida Alberto Lamego 2000
28015-620 Campos dos Goytacazes
Rio de Janeiro, Brazil

Anthony B. Rylands
Center for Applied Biodiversity Science
Conservation International
1919 M Street, NW
Washington, DC 20036, USA

Cristina V. Santos
Departamento de Psicologia
Experimental
Instituto de Psicologia
Universidade de São Paulo
Avenida Professor Mello Moraes 1721
05508-900 São Paulo
São Paulo, Brazil

Gabriel Rodrigues dos Santos
Instituto de Estudos Sócio-Ambientais
do Sul da Bahia
Rua Major Homem Del Rey 147
45650-000 Ilhéus
Bahia, Brazil

Héctor N. Seuánez
Seção de Genética
Instituto Nacional de Câncer
Praça da Cruz Vermelha 23
20230-130 Rio de Janeiro
Rio de Janeiro, Brazil

Faiçal Simon
Fundação Parque Zoológico de São Paulo
Av. Miguel Stefano 4241
Água Funda
04301-905 São Paulo
São Paulo, Brazil

Maria das Graças de Souza
Instituto de Pesquisas Ecológicas (IPÊ)
Caixa Postal 47
12960-000 Nazaré Paulista
São Paulo, Brazil

Tara S. Stoinski
TECHLab
Zoo Atlanta
800 Cherokee Avenue
Atlanta, GA 30315, USA

Suzette D. Tardif
Southwest Regional Primate Research Center
P.O. Box 760549
San Antonio, TX 78245-0549, USA

Cláudio B. Valladares-Padua
Instituto de Pesquisas Ecológicas (IPÊ)
Caixa Postal 47
12960-000 Nazaré Paulista
São Paulo, Brazil

Linda Van Elsacker
Centre for Research and Conservation
Royal Zoological Society of Antwerp
Koningin Astridplein 26
B-2018 Antwerpen, Belgium

FOREWORD

The first field observations of lion tamarins were made by the amazing German naturalist Prince Maximilian zu Wied, who traveled through the Atlantic Forest region of Brazil in the early nineteenth century. At that time he described the golden lion tamarin and golden-headed lion tamarin as common, but unfortunately their numbers in the wild declined dramatically in the second half of the twentieth century. With the rapid economic growth of Brazil beginning in the 1960s, most marked in the Atlantic Forest region, there has been large-scale conversion of forest to cattle pasture, agricultural land, and plantation forestry, and the lion tamarins have been squeezed into smaller and smaller forest patches.

The first to call attention to this in Brazil was Adelmar F. Coimbra-Filho, the pioneer of Brazilian primatology and one of the great leaders of primatology for the entire neotropical region. With his colleague Alceo Magnanini, he carried out fieldwork on the golden lion tamarin in the 1960s and wrote of its plight in the latter half of that decade. In 1969 and 1970, he rediscovered the black, or golden-rumped, lion tamarin in the Morro do Diabo State Reserve in São Paulo, which had not been seen in 65 years. Although his discoveries were published in the latter part of 1970, they received little international attention since they were in Portuguese.

In 1970, I had just finished my undergraduate work at Dartmouth College and had spent 3 months working on the mantled howler monkey on Barro Colorado Island in Panama. Having decided to work on Neotropical primates for my Ph.D., I met with Barbara Harrisson, founder, and then chair, of the IUCN's Species Survival Commission Primate Specialist Group. Barbara encouraged me to go to Brazil and showed me Coimbra's papers on the lion tamarins. Although I could not yet read Portuguese, I struggled through them with my Spanish and wound up fascinated with how little was known about these wonderful little creatures.

About 6 months later, in July 1971, I appeared on Coimbra's doorstep in Rio de Janeiro, speaking the rudimentary Portuguese that I had learned over the past few weeks traveling by boat and bus from the Brazilian Amazon down to Rio. We struck up an immediate friendship that has lasted for three decades and has been one of the most productive of my career. Our discussions of primate conservation laid the groundwork for a series of activities over the next decade, including publication of his early lion tamarin work in English and the need to carry out primate surveys in the Atlantic Forest and Amazonia.

On the international end, there was growing concern about the declining number of captive golden lion tamarins in the United States and Europe following articles published by John Perry (Smithsonian National Zoological Park [SNZP], Washington, D.C.) and Clyde Hill (San Diego Zoo). This led to the historic conference, in February 1972, entitled "Saving the Lion Marmoset," which was organized by the late Don Bridgwater, then of the Minnesota Zoo, Bill Conway of the New York Zoological Society (now the Wildlife Conservation Society), and John Perry. The meeting was held in a backroom behind the cafeteria at the SNZP and laid the groundwork for almost everything else that followed. As a result of this meeting, Devra Kleiman assumed coordination of the captive golden lion tamarin population, which proved amazingly successful, increasing the numbers from about 70 to a controlled 500.

Taking advantage of the international interest in the conference and with some seed money from the World Wildlife Fund (WWF), in 1973, Coimbra and Magnanini, were successful in establishing a captive breeding facility: the Lion Tamarin Biological Bank at the edge of Tijuca National Park in Rio de Janeiro. There, Coimbra began the first colony of black lion tamarins, with six individuals from the Morro do Diabo State Reserve, and the following year, he and Magnanini were also successful in establishing the Poço das Antas Biological Reserve in the lowlands of Rio de Janeiro. International interest was further generated by a second callitrichid conference organized by Devra Kleiman in 1975 (Kleiman, 1978). A third meeting was hosted by Hartmut Rothe, Jurgen Wolters, and John Hearn in 1977 in Göttingen, Germany (Rothe et al. 1978). "The Marmoset Workshop" added an important European dimension, giving the lion tamarins even more status as flagship species for the Atlantic Forest. As a result of these meetings, Coimbra was again able to leverage on-the-ground conservation action, this time resulting in the creation of the Una Biological Reserve for the golden-headed lion tamarin in 1980.

In 1973 and 1974, the first major primate survey of Brazilian Amazonia and a multiyear survey, entitled "Conservation of Eastern Brazilian Primates" (funded by WWF-U.S.), which Coimbra and I had discussed when we first met, became a reality. It had many subcomponents, including the groundbreaking study of the golden-headed lion tamarin by Anthony Rylands in 1980, the initiation of the long-term efforts of Cláudio and Suzana Padua on the black lion tamarin in the mid-1980s, and the first survey of the golden lion tamarin since Coimbra's early research—a 1980 study by Ken Green. The WWF program also helped to lay the groundwork for the SNZP–based field program on the golden lion tamarin (I was able to introduce Devra Kleiman to her first wild golden lion tamarin in the Poço das Antas Biological Reserve in 1982), as well as the field surveys of Ilmar Santos, Cristina Alves, Luiz Paulo Pinto, and Cecília Kierulff. Perhaps most importantly, this program was instrumental in placing the Atlantic Forest high on the agenda of major international conservation organizations—to the point that it is now recognized as one of the top five biodiversity "Hotspots" on Earth (Mittermeier et al. 1999a, 1999b).

Another watershed event was the establishment in 1981 of the first Management Committee for the Golden Lion Tamarin, chaired by Devra Kleiman and initially intended for the management of the captive population. Committees followed for the other species: a golden-headed lion tamarin committee, chaired by Coimbra and Jeremy Mallinson, director emeritus of the Durrell Wildlife Conservation Trust (then the Jersey Wildlife Preservation Trust), and one for black lion tamarins, chaired by Faiçal Simon of the São Paulo Zoo and Devra Kleiman. The Instituto Brasileiro do Meio Ambiente e dos Recursos Naturais Renováveis (IBAMA—Brazilian Institute of the Environment and Renewable Natural Resources) officially recognized the committees in 1990, which became models of international collaboration on behalf of endangered species. Special recognition for making this happen must go to Maria Iolita Bampi, general coordinator for fauna, Directorate of Wildlife and Fisheries of IBAMA, who has long been the government's representative on these key structures. Then, in 1990, came one of the biggest surprises in primatology in the last quarter century: a "newcomer" to the lion tamarin world. The black-faced lion tamarin was discovered by two Brazilian researchers, Maria Lúcia Lorini and Vanessa Persson, on the island of Superagüi in the state of Paraná. This exciting new find amazed all of us, reminding us how little we know of the world's biodiversity.

All in all, I think it fair to say that the international effort of the past 30 years on behalf of the lion tamarins of Brazil represents one of the great success stories in the global conservation movement. Together with the muriquis (*Brachyteles* spp.), they have long been the flagship species for this critically important "Hotspot," and without the lion tamarins it is doubtful that this region would have received the attention that it has. Like the giant panda in China, the orangutan and the Sumatran rhino in Indonesia, and the lemurs in Madagascar, the lion tamarins have been the international ambassadors for their forest home. Although we cannot afford to be complacent, I am very optimistic about the future of the lion tamarins and the Atlantic Forest as a whole. The lion tamarin effort has set high international standards and has been a model for the development of similar efforts around the world. By summarizing all the good work that has taken place, the editors, Devra Kleiman and Anthony Rylands, and all the authors in this book have made a great contribution to biodiversity conservation and have really helped us in the great challenges that we face over the next couple of decades. Without their dedication, their commitment, and their many sacrifices, the lion tamarins would probably be close to extinction today. Instead, they represent one of our greatest reasons for hope in the saving of the earth's biodiversity.

Finally, I am very pleased that this book has been dedicated to one of the world's greatest ecoheroes, Dr. Adelmar F. Coimbra-Filho. Without Coimbra, none of what is recounted in the pages that follow would have been possible. He has been a pioneer, a leader, a visionary, and a great spokesman for conservation for nearly half a century and is deserving of our greatest respect and admiration. I am proud to call him one of my closest friends and am honored that he saw fit to involve me in this noble effort so long ago.

—RUSSELL A. MITTERMEIER
PRESIDENT, CONSERVATION INTERNATIONAL
CHAIRMAN, IUCN/SSC PRIMATE SPECIALIST GROUP

INTRODUCTION AND ACKNOWLEDGMENTS

This book represents the efforts of multinational teams of scientists and conservationists to save from extinction four unique and beautiful primates: the lion tamarins—golden, golden-headed, black-faced, and black. They are one of the more spectacular elements of the extraordinary fauna of the tropical forests in eastern Brazil, once extending over considerably more than a million square kilometers along the Atlantic coast, but now very largely decimated. To save these monkeys, it is necessary to save their forests. Understanding them is vital for informed and wise conservation action, both in the wild and in the maintenance of secure and healthy captive populations—very important when so few of them remain. Unlocking the mysteries of lion tamarin biology has required research in many fields of science: reproductive physiology, genetics, veterinary science, husbandry, all aspects of lion tamarins' behavior in captivity and the wild (communication and their social, mating, and rearing systems), and their ecology and demography. However, braking and reversing their slide into extinction requires so much more. Efforts to protect these animals demand a full knowledge of where they still remain in the wild, a profound understanding of the sociopolitical and economic forces causing the continued loss of their forests, and the creation and management of new protected areas. Very little can be accomplished, however, without the social circumstances and political will, both locally and nationally, to promote their survival. Environmental education, lobbying, and working closely with landowners and local communities are needed to demonstrate that protecting these animals is not only safeguarding the natural heritage of the country but also providing for the protection of its forests and the well-being of the people who live there.

The publication of this book will mark 30 years since the seminal workshop

"Saving the Lion Marmoset," which consolidated the then incipient awareness of the lion tamarins' endangered status and set the stage for the in situ and ex situ efforts that followed (Bridgwater 1972a). The accomplishments of the numerous people who have been involved in the conservation of the lion tamarins both in the wild and in captivity since then have been truly extraordinary. Many of them have devoted a major portion of their careers to this effort. It is entirely likely that at least three of the four species would be effectively extinct today without the motivation, persistence, energy, and passion of those working in laboratories, in zoos, in the forests, with local schools and farmers, and not least in government offices. The tasks of all have had their fascinations, their successes, and their frustrations, such as monitoring a healthy group of reintroduced captive-bred lion tamarins foraging in the forest and then spending days looking for a specific study animal and failing to find it. Having yet another meeting with a newly appointed public official while vital documents languish and then getting the final signature that creates a new protected area. Talking to a large gathering in the local community and feeling the crowd's enthusiasm and support mount and then responding to a call that your study group has been stolen by hunters and seeing that a landowner has cut down his last forest fragment. We salute our colleagues for their commitment and incredible dedication to tasks that all too often involve inordinate and prolonged drudgery and to ideals that sometimes seem impossible to achieve. Looking at the bigger picture and summing the seemingly "small" successes, we hope that this book will maintain optimism for the story it tells of so much progress but so much yet to do.

We have divided the contributions to this book into three parts. Part One covers the history and institutional framework of research and conservation efforts for the four species. This section provides a rare historical look at the circumstances that resulted in the development of diverse fronts of action. It is a story of both individual and institutional feats, ranging from the achievements of scientists to those of civil servants, from the accomplishments of global zoos to those of governments to those of national and international nongovernmental organizations, from independent to cooperative efforts. This history demonstrates how complex and multidimensional conservation work can be.

Part Two covers the principal research fields that have played an important role in contributing directly and indirectly to the management of the species in captivity and the wild. The topnotch research on genetics, population dynamics, husbandry, sociobiology, and behavioral ecology has not only informed conservation actions but also has pushed the envelope within the individual disciplines. Saving the lion tamarins has resulted in scientific advances. There are two take-home messages from this section. First, there is no substitute for long-term data on one or more populations of a species. Conservation action depends upon an understanding of natural variation in numerous life history parameters—something that cannot be obtained without long-term studies that document differences and change. Second, there is no substitute for good science when developing recovery or conservation programs. Suggestions that there is a need for more conservation action and less research arise from a false dichotomy—conservation and research are not antithetical, but supportive and interactive, and we clearly cannot apply adaptive management principles without the broadest and strongest scientific underpinning.

Part Three focuses on direct interventions for the conservation of the wild populations and their habitats. The establishment and sound management of protected areas is obviously the keystone for establishing refuges for wildlife in otherwise transformed landscapes, but in the case of the lion tamarins, most especially the golden lion tamarin in the lowland forests of Rio de Janeiro and the black lion tamarin in the west of the state of São Paulo, this is not enough. Research on the genetics, demography, and ecology of the lion tamarins discussed in the second part of this book clearly demonstrates that the small protected forests that remain are insufficient to guarantee the long-term viability of the species. The populations are too small, too few, and too isolated. Metapopulation management strategies are emerging that promote the mixing of the current captive and wild populations in order to maintain genetic diversity. This section focuses on the implementation of conservation action guided by scientific principles and the role of the conservation educators in realizing and executing the conservation objectives. In this final part, we provide an overview of the accomplishments of lion tamarin conservation and our perspective of its future directions and challenges.

The idea of editing this book arose at a meeting to discuss lion tamarin research and conservation held in Belo Horizonte, Brazil, in 1997, 25 years after "Saving the Lion Marmoset." It was hosted by the Fundação Biodiversitas (Biodiversity Foundation) and sponsored by the Jersey Wildlife Preservation Trust (now the Durrell Wildlife Conservation Trust), Instituto Brasileiro do Meio Ambiente e dos Recursos Naturais Renováveis (IBAMA—Brazilian Institute of the Environment and Renewable Natural Resources), Conservation International, Margot Marsh Biodiversity Foundation, U.S. Fish and Wildlife Service, TransBrasil Airlines, and Smithsonian National Zoological Park (SNZP). We thank, particularly, Maria Inês Castro (then at the SNZP) and Ilmar Santos (executive director of Fundação Biodiversitas) for their tireless work in organizing this meeting. It combined three events: a 2-day symposium reviewing progress in the research and conservation efforts for the lion tamarins; the second Population and Habitat Viability Analysis (PHVA) Workshop for lion tamarins, run by the World Conservation Union/Species Survival Commission (IUCN/SSC) Conservation Breeding Specialist Group (CBSG) (Ballou et al. 1998); and the annual meeting of the International Recovery and Management Committees for each of the species. Besides the scientific presentations, the modeling, the reviewing, the planning, the discussions, and the analysis of the successes and the setbacks, it was a celebration of the extraordinary progress that had been made, and this we decided should be put into print.

We are most grateful to the 48 contributors and the many people who have helped us in the preparation of this book. Stephen Nash, scientific illustrator at the State University of New York, Stony Brook, generously donated the frontispiece and drew the distribution maps for each of the species. Every chapter was reviewed by two or three independent, anonymous scholars; the reviewers' time and expertise were invaluable to us and are most gratefully acknowledged: Gustl Anzenberger, Andrew Baker, Lynne Baker, Jonathan Ballou, Sue Boinski, Hilary Box, Robert Cooper, Anthony Di Fiore, Siân Evans, Anna Feistner, Murray Fowler, Paul Garber, Alexander Harcourt, Eckhard Heymann, Keith Hodges, Robert Horwich, Susan Jacobson, Robert Lacy, Brian Miller, Philip Miller, Martha Monroe, Nicholas Mundy, Elizabeth Nagagata, Michael Phillips, Alfred Rosenberger, Oliver Ryder, Alan Shoemaker, Pritpal Soorae, Miranda Stevenson, Crispen Wilson, Maria Emília Yamamoto, and Tony Ziegler. Kimberley Meek and Glenda Fábregas at the Center for Applied Biodiversity Science, Conservation International, kindly helped us manage and format the graphs and figures.

Anthony Rylands pays special homage to Célio Valle, Ângelo Machado, and Gustavo Fonseca, professors and colleagues at the Federal University of Minas Gerais, and to Ilmar Santos, executive director of the Fundação Biodiversitas, dedicated pioneers of enormous influence in the conservation of the beleaguered fauna and flora of the Atlantic Forest. Devra Kleiman is grateful to the Smithsonian Institution and the SNZP and especially the vision of Ted Reed (former SNZP director) for supporting her efforts in this conservation endeavor. Ted Reed once declared in the mid-1970s that we were breeding golden lion tamarins in order to reintroduce them, an outcome that Kleiman then believed was unattainable. Kleiman also thanks her colleagues in the Golden Lion Tamarin Conservation Program for what has been a stimulating and wonderful collaboration.

We have dedicated this book to Adelmar F. Coimbra-Filho but also recognize and acknowledge the leadership of several other individuals who have made outstanding contributions to advancing the cause of lion tamarin research and conservation in Brazil (in alphabetical order): Maria Iolita Bampi, Anne Binney, Ibsen de Gusmão Câmara, Lou Ann Dietz, the late Gerald Durrell, Alceo Magnanini, Jeremy Mallinson, and Russell Mittermeier. Finally, the leadership and contributions of the Jersey Zoo (Durrell Wildlife Conservation Trust) and the Smithsonian's National Zoological Park have been exceptional.

Lastly, we most sincerely thank Vincent Burke, science acquisitions editor at the Smithsonian Institution Press, who was wise and patient in helping us to deal with the intricacies and details of finalizing the manuscript; an anonymous reviewer, appointed by the Smithsonian Press, who provided an enthusiastic and most insightful critique; and Betsy Hovey, our copy editor, who did a wonderful job of finding and fixing errors and inconsistencies and clarifying confusing text.

ACRONYMS

AMLD—Associação Mico-Leão-Dourado (Golden Lion Tamarin Association)
CPRJ/FEEMA—Centro de Primatologia do Rio de Janeiro (Rio de Janeiro Primate Center)/Fundação Estadual de Engenharia do Meio Ambiente (State Foundation for Environmental Engineering)
DWCT—Durrell Wildlife Conservation Trust (formerly JWPT—the Jersey Wildlife Preservation Trust)
GLTCP—Golden Lion Tamarin Conservation Program
IBAMA—Instituto Brasileiro do Meio Ambiente e dos Recursos Naturais Renováveis (Brazilian Institute of the Environment and Renewable Natural Resources)
ICCM—International Committee for the Conservation and Management of Lion Tamarins (Comitê Internacional para a Conservação e Manejo dos Micos-Leões)
IESB—Instituto de Estudos Sócio-Ambientais do Sul da Bahia (Institute for Social and Environmental Studies of Southern Bahia)
IPÊ—Instituto de Pesquisas Ecológicas (Institute for Ecological Research)
IRMC—International Recovery and Management Committee
IUCN/SSC—World Conservation Union/Species Survival Commission
JWPT (see DWCT)
RPPN—Reserva Particular de Patrimônio Natural (Private Natural Heritage Reserve)
SI—Smithsonian Institution
SNZP—Smithsonian National Zoological Park
WWF—World Wildlife Fund (United States and Canada) or Worldwide Fund for Nature (internationally)

Part One

THE HISTORY AND STATUS OF LION TAMARINS

ANTHONY B. RYLANDS, JEREMY J. C. MALLINSON,
DEVRA G. KLEIMAN, ADELMAR F. COIMBRA-FILHO,
RUSSELL A. MITTERMEIER, IBSEN DE GUSMÃO CÂMARA,
CLÁUDIO B. VALLADARES-PADUA,
AND MARIA IOLITA BAMPI

1
A HISTORY OF LION TAMARIN RESEARCH AND CONSERVATION

Among the most endangered of the New World primates are the beautiful and dramatic lion tamarins (*Leontopithecus* spp.), including four forms: the golden lion tamarin (*L. rosalia*), black lion tamarin (*L. chrysopygus*), golden-headed lion tamarin (*L. chrysomelas*), and black-faced lion tamarin (*L. caissara*). Endemic to the Atlantic Forest of Brazil, they are restricted to forest patches in the states of Bahia, Minas Gerais, Rio de Janeiro, São Paulo, and Paraná, all of which are now densely populated by people. Today, less than 7 percent of the original forest remains (Fundação SOS Mata Atlântica/INPE 1998; Mallinson 1994b, 1995; Mittermeier et al. 1999b). A history of the destruction of the Atlantic Forest and its causes is given by Dean (1995).

Concern about the imminent disappearance of these animals began in the first half of the 1960s when Adelmar F. Coimbra-Filho called attention to the severe plight of the golden lion tamarin in the state of Rio de Janeiro, as well as the golden-headed and black lion tamarins in Bahia and São Paulo (Coimbra-Filho and Magnanini 1968). By the 1960s, golden lion tamarin numbers had decreased dramatically due both to exportation for the pet trade and zoos and to deforestation in the state of Rio de Janeiro (Coimbra-Filho and Mittermeier 1977). Less was known about the golden-headed and black (golden-rumped) lion tamarins. In fact, Coimbra-Filho and Alceo Magnanini (1968) considered that *L. chrysopygus* was probably extinct by 1964, and only in 1970, after an interval of 65 years during which the species had not been seen, did Coimbra-Filho (1970b) rediscover the black lion tamarin in the state of São Paulo.

Through the determination and tenacity of Coimbra-Filho and Magnanini, two reserves were created specifically for protecting lion tamarins, the Poço das Antas

Biological Reserve for *L. rosalia* in Rio de Janeiro (1974) and the Una Biological Reserve for *L. chrysomelas* in Bahia (1980). *Leontopithecus chrysopygus* owes its survival to the fortuitous permanence of the Morro do Diabo Forest Reserve, decreed a forest reserve in 1941 and a state park in 1986, on the west side of the state of São Paulo; all other forests in the region were largely decimated by the 1960s. *Leontopithecus caissara* was discovered in 1990 because of the establishment in the late 1980s of a complex of protected areas created to preserve the lowland and upland rain forests and *restinga* forests (coastal forests on sand) and mangroves on the coast of the state of Paraná, in the region of Guaraqueçaba.

International efforts for the conservation of the lion tamarins centered first on *L. rosalia,* with the key moment being the conference "Saving the Lion Marmoset" (Bridgwater 1972a). The little that was then known of their status was reviewed, and recommendations were developed for their conservation. A captive management and research program was then established for the *L. rosalia* population by Devra Kleiman (see Chapter 4 this volume).

In 1975 and 1977, two further conferences were held to review the research, status, and conservation of all the callitrichids (Kleiman 1978a; Rothe et al. 1978). Both served to provide a solid basis for future research directions. In the mid-1980s, International Recovery and Management Committees (IRMCs) and captive breeding programs were set up for *L. chrysomelas* and *L. chrysopygus* following the model for *L. rosalia* developed in 1981 (Kleiman and Mallinson 1998; see also Chapters 3 and 4 this volume). Thus, by 1990, there were in situ and ex situ conservation programs in place for three of the species. A Population Viability Analysis (PVA) Workshop was held for the lion tamarins in 1990 (Seal et al. 1990); it evaluated the status of the four species in captivity and the wild and included computer simulations of the viability of the wild populations in protected areas. An outcome of the workshop was the formal recognition of the management committees by the Instituto Brasileiro do Meio Ambiente e dos Recursos Naturais Renováveis (IBAMA—Brazilian Institute of the Environment and Renewable Natural Resources) and the expansion of their mandate to include overseeing and advising on conservation efforts and research in situ (Kleiman and Mallinson 1998; Mallinson 1997b; Chapter 3 this volume). The discovery of the fourth lion tamarin, *L. caissara* (Lorini and Persson 1990), was announced at the 1990 PVA Workshop and resulted in the establishment of a fourth committee to oversee research and conservation efforts for this species (Kleiman and Mallinson 1998; Chapter 3 this volume).

A second Population and Habitat Viability Analysis (PHVA) Workshop for the four species was held in 1997 (Ballou et al. 1998; Figure 1.1). In contrast to the 1990 PVA, which emphasized the need for further research on the wild populations, recommendations for future action centered more on the resolution of the problems concerning socioeconomic issues and the various protected areas, along with habitat analysis and forest restoration to increase the area of forest available for the wild populations (Ballou et al. 1998).

Genetic studies comparing the species have been carried out by Forman et al. (1986) and Seuánez et al. (1988), who concluded the lion tamarins should be considered subspecies of a single species, *L. rosalia*. Despite this, conservationists have continued to classify the forms as individual species to ensure their continued legal protection. Chapter 5 reviews our most current understanding of lion tamarin genetics. The taxonomy and taxonomic history of the first three species have been reviewed by Coimbra-Filho (1970a, 1976b), Coimbra-Filho and Mittermeier (1972), Hershkovitz (1977), and Rosenberger and Coimbra-Filho (1984). Groves

Figure 1.1. Maria Iolita Bampi of the Instituto Brasileiro do Meio Ambiente e dos Recursos Naturais Renováveis (IBAMA—Brazilian Institute of the Environment and Renewable Natural Resources) with the first chair of the *caissara* International Recovery and Management Committee (IRMC), Ibsen de Gusmão Câmara, discussing black-faced lion tamarins at the 1997 conference that led to this volume. In the background are Ilmar Santos of Fundação Biodiversitas (Biodiversity Foundation) and Susie Ellis of the Conservation Breeding Specialist Group of the World Conservation Union/Species Survival Commission (IUCN/SSC). (Photo by Devra Kleiman)

(2001), Rylands and Rodríguez-Luna (2000), and Rylands et al. (1993, 2000) have reviewed the taxonomy, distributions, and conservation status of all the callitrichids.

Lion tamarin conservation and research have played an important role in the growth of primatology in Brazil since the late 1970s. Until 1976 no primate field studies (yearlong observations, for example, of behavior and ecology) had been carried out in Brazil, and the research of Coimbra-Filho in the 1960s and, with Russell Mittermeier, in the 1970s and early 1980s was truly pioneering, providing the initial stimulus for preserving not only lion tamarins but also other threatened species and for promoting Brazilian field primatology in general.

Here we present a brief history of the conservation and research efforts for each of the lion tamarin species and discuss the major actions in terms of research, protected areas, surveys, and field studies. Chapter 5 provides further details on the history of the captive breeding programs, and Chapter 3 discusses the IRMCs. Tables 1.1, 1.2, and 1.3 respectively detail the history of the major events involving conservation and ecological research; endangered species and protected area legislation; captive breeding and management; and workshops, conferences, and committees on research and conservation of lion tamarins.

GOLDEN LION TAMARIN (*LEONTOPITHECUS ROSALIA*)

Historical Review

In December 1519, the priest Antonio Pigafetta, chronicler of Magellan's voyage around the world, provided the first known reference to golden lion tamarins after observing them in the wild, calling them "beautiful simian-like cats similar to small lions" (Feio 1953). Perhaps the popularity of marmosets and tamarins with the aristocracy explains their inclusion in some late sixteenth- and early seventeenth-century paintings (Mallinson 1996). The name *golden lion tamarin* derives from M. J. Brisson's reference to "Le petit singe-lion" when describing a specimen brought from Brazil to Paris in 1754 for Madame La Maquise de Pompadour (Brisson 1756). In 1763, the naturalist Compte de Buffon examined a living specimen; 3 years later the Swedish scientist Carolus Linnaeus first described the species fully (Linnaeus 1766). When Prince Maximilian zu Wied traveled through southeastern Brazil in 1816 and 1817, he considered the species common (Hill 1970; Wied-Neuwied 1940).

It was not until the early 1960s that Adelmar F. Coimbra-Filho, the doyen of Brazil's primatologists, was able to focus public attention on the severe plight of the golden lion tamarin in Brazil. While working on the planning of a proposed

Table 1.1
A Chronology of Some of the Principle Events Concerning Lion Tamarin Conservation and Ecological Research

Date	Event	Reference
1519	*L. rosalia* first reference	Feio 1953
1766	*L. rosalia* (Linnaeus 1766) described	Linnaeus 1766
1818–1824	*L. rosalia* in Paris Menagerie	Geoffroy Saint-Hilaire 1827
1820	*L. chrysomelas* (Kuhl 1820) described	Kuhl 1820
1822	*L. chrysopygus* discovered by Johann Natterer, near Ipanema, São Paulo	Hill 1957
1823	*L. chrysopygus* (Mikan 1823) described	Mikan 1823
1869	*L. chrysomelas* at London Zoo	
1905	A skin of *L. chrysopygus* (from Bauru, São Paulo) donated by O. Hume to the São Paulo Zoology Museum (MZSP). The last record of the species till 1970	Coimbra-Filho 1976a, 1976b
1967	First observation of *L. rosalia* using a tree hole as a sleeping site	Coimbra-Filho 1978
1968–1969	Botanical inventory of the Morro do Diabo Forest Reserve	Campos and Heinsdijk 1970
1969	Survey for *L. chrysomelas* in southern Bahia. Proposal for a reserve in the region of Una	Coimbra-Filho 1970a
1970	Survey and rediscovery of *L. chrysopygus* in the Morro do Diabo Forest Reserve by Adelmar F. Coimbra-Filho	Coimbra-Filho 1970a, 1970b
1976	*L. chrysopygus* discovered in the Fazenda Paraíso, Gália, (later a state reserve) by Olav Mielke, entomologist, Federal University of Paraná	Coimbra-Filho 1976b
1980	Survey of the vegetation of the Poço das Antas Biological Reserve and first estimate of the population of *L. rosalia*	Green 1980
1980	First field study of *L. chrysomelas* at the Lemos Maia Experimental Station (Centro de Pesquisas de Cacau [CEPEC—Cocoa Research Center]/ Comissão Executiva do Plano da Lavoura Cacaueira [CEPLAC—Regional Cocoa Growing Authority for Bahia]), Una, Bahia	Rylands 1982, 1989b, 1993
1980–1981	WWF-U.S. Primate Program Survey of Atlantic Forest primates and protected areas	Mittermeier et al. 1981, 1982; Rylands et al. 1991/1992; Santos et al. 1987

Continued on next page

Table 1.1 continued

Date	Event	Reference
1982	Survey of the vegetation of the Poço das Antas Biological Reserve by the Rio de Janeiro Botanical Garden (Instituto Brasileiro de Desenvolvimento Florestal [IBDF—Brazilian Forestry Development Institute]/JBRJ)	Guimarães et al. 1985
1983	Initiation of reintroduction program for captive-born *L. rosalia*	Beck et al. 1986
1983–1984	WWF-U.S. Primate Program Survey of Atlantic Forest primates and protected areas	Rylands et al. 1991/1992; Santos 1983, 1984; Santos et al. 1987
1983–1984	First field study of *L. rosalia* at the Poço das Antas Biological Reserve	Dietz et al. 1997; Peres 1986a, 1986b, 1989a, 1989b
1983–present	Initiation of the Golden Lion Tamarin Conservation Program, National Zoological Park, Washington, D.C.	Dietz et al. 1986, 1994b; Kleiman 1984a, 1984b; Kleiman et al. 1985, 1986, 1990a, 1991
1983–present	Studies of the status, behavioral ecology, and demography of *L. rosalia* at the Poço das Antas Biological Reserve	Baker 1991; Baker et al. 1993; Dietz and Baker 1993; Dietz et al. 1994a, 1995, 1997
1983–present	Environmental education program for *L. rosalia* and the Poço das Antas Biological Reserve	Dietz 1985; Dietz and Nagagata 1986, 1995, 1997
1984	First captive-born *L. rosalia* introduced to the Poço das Antas Biological Reserve	Beck et al. 1986a, 1991, 1994
1984–present	Survey of the status and distribution of *L. chrysopygus*—"The Black Lion Tamarin Conservation Biology Project"	Valladares-Padua 1987, 1993; Valladares-Padua and Cullen 1994
1985–1986	Rescue operation for *L. chrysopygus* groups in the inundation area of the Rosana hydroelectric dam at the Morro do Diabo State Reserve (Instituto Florestal de São Paulo [IF/SP—São Paulo Forestry Institute], WWF, Companhia Energética de São Paulo [CESP—Sao Paulo Electricity Company], Fundação Brasileira para a Conservação da Natureza [FBCN—Brazilian Foundation for the Conservation of Nature], and Universidade Federal de Minas Gerais [UFMG— Federal University of Minas Gerais])	Carvalho et al. 1989; Carvalho and Carvalho 1989; Valle and Rylands 1986

Table 1.1 continued

Date	Event	Reference
1986–1987	Survey of the status and distribution of endemic arboreal mammals of the Atlantic Forest of southern Bahia and Espírito Santo	Oliver and Santos 1991
1987	Survey of the primates of the Rio Jequitinhonha valley, Minas Gerais	Rylands et al. 1988
1987	Studies of *L. rosalia* vocalizations initiated at the Poço das Antas Biological Reserve	Halloy and Kleiman 1994
1987–1988	Survey of the conservation status of the primates of southern Bahia, two expeditions, Federal University of Paraíba	Lima 1990
1988	Environmental education program for *L. chrysopygus* and the Morro do Diabo State Park	Padua 1991, 1994a, 1994b, 1997; Padua and Jacobson 1993; Padua and Valladares-Padua 1997
1988	Founding of the NGO Fundação Biodiversitas (Biodiversity Foundation), Belo Horizonte, Minas Gerais	
1988	Field study of *L. chrysopygus* in the Caetetus State Ecological Station	Keuroghlian 1990
1989–1991	Field study of *L. chrysopygus* in the Caetetus State Ecological Station	Passos 1991, 1992, 1997a
1989–1992	Studies on forest regeneration in the Poço das Antas Biological Reserve	Kolb 1993; Pessamílio 1994
1989–1993	Studies of small mammals and the effects of fires in the Poço das Antas Biological Reserve	Procópio de Oliveira 1993
1990	*L. caissara* first described	Lorini and Persson 1990
1990–1991	Field surveys for *L. caissara* in the southeast of the state of São Paulo	Martuscelli and Rodrigues 1992; Rodrigues et al. 1992
1990–1995	Distribution survey of *L. caissara* in Paraná	Lorini and Persson 1994a, 1994b
1990–1996	Community conservation education program for *L. chrysomelas*—Projeto Mico-Leão-Baiano	Alves 1991, 1992; Konstant 1990; Nagagata 1994a, 1994b
1991	Studies of *L. rosalia* vocalizations at the Poço das Antas Biological Reserve	Boinski et al. 1994
1991–1992	Distribution survey of *L. rosalia* in Rio de Janeiro	Kierulff 1993a, 1993b; Kierulff and Stallings 1991; Kierulff et al. 1997

Continued on next page

Table 1.1 continued

Date	Event	Reference
1991–1993	Distribution survey of *L. chrysomelas* in southern Bahia	Pinto 1994; Pinto and Rylands 1997; Pinto and Tavares 1994
1991–present	Field study of the demography and ecology of *L. chrysomelas* in the Una Biological Reserve	Dietz et al. 1994c and 1996
1992	Environmental education program for *L. chrysopygus* and the Caetetus State Ecological Station	Padua 1994a, 1994b; Padua and Jacobson 1993
1992	Founding of the NGO Instituto de Pesquisas Ecológicas (IPÊ—Institute for Ecological Research), São Paulo, centered on conservation and research on *L. chrysopygus*	IPÊ 1995/1996
1992	Founding of the NGO Associação Mico-Leão-Dourado (AMLD—Golden Lion Tamarin Association), Rio de Janeiro, for conservation and research on *L. rosalia*	
1993–1996	Study of the arboreal vegetation of the Caetetus State Ecological Station	Kim and Passos 1994
1993–1996	Field study of *L. chrysopygus* in the Caetetus State Ecological Station	Passos 1994
1994	Studies of locomotion and positional behavior of *L. rosalia* at the Poço das Antas Biological Reserve	Stafford and Ferreira 1995
1994	Founding of the NGO Instituto de Estudos Sócio-Ambientais do Sul da Bahia (IESB—Institute for Social and Environmental Studies of Southern Bahia) in the region of Una, Bahia	IESB 1996
1994–1997	Translocation of isolated groups of *L. rosalia* to the Fazenda União (2,400 ha), Rio de Janeiro	Kierulff and Procópio de Oliveira 1994, 1996
1994–present	Botanical surveys and research in the Poço das Antas Biological Reserve by the Rio de Janeiro Botanical Garden (JBRJ/IBAMA)	
1995	First translocation of a black lion tamarin group, initiating the metapopulation management program for the species	Medici 2001; Chapter 14 this volume
1995	Start of the environmental education program for the Una Biological Reserve, Bahia	Santos 1995; Santos and Blanes 1997, 1999
1995	Translocation of a group of *L. chrysopygus* from Fazenda Rio Claro, Lençois Paulista, to a forest (2,000 ha) in the Fazenda Mosquito, Narandiba, São Paulo (May 1995)	IPÊ 1995/1996

Table 1.1 continued

Date	Event	Reference
1995–1997	First field study of *L. caissara,* in the Superagüi National Park	Prado 1999; Prado and Valladares-Padua 1997; Valladares-Padua and Prado 1996
1999	Translocation of a second black lion tamarin group from the Fazenda Rio Claro to the Fazenda Mosquito, São Paulo (July 1999)	Medici 2001; Chapter 14 this volume
1999–present	Distribution survey to examine the northern limits to the range of *L. caissara*	Valladares-Padua et al. 2000a
2000	Golden lion tamarins reintroduced reached 153, and the total reintroduced population reached 359 (50 groups)	Beck and Martins 2001
2000	First mixed, captive-/wild-born group of *L. chrysopygus* introduced into the Morro do Diabo State Park	Chapter 14 this volume
2001	Translocated golden lion tamarin population successfully established in the União Biological Reserve	Kierulff 2000
2001	10 years of the International Research and Management Committees for the genus *Leontopithecus* marked by IBAMA presenting commemorative plaques to Adelmar F. Coimbra-Filho, Jeremy Mallinson, and Devra Kleiman, Ilhéus, Bahia (May 2001)	
2001	A golden lion tamarin birth resulted in the known population in the wild reaching 1,000 individuals	
2001	Adelmar F. Coimbra-Filho honored by the minister of the environment, José Sarney Filho, for his dedication to lion tamarin research and conservation	

Note: IBAMA = Instituto Brasileiro do Meio Ambiente e dos Recursos Naturais Renováveis (Brazilian Institute of the Environment and Renewable Natural Resources), JBRJ = Jardim Botânico do Rio de Janeiro (Rio de Janeiro Botanical Garden), NGO = nongovernmental organization, and WWF = World Wildlife Fund.

reserve at Jacarepaguá, in the state of Guanabara (now Rio de Janeiro) in 1962, Alceo Magnanini and Coimbra-Filho began their first attempts at breeding a pair of *L. rosalia* in the hope of setting up a reintroduction program, lion tamarins already being extinct in the area (Coimbra-Filho and Magnanini 1962). Unfortunately, attempts to establish the reserve were unsuccessful due to lack of interest from the government, and the lion tamarins were consequently moved to the Rio de Janeiro Zoo in 1963 (Magnanini and Coimbra-Filho 1972; Magnanini et al. 1975). Coimbra-Filho was then head of the research department at the zoo (Coimbra-Filho 1965). In 1972 through 1974, Coimbra-Filho and Magnanini set up the Tijuca Biological Bank, in the Tijuca National Park, specifically for the conservation and captive breeding of lion tamarins (Coimbra-Filho and Mittermeier 1977; Coimbra-Filho et al. 1986b). Its success resulted in the transfer of the lion tamarins, and other threatened Atlantic Forest callitrichids, to the Centro de Primatologia do Rio de Janeiro (CPRJ—Rio de Janeiro Primate Center), inaugurated in 1979 in the Serra dos Órgãos, northeast of Rio de Janeiro.

During the 1960s and 1970s, Coimbra-Filho traveled to many of the municipalities in the state of Guanabara (now Rio de Janeiro) in search of remnant populations of golden lion tamarins and witnessed year-by-year the destruction of their forests. He received very small grants from 1965 to 1968 from the Conselho Nacional de Desenvolvimento Científico e Tecnológico (CNPq—Brazilian National Science Council) and the International Biological Program (IBP), but most of his expeditions, visiting remnant forest patches and interviewing hunters and local people, were by train and by foot and of his own initiative.

The results of Coimbra-Filho's pioneering research on *L. rosalia,* both in captivity and in the wild, were presented at the Third Brazilian Zoology Congress at the National Museum, Rio de Janeiro, in July 1968. Through Coimbra-Filho's insistence, the golden lion tamarin was chosen as the symbol of the congress, and the paper he presented was subsequently published in a supplement of the *Anais da Academia Brasileira de Ciências* in 1969 (Coimbra-Filho 1969). This classic paper describes aspects of the golden lion tamarin's morphology and karyotype, habitat, behavior, and reproduction and management in captivity, and it provides the first distribution map for the species. Several other papers of that period also describe his experiences with *L. rosalia* in captivity (Coimbra Filho 1965; Coimbra Filho and Magnanini 1972). In 1971, Russell Mittermeier, then a Harvard University graduate student, teamed up with Coimbra-Filho, which resulted in additional important publications reporting on the taxonomic status, ecology, and conservation of the lion tamarins (Coimbra-Filho and Mittermeier 1972, 1973, 1976, 1977, 1978, 1982; Coimbra-Filho et al. 1975; see also Kleiman 1981).

In 1964, Coimbra-Filho and Magnanini outlined the threatened status of numerous vertebrate species in Brazil, and their work formed the basis for Brazil's

Table 1.2

A Chronology of Principle Events Concerning Legislation, Endangered Species Lists, and Protected Areas

Date	Event	Reference
1941	Morro do Diabo Forest Reserve (37,157 ha) created. Administered by the secretary of agriculture, São Paulo	State Decree Law No. 12.279/29 October 1941
1942	Lagoa São Paulo Forest Reserve (13,343 ha) created. Administered by the secretary of agriculture, São Paulo	State Decree Law No. 13.049/6 November 1942
1942	Reserva do Pontal (246,840 ha) created. Administered by the secretary of agriculture, São Paulo	State Decree Law No. 13.075/25 November 1942
1956–1957	Decrees reinforcing the protection of the three Pontal forest reserves	State Decrees Nos. 25.363 and 25.364/ November 1956; State Decree No. 28.338/ 8 May 1957
1957	Illegal squatters occupying the Morro do Diabo Forest Reserve in the 1940s and 1950s removed	Leite 1979
1965	Brazilian Forest Code	Law No. 4.771/ 15 September 1965
1966	State Decrees Nos. 25.363 and 25.364/November 1956 and 28.338/8 May 1957 for the protection of the forests of the Pontal region, São Paulo, revoked	State Decree 45.897/ 12 January 1966
1966	*L. rosalia* blacklisted by the American Association of Zoological Parks and Aquariums (AAZPA)	Bridgwater 1972b
1966	*L. rosalia* included in the *IUCN Red Data Book*	
1967	Brazilian Fauna Protection Law	Law No. 5.197/3 January 1967
1968	First Brazilian Official List of Species Threatened with Extinction. Gave legal protection to *L. rosalia, L. chrysomelas,* and *L. chrysopygus*	Edict No. 303/20 May 1968; Carvalho 1968; Coimbra-Filho 1972; Coimbra-Filho and Magnanini 1968
1968	The establishment of reserves proposed by Adelmar F. Coimbra-Filho for *L. rosalia* in the Rio São João basin and area of Poço das Antas, Rio de Janeiro	Brazil, MA/IBDF/FBCN 1981; Coimbra-Filho 1976a
1969	A reserve for *L. chrysomelas* in the municipality of Una proposed by Adelmar F. Coimbra-Filho	Coimbra-Filho 1970a; Magnanini 1978
1969	Construction of a highway bisecting the Morro do Diabo Forest Reserve	Leite 1979

Continued on next page

Table 1.2 continued

Date	Event	Reference
1969	A ban on importation of *L. rosalia* pledged by the International Union of Directors of Zoological Gardens (IUDZG)	J. Perry (personal communication in Coimbra-Filho and Mittermeier 1977)
1969	U.S. Rare and Endangered Species Act	U.S. Public Law 90-135
1969	Jacupiranga State Park (150,000 ha) created. Administered by IF/SP	State Decree Law No. 145/8 August 1969
1973	Una, Bahia, visited by Adelmar F. Coimbra-Filho, who was formally invited by IBDF to select a site for a reserve for *L. chrysomelas*	
1973	First management plan for the Morro do Diabo Forest Reserve	Deshler 1975
1973	Part of the forest of the Morro do Diabo Forest Reserve destroyed by defoliants 2,4,5T and 2,4D	Coimbra-Filho 1976b; Coimbra-Filho and Mittermeier 1977
1974	Creation of Poço das Antas Biological Reserve, Rio de Janeiro. Administered by IBAMA, Brasília	Decree No. 73.791/ 11 March 1974; Magnanini 1978
1975	Brazil a signatory of the Washington Convention on International Trade in Endangered Species of Wild Flora and Fauna (CITES)	Legislative Decree No. 54/24 July 1975; Decree No. 76.623/ 17 November 1975
1975	The limits of the Poço das Antas Biological Reserve altered, and its size increased to 5,500 ha	Decree No. 76.534/ 3 November 1975. Decree No. 76.583/ 3 November 1975 authorized expropriation of the land.
1976	Gália State Reserve (2,179 ha) created. Administered by IF/SP	State Decree No. 8.346/ 9 August 1976
1976	Purchase of 5,342 ha by IBDF (later IBAMA) in the region of Una, Bahia, for the creation of a biological reserve	Mallinson 1984
1979–1981	Construction of the Rio São João dam, affecting the southern portion of the Poço das Antas Biological Reserve	Brazil, MA/IBDF/FBCN 1981; Magnanini 1978
1980	Creation of the Una Biological Reserve, Bahia. Administered by IBAMA, Brasília	Decree No. 85.463/ 10 December 1980
1980–1981	Management plan published for the Poço das Antas Biological Reserve	Brazil, MA/IBDF/FBCN 1981
1982	Creation of the Guaraqueçaba Ecological Station (14,000 ha), Paraná, protecting mangroves and *restinga*. Administered by IBAMA	Decree No. 87.222/ 31 May 1982

Table 1.2 continued

Date	Event	Reference
1983	Management plan for the Morro do Diabo Forest Reserve	Guillaumon et al. 1983
1984	Creation of the Cananéia-Iguape-Peruíbe Environmental Protection Area, São Paulo. Administered by IBAMA	Decrees No. 90.347/ 23 October 1984 and No. 91.892/ 6 November 1985
1985	Creation of the Guaraqueçaba Environmental Protection Area (313,400 ha). Administered by IBAMA	Decree No. 90.883/ 31 January 1985
1985–2001	Removal of squatters from the Una Biological Reserve	Coimbra-Filho et al. 1993
1986	Inundation of about 3,000 ha of the Morro do Diabo State Park by the Rosana hydroelectric dam (27,614 ha) of the Companhia Energética de São Paulo (CESP—São Paulo Electricity Company), Rio Paranapanema	Audi 1986; Valle and Rylands 1986
1986	Morro do Diabo Forest Reserve changed to a state park (34,441 ha). Administered by IF/SP	State Decrees No. 25.342/ 4 July 1986 and No. 28.169/21 January 1988
1987	Gália State Reserve changed to the Caetetus State Ecological Station (2,179 ha). Administered by the IF/SP	State Decree No. 26.718/6 February 1987
1987	Creation of the Paraíso State Ecological Station (4,920 ha), Serra dos Órgãos, around the Centro de Primatologia do Rio de Janeiro (CPRJ—Rio de Janeiro Primate Center), Rio de Janeiro	State Decree No. 9.803/12 March 1987; Brazil, FEEMA 1989
1988	The Atlantic Forest and the Serra do Mar—with Amazonia, the Pantanal of Mato Grosso, and the coastal zone—declared a national heritage in the Brazilian Constitution of 1988. Chapter VI, article 4	Brazil 1988
1989	Superagüi National Park (21,400 ha) created. Administered by IBAMA, Brasília	Decree No. 97.688/ 25 April 1989
1989	Brazilian Official List of Species Threatened with Extinction revised to include *L. rosalia, L. chrysomelas,* and *L. chrysopygus*	Edict No. 1.522/ 19 December 1989; Bernardes et al. 1990; Fonseca et al. 1994
1990	Creation of the protected area category RPPN	Decree Law 98.194/ 31 January 1990
1990	International Management Committee for *L. chrysopygus* officially recognized by the Brazilian government	Edict No. 1.203/18 July 1990

Continued on next page

Table 1.2 continued

Date	Event	Reference
1990	International Management Committee for *L. chrysomelas* officially recognized by the Brazilian government	Edict No. 1.204/18 July 1990
1990	International Management Committee for *L. rosalia* officially recognized by the Brazilian government	Edict No. 2.342/ 28 November 1990
1990	Presidential decree protecting the Atlantic Forest	Decree No. 99.547/ 25 September 1990
1990–1991	Una Biological Reserve increased in size by 659 ha	Coimbra-Filho et al. 1993
1991	Official donation to IBAMA of 659 ha of land for the Una Biological Reserve by H.R.H. the Duke of Edinburgh, president of Worldwide Fund for Nature (WWF) International (March 1991)	Coimbra-Filho et al. 1993
1992	*L. caissara* included on the Brazilian Official List of Species Threatened with Extinction, Edict No. 1.522/19 December 1989	Edict No. 045/92-N/ 27 April 1992
1992	International Management Committee for *L. caissara* officially recognized by the Brazilian government	Edict No. 106-N/ 30 September 1992
1993	Decree protecting the Atlantic Forest (replaced Decree No. 99.547/25 September 1990)	Decree No. 750/10 February 1993 (published 11 February 1993)
1993	Una Biological Reserve increased in size by 1,058 ha	Coimbra-Filho et al. 1993
1995	*L. caissara* placed on the threatened species list for the state of Paraná	State Law 11.067/17 February 1995 Brazil, Paraná, SEMA 1995
1996	Regulations concerning the creation of RPPNs	Decree No. 1.922/ 5 June 1996
1996	*L. rosalia, L. chrysopygus,* and *L. caissara* listed as "critically endangered" and *L. chrysomelas* listed as "endangered" in the 1996 *IUCN Red List of Threatened Animals,* using the Mace-Lande categories	IUCN 1994, 1996
1996	*L. chrysomelas* placed on the threatened species list for the state of Minas Gerais	Deliberation of the State Council for Environmental Policy (COPAM), 20 January 041/95; *Minas Gerais, Orgão Oficial dos Poderes do Estado, Diário Executivo, Legislativo e Publicações de Terceiros* (14, part 1): 1–4, 20 January 1996; Machado et al. 1998

Table 1.2 continued

Date	Event	Reference
1997	Preliminary version of a management plan for the Una Biological Reserve	Saracura 1997
1997	The boundaries of the Superagüi National Park altered, increasing its size from 21,400 ha to 34,254 ha	Law No. 9.513/ 20 November 1997; Capobianco 1998
1998	Creation of the União Biological Reserve (3,200 ha), Rio de Janeiro	Decree, 22 April 1998; Kierulff 1999; Kierulff and Procópio de Oliveira 1996
1998	*L. chrysopygus* placed on the threatened species list for the state of São Paulo	State Decree No. 42.838/4 February 1998; *Diário Oficial do Estado de São Paulo* 108(25):1–7, 5 February 1998; Brazil, São Paulo, SMA 1998
1999	Superagüi National Park declared a World Heritage Site (December 1999)	
1999	Committees for the four species combined and renamed the International Committee for the Conservation and Management of Lion Tamarins (ICCM)	Instituto Brasileiro do Meio Ambiente e dos Recursos Naturais Renováveis (IBAMA), Edict 764, *Diário Oficial da União,* 15 December 1999
2000	*L. rosalia* placed on the threatened species list for the state of Rio de Janeiro	Bergallo et al. 2000
2001	Last squatters removed from the Una Biological Reserve	S. de Sousa, personal communication

Note: IBAMA = Instituto Brasileiro do Meio Ambiente e dos Recursos Naturais Renováveis (Brazilian Institute of the Environment and Renewable Natural Resources), IBDF = Instituto Brasileiro de Desenvolvimento Florestal (Brazilian Forestry Development Institute), IF/SP = Instituto Florestal de São Paulo (São Paulo Forestry Institute), and RPPN = Private Natural Heritage Reserve.

first threatened species' list, including the three lion tamarins already described (*L. caissara* was not discovered until 1990), prepared by the Fundação Brasileira para a Conservação da Natureza (FBCN—Brazilian Foundation for the Conservation of Nature) (Carvalho 1968; Coimbra-Filho 1972; Coimbra-Filho and Magnanini 1968; Magnanini 1975). The Brazilian Fauna Protection Law of 1967, with the Brazilian Official List of Species Threatened with Extinction of 1968, forbade the capture, hunting, purchase, sale, and exportation of all endangered species and any products made from them. Between 1965 and 1967, the international zoo community was also moving to prevent the importation of golden lion tamarins (Hill 1970). Besides the Brazilian Fauna Protection Law regulating the export of animals from Brazil (1967), measures included the 1969 U.S. Rare and Endangered Species Act, which effectively prevented further lion tamarin acquisitions by zoos in the United States, and the "hands off" importation policy adopted by the international zoo community; these helped to end the legal importation of this species (Bridgwater 1972b; Coimbra-Filho and Mittermeier 1972, 1977; Perry 1971). Although they were significant steps, they unfortunately did nothing to address problems of habitat destruction, which in reality was the major threat to these monkeys. From 1969 to 1979, Magnanini and Coimbra-Filho's efforts for the conservation of *L. rosalia* in the wild centered around the slow and difficult process of setting up a protected area, the Poço das Antas Biological Reserve, within the species' already restricted range in Rio de Janeiro.

In 1981, the Smithsonian National Zoological Park Golden Lion Tamarin Conservation Program (SNZP/GLTCP) held the first negotiations with the Instituto Brasileiro de Desenvolvimento Florestal (IBDF—Brazilian Forestry Development Institute, now IBAMA) to initiate a long-term research program based initially in the Poço das Antas Biological Reserve (Kleiman et al. 1985, 1986, 1988, 1990a, 1991). The aim was an integrated in situ and ex situ conservation effort for the species (see Chapter 3 this volume for the list of missions of the GLTCP). From 1983, the GLTCP was administered through the FBCN in Rio de Janeiro, but in 1992 the Associação Mico-Leão-Dourado (AMLD—Golden Lion Tamarin Association) was set up in Casimiro de Abreu, with Alcides Pissinatti as president and Denise Rambaldi as executive director (Chapter 3 this volume). The AMLD has the mandate of fulfilling the original missions of the GLTCP in the broadest sense.

Results of the field and captive research on the golden lion tamarin provided the baseline data for the PVA Workshop held in Belo Horizonte, Brazil, in June 1990 (Seal et al. 1990). This important exercise resulted in recommendations for in situ and ex situ action for the lion tamarins for the following 7 years.

A key question, still unresolved by 1990, was the status of *L. rosalia* outside of

Poço das Antas. In 1991, Maria Cecília Kierulff initiated a major survey of the forests throughout the original known range of the species (Kierulff 1993a, 1993b; Kierulff and Stallings 1991; Chapter 2 this volume). In 1994, with Paula Procópio de Oliveira, she subsequently rescued some of the most threatened groups and translocated them to a well-preserved forest of 2,400 ha in the Fazenda União (União Ranch), about 20 km northeast of the Poço das Antas Biological Reserve (Kierulff 1999; Kierulff and Procópio de Oliveira 1994; see Chapter 12 this volume).

Captive Breeding and Research

Research on *L. rosalia* in captivity prior to 1972 was reviewed by Bridgwater (1972a), who provided a full bibliography to that date. Snyder (1974) also reviewed what was known of this species' behavior. Since then, Adelmar F. Coimbra-Filho, Alcides Pissinatti, and Roberto da Rocha e Silva and collaborators at the CPRJ have carried out and published numerous studies on the breeding, management, and morphology, pathology, and biology of *L. rosalia,* as well as *L. chrysomelas* and *L. chrysopygus* (Burity et al. 1997a, 1997b, 1997c, 1997d, 1999; Castro 1990; Coimbra-Filho 1981; Coimbra-Filho and Maia 1977, 1979a, 1979b; Coimbra-Filho and Rocha 1978; Coimbra-Filho et al. 1980, 1981, 1984a, 1984b, 1986a, 1991; Ferreira et al. 1997; French et al. 1996b; Marques et al. 1997; Pinder and Pissinatti 1991; Pissinatti 2001; Pissinatti and Tortelly 1984; Pissinatti et al. 1981, 1984a, 1984b, 1992, 1993, 2000; Rocha e Silva 1984, 1986, 2001; Rocha e Silva et al. 1991; Rosenberger and Coimbra-Filho 1984; Chapter 11 this volume). Valladares-Padua (1986) compared the survivorship of *L. rosalia* in the CPRJ with that in the Brookfield Zoo, Chicago.

Reproductive, social, and communication behavior, as well as physiology, were researched extensively by Devra Kleiman and coworkers at the SNZP (Green 1979; Halloy and Kleiman 1994; Hoage 1977, 1978, 1982; Kleiman 1978b, 1978c, 1979, 1983; Kleiman and Mack 1977, 1980; Mack and Kleiman 1978; McLanahan and Green 1978; Rathbun 1979; Ruiz 1990; Thompson et al. 1994) and by Jeffrey French and his students and collaborators at the University of Nebraska at Omaha (Benz et al. 1990, 1992; French and Inglett 1989; French and Stribley 1985, 1987; French et al. 1989, 1992; Inglett et al. 1989, 1990), as well as at CPRJ during 6 months in 1993 (French et al. 1996b). Snowdon et al. (1986) carried out a comparative study of long calls of the three species at the Monkey Jungle, Florida, and at CPRJ. More recently, studies on the locomotion of lion tamarins have been carried out in SNZP by Stafford and Rosenberger (Rosenberger and Stafford 1994; Stafford et al. 1994), with comparative data being obtained from wild groups (Stafford and Ferreira 1995). Rapaport (1997, 1999) and Ruiz-Miranda et al. (1999) have studied the behavior involved in food transfers in *L. rosalia* at SNZP and at the Poço das Antas Biological Reserve.

Pathology and diseases of *L. rosalia* have been researched at the SNZP as well as by Pissinatti at the CPRJ (Bush et al. 1980, 1993, 1996; Goff et al. 1986, 1987; Montali 1993a, 1993b; Montali and Bush 1981, 1999; Montali et al. 1980, 1983, 1993, 1995a, 1995b; Pissinatti and Tortelly 1984; Randolph et al. 1981; Scanga et al. 1993; Wilson et al. 1989).

Protected Areas: The Poço das Antas Biological Reserve and the União Biological Reserve

During the 1960s, Coimbra-Filho identified important areas where golden lion tamarins still survived in the municipalities of Cabo Frio, Casimiro de Abreu, and Silva Jardim, mainly along the tributaries of the Rio São João (Coimbra-Filho 1969). The best location in terms of the state of the forest was in the vicinity of the Rio Iguape at Poço d'Anta, although he had found golden lion tamarins to be scarce there. Coimbra-Filho argued for a minimum size of 40,000 ha in order to maintain a viable population, and in 1969, with Alceo Magnanini and José Candido de Melo Carvalho, then president of the FBCN, he selected two excellent sites for the establishment of a reserve in the basin of the São João. Both were unfortunately destroyed the year after (Magnanini and Coimbra-Filho 1972), and a third site was chosen, of 3,000 ha, in the municipality of Silva Jardim. The decree creating the reserve was submitted to the Ministry of Agriculture in 1971, but only in 1974 were decrees signed creating the reserve and authorizing expropriation. Poço das Antas was the first biological reserve created in Brazil.

By April 1975, much of the forest had been destroyed, with dense forest covering an estimated 10 percent of the area and degraded forests another 30 percent. During this time, a major socioeconomic development program was in progress for the São João valley, including grandiose drainage schemes and the proposed construction of a 2.5 km dam, which would have a direct impact on the reserve. Magnanini and Coimbra-Filho subsequently proposed a decree altering the limits of the reserve and increasing its size to 5,500 ha (Coimbra-Filho and Mittermeier 1977; Magnanini 1978). They launched a major campaign to solicit support for this, and due in large part to the insistence of then state governor Floriano Faria Lima, the decree was signed in November 1975. In January 1980, FBCN, in collaboration with the Fundação Estadual de Engenharia do Meio Ambiente (FEEMA—State Foundation for Environmental Engineering) and IBDF, initiated studies for the elaboration of a management plan (Brazil, MA/IBDF/FBCN 1981).

Brief vegetation surveys of the reserve were made in 1980 (Green 1980; Kleiman et al. 1988) and 1982 by the Jardim Botânico do Rio de Janeiro (JBRJ—

Table 1.3
A Chronology of Some Important Events Regarding Captive Breeding and Management and Committees, Meetings, and Workshops for the Genus *Leontopithecus*

Date	Event	Reference
1962	First attempts to breed *L. rosalia* at the proposed Jacarepaguá Biological Reserve (the reserve was never established), with a view to reintroduction	Coimbra-Filho and Magnanini 1972
1963	Initiation of breeding program at the Rio de Janeiro Zoo	Magnanini and Coimbra-Filho 1972
1964	Paper on rare Brazilian animals threatened with extinction submitted by Adelmar F. Coimbra-Filho and Alceo Magnanini to the Goeldi Museum, Belém, Pará. It warned of the threatened status of the three lion tamarin species (published in 1968).	Coimbra-Filho and Magnanini 1968
1965	First publication on the management of *L. rosalia* in captivity	Coimbra-Filho 1965
1965	Appeal by Clyde Hill, San Diego Zoological Society, to the American Association of Zoological Parks and Aquariums (AAZPA) to blacklist *L. rosalia* for international trade	Hill 1970
1966	Wild Animal Propagation Trust Committee formed to review captive status	Bridgwater 1972b
1966–1968	International Biological Program (IBP). Project on the status of *L. rosalia* conducted by Adelmar F. Coimbra-Filho	Magnanini and Coimbra-Filho 1972
1969	First evaluation of the status of *L. rosalia* in captivity. Provided by the Wild Animal Propagation Trust Committee	Bridgwater 1972b
1972	Conference "Saving the Lion Marmoset," National Zoological Park, Smithsonian Institution, Washington, D.C.	Bridgwater 1972a
1972	First guidelines for captive management. Established at the conference "Saving the Lion Marmoset"	DuMond 1972
1972	Guidelines for research priorities to improve captive breeding. Established at the conference "Saving the Lion Marmoset"	Kleiman 1972
1972	First published compilation of the status of *L. rosalia* in captivity 1966–1971 in international studbook	Bridgwater 1972b; Jones 1973; Kleiman 1977a, 1977b; Kleiman and Jones 1978

Continued on next page

Table 1.3 continued

Date	Event	Reference
1972–1974	Establishment of the Tijuca Biological Bank for breeding *Leontopithecus* species for research and reintroduction	Coimbra-Filho and Mittermeier 1977; Magnanini and Coimbra-Filho 1972
1973	First captive colony of *L. chrysopygus,* Tijuca Biological Bank, Rio de Janeiro	Coimbra-Filho 1976a, 1976b
1975	Conference "The Biology and Conservation of the Callitrichidae," National Zoological Park, Washington, D.C.	Kleiman 1978a
1977	"The Marmoset Workshop," Göttingen, Germany	Rothe et al. 1978
1978	Global Primate Action Plan of the International Union for the Conservation of Nature and Natural Resources/SSC Primate Specialist Group	Mittermeier 1978
1979	Inauguration of the Centro de Primatologia do Rio de Janeiro (CPRJ/FEEMA)	Coimbra-Filho et al. 1986b
1979–1990	Rapid growth of the captive population of *L. rosalia*	Ballou and Sherr 1997
1981	Cooperative Research and Management Agreement for captive *L. rosalia*. Formation of international committee to manage the captive population	Kleiman 1984a
1983	*L. chrysomelas* illegally exported from Bolivia and Guyana to Belgium, France, Hong Kong, and Japan	Konstant 1986; Mallinson 1984, 1987b
1985	Formation of International Recovery and Management Committee for *L. chrysomelas* and first meeting in San Diego	Mallinson 1989
1986	*L. chrysopygus* received by São Paulo Zoo. Captured from the area of inundation of the Rosana hydroelectric dam, São Paulo	Simon 1988
1987	Formation of the International Committee for the Preservation and Management of the Black Lion Tamarin	
1987	Preliminary international studbook for *L. chrysomelas*	Ballou 1989; Mallinson 1987a
1988	Preliminary studbook for *L. chrysopygus*	Simon 1988
1989	First international studbook for *L. chrysopygus*	Simon 1989
1990	First Population Viability Analysis (PVA) Workshop, Belo Horizonte, Brazil, in collaboration with the IUCN/SSC/CBSG	Rylands 1993/1994; Seal et al. 1990
1991	Title of captive population of *L. rosalia* returned to IBAMA by global zoos owning the species	
1991	Initiation of Lion Tamarins of Brazil Fund	Mallinson 1994a

Table 1.3 continued

Date	Event	Reference
1991–1992	Publication of the *Global Captive Action Plan for Primates,* IUCN/SSC/CBSG/PSG and CI	Stevenson et al. 1991, 1992
1991–1996	Reduction and stabilization of the captive population of *L. rosalia* (476 animals on 31 December 1996)	Ballou and Sherr 1997
1992	First Annual Meeting of the International Management Committees with IBAMA, Rio de Janeiro	
1993	Second Annual Meeting of the International Management Committees and First Symposium on *Leontopithecus,* Poço das Antas Biological Reserve, Rio de Janeiro	Rylands and Rodríguez-Luna 1994
1993	Emergency action plan drawn up for *L. caissara* and presented to IBAMA (June 1993)	Câmara 1993, 1994
1994	Third Annual Meeting of the International Management Committees and Second Symposium on *Leontopithecus,* Ilhéus, Bahia	
1995	Fourth Annual Meeting of the International Management Committees, Guaraqueçaba, Paraná	
1996	Fifth Annual Meeting of the International Management Committees, Brasília	
1996	Draft action plan drawn up for the captive management of *L. chrysopygus*	Valladares-Padua and Ballou 1998
1997	Sixth Annual Meeting of the International Management Committees, Third Symposium on *Leontopithecus,* and second Population and Habitat Viability Analysis (PHVA) Workshop, Belo Horizonte, Brazil, in collaboration with the CBSG	Ballou et al. 1998
1997	Metapopulation management plan for *L. chrysopygus* approved by the International Management Committee	Valladares-Padua and Martins 1996
1997	First issue of *Tamarin Tales,* newsletter of the International Committees for Recovery and Management of *L. rosalia, L. chrysopygus, L. chrysomelas,* and *L. caissara*	
1998	Seventh Annual Meeting of the International Management Committees, Instituto de Pesquisas Ecológicas (IPÊ—Institute for Ecological Research), Nazaré Paulista, São Paulo	

Continued on next page

Table 1.3 continued

Date	Event	Reference
1999	Eighth Annual Meeting of the International Management Committees, Museu de Biologia Melo Leitão (MBML—Melo Leitão Biological Museum), Santa Teresa, Espírito Santo	
2000	Ninth Annual Meeting of the International Management Committees, Belo Horizonte, Minas Gerais	
2001	Tenth Annual Meeting of the International Committee for Conservation and Management of the genus *Leontopithecus* (ICCM), Ilhéus, Bahia	

Note: CBSG = IUCN/SSC Conservation Breeding Specialist Group, CI = Conservation International, CPRJ = Centro de Primatologia do Rio Janeiro (Rio Janeiro Primate Center), FEEMA = Fundação Estadual de Engenharia do Meio Ambiente, Rio de Janeiro (State Foundation for Environmental Engineering), IBAMA = Instituto Brasileiro do Meio Ambiente e dos Recursos Naturais Renováveis (Brazilian Institute of the Environment and Renewable Natural Resources), IUCN = World Conservation Union, PSG = IUCN/SSC Primate Specialist Group, and SSC = IUCN Species Survival Commission.

Rio de Janeiro Botanical Garden) (Guimarães et al. 1985). Green (1980) estimated the area of forest to be about 2,000 ha, highly fragmented, with only about 500 ha of mature forest. From this, he calculated the density of golden lion tamarins to be about 0.05/km^2, or 75 golden lion tamarins in the entire reserve (Kleiman et al. 1988). The areas of grassland within the reserve were a constant source of threat to the forests, in dry years being subject to potentially catastrophic fires. From the earliest days, reforestation of these open areas was seen as a vital step to increase the carrying capacity of the reserve for the golden lion tamarins, as well as to reduce the threat of fires (Kleiman et al. 1988; Pessamílio 1994). This still remains a major challenge.

In 1983, the SNZP/GLTCP established long-term projects on the demography and socioecology of the lion tamarins (see Chapter 8 this volume), a reintroduction program (initially for the Poço das Antas Biological Reserve but later for neighboring forest patches—see Chapter 13), and a community environmental education program (Chapter 15). The Jardim Botânico do Rio de Janeiro resumed its studies of the flora and vegetation of the reserve in 1994. In 1997, the population of lion tamarins in the Poço das Antas Biological Reserve was estimated to be 347 animals and believed to be at carrying capacity (Ballou et al. 1998; J. M. Dietz, unpublished data). However, since 1997 an increase in predation (five entire groups were predated in 2000) resulted in a population decline to about 220 golden lion tamarins, with the average group size in the reserve falling from 5.6 in 1997 to 3.4 in 2000 (Franklin and Deitz 2001). The predators involved have not yet been identified, although coatimundis (*Nasua* spp.) or tayras (*Eira barbara*) are suspected.

The reintroduction program for golden lion tamarins in and around the Poço das Antas Biological Reserve began in 1983, with the first releases in 1984. Currently, Benjamin Beck of SNZP and Andréia Martins of AMLD jointly coordinate the project (Beck and Martins 2001; Beck et al. 1986a, 1991, 1994; Bush et al. 1993; Castro et al. 1998; Dietz 1985; Kleiman et al. 1986; Pinder 1986a, 1986b; Chapters 12 and 13 this volume). Although six zoo-born golden lion tamarins were released in 2000, researchers estimated that the entire habitat within practical commuting distance for the reintroduction team was at carrying capacity in 2001. In 2000 the population of reintroduced golden lion tamarins had reached 359 (95 percent wild born), in 50 groups, on about 3,200 ha of forest (Beck and Martins 2001; Chapters 7 and 12 this volume). One of the key results of this program is the protection of golden lion tamarin habitats on private lands (over 20 privately owned ranches are collaborating). Stoinski (2000) carried out a study on the underlying differences in survival-critical behaviors between zoo-born reintroduced golden lion tamarins and their wild-born offspring (see Chapter 13 this volume).

Laurenz Pinder translocated the first groups of golden lion tamarins in 1983 into the Poço das Antas Biological Reserve (Pinder 1986a, 1986b). As a result of the distribution surveys of Kierulff and Stallings in 1991 to 1992 (Kierulff 1993a, 1993b; Kierulff and Procópio de Oliveira 1996; Kierulff and Stallings 1991), 60 individuals in 12 groups were found surviving in nine small and isolated forest fragments. Discussions commenced concerning translocating them to a forest of 2,400 ha, then a reserve of the Rede Ferroviaria Federal S.A. (RFFSA—Brazilian Railway Company) in the Fazenda União, municipality of Rio das Ostras, about 20 km to the north of the Poço das Antas Biological Reserve (Kierulff and Procópio de Oliveira 1994). Between 1994 and 1997, 43 golden lion tamarins (from six of the isolated groups) were translocated to União, and by 2001 a population of more than 120 was surviving there (see Chapter 12 this volume). Five years of negotiations involving IBAMA, RFFSA, and AMLD led to the creation of the União Biological Reserve in April 1998 (Kierulff 1999).

Field Research and Environmental Education

Coimbra-Filho first described aspects of the ecology and behavior of golden lion tamarins (e.g., their use of nest holes for sleeping) (Coimbra-Filho 1978). In 1983, Coimbra-Filho and Kleiman jointly organized the development of long-term field studies on the golden lion tamarin in the Poço das Antas Biological Reserve

(Figure 1.2). They have included ecological and behavioral studies (the coordinators are James Dietz and Andrew Baker) of the golden lion tamarins both in and around the Poço das Antas Biological Reserve and studies on feeding and ranging behavior (Dietz et al. 1997; Peres 1986a, 1986b, 1989a, 1989b), demography, mating systems and infant care (Baker 1991; Baker et al. 1993; Bales et al. 2000; Dietz and Baker 1993; Dietz et al. 1994a, 1995; Chapter 8 this volume), food transfers (Ruiz-Miranda et al. 1999), locomotion and posture (Stafford and Ferreira 1995), and vocalizations (Boinski et al. 1994; Halloy and Kleiman 1994). Long-term research on the feeding ecology, ranging behavior, and demography of the lion tamarins translocated to the União Biological Reserve has been carried out since 1994 by Kierulff and Procópio de Oliveira (1994, 1996). More recently, Rapaport (1997, 1999) and Ruiz-Miranda et al. (1999; Ruiz-Miranda and Rapaport, in press) have been studying the ontogeny of foraging and feeding in lion tamarins in both reserves.

Conservation education programs for the local communities and for professional training were developed by Lou Ann Dietz and Elizabeth Nagagata, beginning in

Figure 1.2. Adelmar F. Coimbra-Filho and James Dietz discuss trapping techniques for golden lion tamarins in 1984 at the beginning of the long-term field studies in the Poço das Antas Biological Reserve. Cláudio Valladares-Padua is in the background. (Photo by Devra Kleiman)

1983 (Dietz and Nagagata 1986, 1995, 1997; Fernandes et al. 2000; Chapter 15 this volume). The programs are currently overseen by Denise Rambaldi, executive director of the AMLD. The community focus includes the establishment of organic and sustainable agroforestry methods with "landless" people who have settled near the Poço das Antas Biological Reserve and have had a negative impact on the forest. A forest restoration program, formerly coordinated by Dionízio Pessamílio, then director of the reserve (Kolb 1993; Pessamílio 1994), has now expanded to involve staff of the reserve, the Jardim Botânico do Rio de Janeiro, and the AMLD. Current activities involve habitat restoration and reforestation within reserve areas as well as the building of corridors between forest fragments both within and outside reserve areas. The reserve's staff and AMLD are also assisting and encouraging local landowners, especially those with reintroduced golden lion tamarins on their property, in the formal protection of their land through the establishment of a Reserva Particular de Patrimônio Natural (RPPN—Private Natural Heritage Reserve) (D. Rambaldi, personal communication).

GOLDEN-HEADED LION TAMARIN (*LEONTOPITHECUS CHRYSOMELAS*)

Historical Review

Very little was known of *L. chrysomelas* prior to the studies of Coimbra-Filho (1970a). There were only about 20 museum specimens worldwide, 11 of them, in the Museu Nacional, Rio de Janeiro (National Museum), resulting from the research of Laemmert et al. (1946) on the occurrence of yellow fever in both the lion tamarins and the marmoset (*Callithrix kuhlii*) in southern Bahia. Coimbra-Filho (1970a) reviewed the taxonomic history, geographic distribution, and habitat of *L. chrysomelas* and also reported his observations of three captive individuals at the Rio de Janeiro Zoo. He pointed out that the golden-headed and the black lion tamarins were in fact lion tamarins and very closely related to *L. rosalia,* which may seem obvious today but was not clear to some earlier authors, for example, Elliot (1913) and Thomas (1922). He was the first to set up a breeding colony of golden-headed lion tamarins, initially at the Rio de Janeiro Zoo and, from 1972, the Tijuca Biological Bank.

Coimbra-Filho visited southern Bahia for the first time in 1969 and observed firsthand the habitat of the species and the remaining forest that was relatively widespread and intact when compared with the habitat of *L. rosalia* (see Coimbra-Filho 1970a). However, the degree of deforestation already underway alarmed

Coimbra-Filho to the extent that he argued that, of the species *L. chrysomelas, L. chrysopygus,* and *L. rosalia, L. chrysomelas* was in the greatest danger of extinction due to the "devastation and degradation of the primitive forests by lumber and agricultural interests (especially cacao planters)" (Coimbra-Filho and Mittermeier 1972, p. 60). In 1973, he surveyed the region again to locate a suitable area to establish a reserve, and in Una he found the last significant tract of well-preserved forest within the species' range. It was not until 1976, however, that some of the land was purchased, and a further 4 years passed before the Una Biological Reserve was decreed in December 1980.

From 1979 to 1985, Russell Mittermeier coordinated a survey of the eastern Brazilian primates sponsored by the World Wildlife Fund (WWF) Primate Program (Mittermeier et al. 1981, 1982; Santos 1983, 1984; Santos et al. 1987). Findings in southern Bahia indicated that isolated and small populations remained throughout a large part of the species' known range. A report with specific recommendations regarding the protection of the species and of the Una Biological Reserve, based on these surveys, was submitted to the Brazilian government (Santos 1983, 1984; Santos et al. 1987). From May to November of 1980, Rylands carried out the first field study of the ecology and behavior of one of four groups of *L. chrysomelas* occupying a small part of the Lemos Maia Experimental Station (a forest of 240 ha) of the Comissão Executiva do Plano da Lavoura Cacaueira (CEPLAC—Regional Cocoa Growing Authority for Bahia) (Rylands 1980, 1982, 1983, 1989b). Further surveys were carried out within the range of *L. chrysomelas* in the late 1980s by Oliver and Santos (1991) and Lima (1990).

In 1983 to 1984, large numbers of golden-headed lion tamarins were illegally exported from Brazil, via Bolivia and Guyana, and found their way into the hands of animal dealers and private animal collections, mainly in Belgium and Japan (Konstant 1986; Mallinson 1984, 1987b). As a result, IBAMA formed the International Recovery and Management Committee (IRMC) in March 1985, with Jeremy Mallinson and Adelmar F. Coimbra-Filho as cochairs. With IBAMA, the committee was successful in having the majority of the illegally exported lion tamarins returned to Brazil in 1986 (Mallinson 1987a, 1987b).

The PVA Workshop held in 1990 examined the status of the Una Biological Reserve and the situation concerning the, by then, very rapid growth of the captive population (Rylands 1993/1994; Seal et al. 1990). The workshop resulted in 21 specific recommendations in five major categories: protected areas, potential protected areas, environmental education, captive breeding, and reintroduction. Eventually, a reintroduction program was considered to be unnecessary, mainly due to the fact that there were still sufficient numbers in the wild (despite continuing forest destruction), there was not sufficient protected habitat, and the causes of the decline of the lion tamarins were still not controlled (Kleiman 1990b; Kleiman et al. 1994).

There have been only two field projects for *L. chrysomelas* since 1991: (1) a survey of the distribution and status of *L. chrysomelas* (Pinto 1994; Pinto and Rylands 1997; Pinto and Tavares 1994) and (2) a study of the demography and ecology of the lion tamarins in the Una Biological Reserve (Dietz 1993; Dietz et al. 1994c, 1996; Chapter 7 this volume).

The western half of the range of *L. chrysomelas* is dominated by cattle ranching, and the scarce populations surviving are found only in small and isolated forest fragments (Pinto and Rylands 1997). The majority of the remaining populations survive in the eastern half, which is the cocoa-growing region of southern Bahia (Pinto and Rylands 1997). The permanence of the forests and the continued survival of *L. chrysomelas* are due to the fact that until the 1960s many large landowners still maintained more old growth forest on their land than cocoa plantations (Alger and Caldas 1994). Over the last three decades, however, the social and economic scenario of the region has been changing dramatically, resulting in a rapid loss of enormous areas of forest. In the late 1970s and the first half of the 1980s, high prices for cocoa resulted in an increase in the area planted, but a significant drop in prices from 1986 to 1992 and the appearance of witch's broom disease (combated only by pruning the affected branches or burning the entire tree) in 1989 have resulted in farm owners abandoning their plantations, logging, and switching to alternative crops (coconut palms, African oil palm, cloves, and peppers) and cattle ranching. There has consequently been an upsurge of forest cutting (Alger and Caldas 1994). The importance of the Una Biological Reserve, not only for the golden-headed lion tamarins but also as one of the last forests protecting the extraordinary biodiversity of the region, increased dramatically with the disappearance of the forests around the reserve. In 1993, Russell Mittermeier (through the Brazil Program of Conservation International) with a partner nongovernmental organization (NGO), the Instituto de Estudos Sócio-Ambientais do Sul da Bahia (IESB—Institute for Social and Environmental Studies of Southern Bahia), formally founded in 1994, began to assess the economic threats and opportunities for conservation in the buffer zone around the Una Biological Reserve. Besides the detailed mapping of the remaining forests, and the assessment of the land ownership (Alger and Araújo 1996; Alger et al. 1996), the project deals directly with the economic activities of the region: logging (Mesquita 1996), cocoa cultivation (Alger and Caldas 1994; Hardner 1996), cattle farming (Reid and Blanes 1996), *piaçava* palm fiber extraction (Moreau et al. 1996), and ecotourism (Mourão 1996). It assesses threats and tendencies and finds economic and political alternatives that will reduce or halt forest cutting (Conservation International 1997).

Captive Breeding and Research

Until the early 1980s, the only colony of golden-headed lion tamarins in captivity was in the Tijuca Biological Bank (up to 1979) and subsequently in CPRJ in Brazil (Coimbra-Filho 1985a). The appearance in 1983 to 1984 of large numbers of illegally exported golden-headed lion tamarins in Belgium, Japan, and French Guiana resulted in the establishment of breeding colonies in Europe, Japan, the United States, and Brazil established with the confiscated animals (Mallinson 1987b, 1989). The development of the captive population is described in Ballou 1989 and Chapter 4, and the work of the IRMC for golden-headed lion tamarins is documented by Mallinson (1984, 1986, 1987b, 1989) and Kleiman and Mallinson (1998). Research on captive social, reproductive, and parental care behavior has been conducted mainly at the Antwerp Zoo (De Vleeschouwer 2000; Van Elsacker et al. 1992), the Durrell Wildlife Conservation Trust (DWCT) (Chaoui and Hasler-Gallusser 1999; Feistner and Price 1999; Moore 1997; Price and Feistner 1993), and the federal universities of Paraíba and Natal (Alonso et al. 1997a, 1997b; Moura and Langguth 1997; Moura et al. 1997; Oliveira et al. 1999).

Protected Area: The Una Biological Reserve

In 1973, Adelmar F. Coimbra-Filho surveyed the region between the Rio Pardo and the Rio de Contas to search for potential sites for a reserve for *L. chrysomelas*. He located an area of 15,000 ha in the municipality of Una, which was privately owned and for sale at $160 (U.S.) per ha. In 1976, IBDF (subsequently IBAMA) purchased a total of 5,342 ha, but the remainder of the 12,470 ha proposed was complicated by legal problems of land title, deeds, and indemnities. It was only in December 1980 that the decree for the creation of the Una Biological Reserve was signed with an area of 11,400 ha.

During this interval, part of the area identified by Coimbra-Filho had been severely degraded and had also been invaded by squatters and the landless (estimated at 100 families at the time [Saracura 1997]), many of them timber merchants from the neighboring state of Espírito Santo. Thus, during the late 1970s, parts of the reserve and the surrounding forest continued to be destroyed, with the tire company Pirelli, for example, planting 2,500 ha of rubber trees immediately to the south. The land acquired was in two blocks, barely connected: the northwestern Piedade zone (named for the adjacent farm) of 2,607 ha and the southeastern Maruim zone (bordered by the Rio Maruim) of 2,735 ha.

Until 1982, the reserve remained without any effective administration or infrastructure. In that year, some $45,000 (U.S.) of Brazilian currency was made available to IBDF for Una through the WWF Tropical Forests and Primates Campaign, but for various bureaucratic reasons little of the money was used, and most of it was lost due to the high inflation rate at the time. Visits to the Una Biological Reserve with the specific aim of evaluating the threats to its integrity, and the status of the lion tamarin population there, were carried out by numerous teams between 1980 and 1985 (Dietz et al. 1985; Lindbergh 1986; Mittermeier et al. 1981, 1982; Santos 1983, 1984; Santos et al. 1987). The resulting reports included a series of recommendations and proposals regarding land tenure and acquisition (particularly the establishment of an effective corridor between the two forest blocks comprising the reserve), the removal of squatters, the removal of a colony of confiscated golden-headed lion tamarins being maintained in the Una Biological Reserve, and the establishment of research and environmental education programs, as well as surveys outside the reserve. A grant proposal of $337,500 (U.S.) was developed and eventually funded to resolve the problems regarding squatters and land title, the acquisition of 1,300 ha (to provide the corridor), and fencing (Sousa et al. 1988).

Estimates of the number of squatter families in the Una Biological Reserve were probably exaggerated until 1984, when Saturnino Neto de Sousa, newly appointed director, estimated 80 in the northwest (Piedade) zone but found none residing in the southeast (Maruim) zone. Sousa has carried out a gradual and extremely successful campaign to remove the squatters, and by June 2001 sufficient funds had been provided by IBAMA and WWF to facilitate the removal of the last families of squatters from the Una Biological Reserve (S. N. F. de Sousa, personal communication).

In 1989, there was a fundraising campaign for the purchase of the corridor to connect the separate forest blocks. In 1991, an area of 659 ha was donated to IBAMA in an official ceremony in the presence of then WWF president H.R.H. the Duke of Edinburgh. The next steps planned to consolidate the reserve were (1) the acquisition of land to the southwest of the Maruim zone and other areas of intact contiguous forest within the decreed area to the north and (2) the redefinition of the reserve limits to exclude the totally destroyed areas in the Piedade zone. In 1992, the Fazenda Eldorado, abutting the northern limits of the Maruim zone, was put up for sale, and a major fundraising campaign resulted in the purchase of 1,058 ha in 1993, the title to which was immediately transferred to IBAMA, increasing the area of the reserve to 7,059 ha (Coimbra-Filho et al. 1993). The Jersey Wildlife Preservation Trust contributed substantially to these efforts, as did WWF, Conservation International, and the Chicago Rainforest Action Group through the Brookfield Zoo.

At the 1990 PVA Workshop, results of Vortex simulations to examine the viability of the population in the Una Biological Reserve indicated that *L. chrysome-*

las was the only wild population of any of the lion tamarin species that had some guarantee of survival over 100 years (Seal et al. 1990). However, the population estimate was based on projections using group sizes that were larger and home ranges that were smaller than demonstrated by later work by Dietz et al. (1994c), who found that the resident population of the Una Biological Reserve was less than half that in the 1990 analyses. This placed the population into the category of doomed over the next 100 years, which was given as an argument that the Una Biological Reserve needed to be increased in size, at least to the 11,400 ha of its original decree (Ballou et al. 1998).

Golden-headed lion tamarins are frequently kept as pets, and over the years the Una Biological Reserve acted as a holding facility for confiscated animals. The IRMC determined that the colony should be removed, and in 1992 Cristina Alves organized the transfer, to a number of Brazilian zoos, of the 25 lion tamarins held captive at the time.

Field Research and Environmental Education

The first field study of *L. chrysomelas* involved a short study of the species' feeding and ranging behavior compared with that of the sympatric marmoset (*Callithrix kuhlii*) (Rylands 1982). In 1980 to 1981, and again in 1991, surveys by different researchers resulted in an expansion of the known distribution of *L. chrysomelas* (Mittermeier et al. 1981, 1982; Oliver and Santos 1991; Rylands et al. 1988, 1991/1992; Santos et al. 1987). Luiz Paulo Pinto carried out a thorough survey (1992–1993) of the distribution and populations of *L. chrysomelas* outside the Una Biological Reserve (Pinto 1994; Pinto and Rylands 1997; Pinto and Tavares 1994). He identified a geographic distribution of 37,500 km^2 and demonstrated a highly fragmented population, evidently under severe decline due to ongoing and widespread deforestation. The population was estimated to be between 6,000 and 14,000 individuals. Pinto (1994) also identified home ranges that were larger than those used in the 1990 Vortex simulation.

Alves (1988, 1990) investigated the role of *cabruca* cocoa plantations (where cocoa is grown beneath shade from remnants of the original forest canopy) in helping to protect lion tamarins. She confirmed the findings of Coimbra-Filho (1970a) and Rylands (1980) that *cabruca* is used by *L. chrysomelas* and can act as an important corridor between forests, but always with intact forest nearby.

In 1993, James Dietz initiated long-term research on the ecology and demography of the population in the Una Biological Reserve (Dietz 1993; Dietz et al. 1994c, 1996). The main objectives were to obtain the necessary demographic data for an understanding of the carrying capacity and the probabilities of the survival of the population, to compare its biology with that of *L. rosalia*, and to expand the role of the Una Biological Reserve as a training and research center, attracting human and financial resources to promote a local, national, and international commitment to the conservation of the region (Dietz 1993). Sufficient data, covering nearly a decade, are now available for demographic analyses. Raboy has studied the behavioral ecology of *L. chrysomelas* (see Chapter 7 this volume) and has done an analysis of birth seasonality and infant survival (Raboy et al. 2001).

The first environmental education program for *L. chrysomelas,* Projeto Mico-Leão-Baiano, coordinated by Cristina Alves, was carried out from 1990 to 1996 and involved numerous educational activities, including, especially, local schools and the establishment of the Nature Education Center in Itabuna, inaugurated in 1994 (Alves 1991, 1992; Konstant 1990). Nagagata (1994a, 1994b) reviewed the environmental education activities of the Projecto Mico-Leão-Baiano (Bahian Lion Tamarin Project) and their effectiveness in changing the knowledge, attitudes, and behavior of the people in the region.

Although relatively numerous and still occurring over a large part of what is believed to be their original geographic range, the populations of golden-headed lion tamarins are extremely fragmented. In November 1995, the Landowners' Environmental Education Programme was set up specifically for landowners and farm workers, not only to encourage the protection of forests on private lands but also to change attitudes toward the Una Biological Reserve (Blanes and Mallinson 1997; Santos 1995; Santos and Blanes 1997, 1999; Chapter 12 this volume).

BLACK LION TAMARIN (*LEONTOPITHECUS CHRYSOPYGUS*)

Historical Review

The black lion tamarin was first discovered by Johann Natterer, who collected the type series (eight specimens) in 1822 from Ipanema (today Varnhagem), near Sorocaba, São Paulo. Mikan described the species a year later, in 1823. In 1902, Ernst Garbe, director of the Zoology Museum, São Paulo, collected three specimens from Vitória, municipality of Botucatu, São Paulo, and in 1905 O. Hume donated a specimen to the museum, collected from Bauru (Coimbra-Filho 1970b, 1976a). These were the only known records until two specimens mounted by a taxidermist in Presidente Wenceslau were discovered by Coimbra-Filho in 1970. The specimens were reported to have been obtained from a forest patch in a nearby ranch, the Fazenda Kitayama; however, the forest had already been cut down (Figure 1.3). Coimbra-Filho subsequently carried out expeditions to the municipalities of Presidente Epitácio and Teodoro Sampaio and on 14 May 1970 observed three individuals in the Morro do Diabo Forest Reserve, on the banks of the Rio Paranapanema, in the far west of the state of São Paulo.

In 1973, Coimbra-Filho captured seven black lion tamarins from Morro do Diabo and started the first breeding colony for the species. He also described the

Figure 1.3. Black lion tamarins were thought to be extinct until 1970, when they were rediscovered by Adelmar F. Coimbra-Filho in the Morro do Diabo State Park. (Photo by Russell A. Mittermeier/Consevation International)

tree holes they were using as sleeping sites (Coimbra-Filho 1976a, 1976b, 1978). The Tijuca Biological Bank and later the CPRJ were the only captive colonies for the species until 14 black lion tamarins, rescued from the inundation area of the Rosana hydroelectric dam, were sent to the São Paulo Zoo in 1986 (see Chapter 4 this volume).

In 1976 Olav Mielke, an entomologist from the Federal University of Paraná, observed and photographed black lion tamarins in an isolated forest patch in the Fazenda Paraíso, municipality of Gália, São Paulo (Coimbra-Filho 1976b). He alerted Coimbra-Filho, and in that same year the forest was decreed a state reserve of 2,179 ha and later, in 1987, a state ecological station (Caetetus State Ecological Station).

National and international action on behalf of the black lion tamarin was galvanized with the imminent threat to part of the population in the Morro do Diabo Forest Reserve by the construction of the Rosana dam in 1983 to 1986 (Audi 1986). A historic agreement was made between the Companhia Energética de São Paulo (CESP—São Paulo Electricity Company) and the Instituto Florestal de São Paulo (IF/SP—São Paulo Forestry Institute) of São Paulo, with the compensatory measures for the flooding of Morro do Diabo being targeted at the implementation of a management plan for the park (Mittermeier et al. 1985).

In 1987 an IRMC for black lion tamarins, chaired by Faiçal Simon of the São Paulo Zoo, was set up on the initiative of the IF/SP and IBAMA to enable all involved parties to contribute to decision making regarding the rescue operation, field research, and the expanded captive population (Chapter 4 this volume).

Fieldwork and conservation efforts have continued and increased over the years since the Morro do Diabo rescue operation. Cláudio Valladares-Padua carried out ecological, behavioral, and genetic studies in the Morro do Diabo State Park from 1984 (before it was a state park), and in 1989, Suzana Padua set up a community environmental education program centered on the Morro do Diabo State Park. The program was extended to the Caetetus State Ecological Station in 1992 (Padua and Valladares-Padua 1997; see Chapter 15 this volume).

In 1991 to 1992, a search for further populations in forest fragments on private lands in the region resulted in Valladares-Padua and his team locating new populations in 5 of 10 localities they visited (Valladares-Padua and Cullen 1994; Valladares-Padua et al. 1994). One of the forests, at the Fazenda Ribeirão Bonito, was in the process of being cut down, and the other four, totaling only 2,800 ha, contained an estimated 114 black lion tamarins. Although small, these populations are important in genetic terms and formed the basis for a metapopulation management plan drawn up by Valladares-Padua and Martins (1996), which was approved by the IRMC in 1996 and resulted in a draft action plan for the species presented at the PHVA Workshop held in 1997 (Valladares-Padua and Ballou 1998). The metapopulation management program, involving translocations, reintroductions of captive-bred lion tamarins, and artificial "dispersal," was begun in 1995 (Valladares-Padua et al. 1994; Chapter 14 this volume).

Ecological studies on black lion tamarins have also been carried out at the Fazenda Rio Claro since 1991 (Mamede-Costa 1997; Mamede-Costa and Gobbi 1998), and one group captured there was translocated in 1995 to one of the forests where the species was found to be absent, the Fazenda Mosquito (Mamede-Costa and Gobbi 1998).

One of the key problems, highlighted in the 1997 PHVA Workshop, was the need to deal with social unrest linked with agrarian reform in the Pontal region, which was potentially a serious threat to the future of the Morro do Diabo State Park and the surrounding forest patches (Ballou et al. 1998; Valladares-Padua et al. 1997). Through the creativity and initiative of the NGO Instituto de Pesquisas Ecológicas (IPÊ—Institute for Ecological Research), established by Suzana Padua and Cláudio Valladares-Padua and based in Nazaré Paulista, São Paulo, and through the environmental education program centered on the park, the threat is being averted, successfully co-opting the collaboration of the landless reform movement and settlers in the region (Padua and Tabanez 1997b; Padua and Valladares-Padua 1997; Valladares-Padua et al. 1997; Chapter 15 this volume).

Captive Breeding and Research

Chapter 4 summarizes the history of the captive black lion tamarin population. Valladares-Padua and Ballou proposed the maintenance of 150 to 200 black lion tamarins in captivity at 95 percent genetic diversity, acting as a subpopulation within a metapopulation management strategy including the populations in the wild (Valladares-Padua and Ballou 1998). Research in captivity was for many years limited to the studies in CPRJ by Coimbra-Filho and his colleagues, including general husbandry and diet (Coimbra-Filho 1976a, 1976b, 1981, 1985b; Coimbra-Filho et al. 1981), reproductive seasonality, infant survival and sex ratios (Coimbra-Filho and Maia 1979b; French et al. 1996b), pathology (Burity et al. 1997c, 1997d, 1999; Ferreira et al. 1997; Gonçalves et al. 1997; Pissinatti et al. 1984b; Pissinatti 2001), and morphology (Burity et al. 1997a, 1997b; Pissinatti et al. 1992, 2000). Since the captive program expanded to include further zoos in the late 1980s, more research is beginning to appear from colonies in São Paulo and the Jersey Zoo on, for example, food sharing (Feistner and Price 2000).

Protected Areas: The Morro do Diabo State Park and the Caetetus State Ecological Station

Serious destruction of the forests of the region known in the past as the Alta Sorocabana in the west of the state of São Paulo began in the 1920s for timber extraction and coffee plantations and later (from 1929/1930) for cotton and peanut farming. Today, the predominant industry in the region is cattle ranching. The timber industry thrived from the 1930s. The town of Presidente Prudente, for example, registered 36 sawmills in 1935. The timber was transported by the Sorocaba railway line (today abandoned), which reached the town of Presidente Prudente in 1917 and passed through the Morro do Diabo State Park. The almost total devastation of the forests of the region was evidenced by the collapse of the timber industry in 1960. Three forest reserves were decreed for the area in 1941 to 1942: Morro do Diabo (37,156 ha), Lagoa São Paulo in the north (13,343 ha), and the immense Reserva do Pontal (246,840 ha). However, the destruction of the forests continued unabated. In 1956, decrees by the state governor Jânio Quadros, along with reinforcement for the policing of the area, slowed down the destruction, but the decrees were revoked 10 years later and resulted in the almost total destruction of all the forests except those in the Morro do Diabo Reserve, although they were extensively logged, hunted, and degraded (Coimbra-Filho 1976a, 1976b; Leite 1979).

It is important to emphasize that the destruction of the São Paulo Atlantic Forest is recent (Dean 1995). In 1900 approximately 70 percent of the forests of the state were intact, by 1920 the figure was reduced to 45 percent, by 1950 to 29 percent, and by 1973 to only 8 percent (Leite 1979). The construction of the Porto Primavera hydroelectric dam, Rio Paraná, begun in 1995, flooded what remained of the Lagoa São Paulo Reserve, and the Rosana dam, completed in 1986, flooded remnant forests along the Rio Paranapanema along with 3,000 ha of the Morro do Diabo Forest Reserve. During its time as a forest reserve, the Morro do Diabo suffered from the lack of any supplementary legislation for its protection. The initial decree simply declared the area to be government land and a reserve for the protection of the fauna and flora (Leite 1979).

The future of the black lion tamarin in the wild depends entirely on the effective protection of the Morro do Diabo State Park (Coimbra-Filho and Mittermeier 1977). It was only in 1983 to 1985, with the construction of the Rosana hydroelectric dam, however, that consideration was given to the effective protection of the park, with the imminent inundation of 10 percent of its forests. On the initiative of a number of institutions—besides the CESP, FBCN, IF/SP, CPRJ, WWF-U.S. Primate Program (see Mittermeier 1982), and Universidade Federal de Minas Gerais (UFMG—Federal University of Minas Gerais)—a program was set up in 1983 (1) to carry out surveys of the fauna and vegetation of the park (then still a forest reserve) and of the black lion tamarins in the inundation area and (2) to rescue the groups during the cutting of the forest prior to the closure of the dam in 1986 (Sério 1986; Valle and Rylands 1986). Eight groups were located within the inundation area and brought into the captive population (see Chapter 4 this volume). The Rosana dam was closed in October 1986, and the area of the Morro do Diabo Forest Reserve was reduced to 34,441 ha. The reserve, which had benefited from considerable investment on the part of CESP, was changed to a state park in July 1986.

The census of the groups affected by the Rosana dam, and a study of a radio-collared group in the reserve (Carvalho and Carvalho 1989; Carvalho et al. 1989), provided the first field data for the species since Coimbra-Filho's observations in the 1970s (1970a, 1970b, 1976a, 1976b). A long-term field study was also set up in the Morro do Diabo State Park by Valladares-Padua starting in 1984, and it included studies of the behavior, ecology, and genetics of wild groups along with vegetation analyses based on the descriptions of Campos and Heinsdijk (1970) (Valladares-Padua 1997).

Although small, the Caetetus State Ecological Station is significant in being well east of the Morro do Diabo State Park and for many years was believed to be the only other locality where black lion tamarins were surviving in the wild. Alexine Keuroghlian carried out a field study there in 1988 (Keuroghlian 1990; Keuroghlian and Passos 2001) and broke the ground for ecological and behavioral studies by Fernando Passos (1991, 1992, 1994, 1997a, 1997b, 1997c, 1999; Passos and Alho 2001; Passos and Keuroghlian 1999; Passos and Kim 1999) from 1991 to 1996. Kim carried out a detailed study of the phytosociology of Caetetus from 1993 to 1996 (Kim and Passos 1994).

Field Research and Environmental Education

In the mid-1980s, Carvalho and colleagues initiated field studies of the ecology of wild black lion tamarins in the Morro do Diabo State Park (Carvalho and Carvalho 1989; Carvalho et al. 1989) and the Caetetus State Ecological Station (see references for Keuroghlian and Passos in the preceding section). In 1984, Valladares-Padua initiated in situ research and conservation efforts, centered on the Morro do Diabo State Park, and included distribution surveys and a research program in the Fazenda Rio Claro, near the Caetetus State Ecological Station (Valladares-Padua 1993; Valladares-Padua and Cullen 1994; Valladares-Padua and Mamede 1996b; Mamede-Costa 1997; Mamede-Costa and Gobbi 1998). The latter studies are ongoing.

An environmental education program specifically for the Morro do Diabo State Park was set up in 1988 by Suzana Padua (Padua 1991, 1994a, 1994b, 1997; Padua and Jacobsen 1993; Padua and Valladares-Padua 1997; Chapter 15 this volume).

BLACK-FACED LION TAMARIN (*LEONTOPITHECUS CAISSARA*)

Historical Review

The Brazilian NGO Sociedade de Pesquisa em Vida Selvagem e Educação Ambiental (SPVS—Society for Wildlife Research and Environmental Education), based in Curitiba, Paraná, set up a project in 1987 for a primate survey of the Guaraqueçaba Environmental Protection Area (313,400 ha), created 2 years previously on the coast of the state of Paraná. Reporting on their results at the XVII Brazilian Zoology Congress in 1990, Oliveira and Pereira (1990) registered the first record for the state of a titi monkey (*Callicebus personatus*). The photo they displayed at the poster session, however, was recognized by others as a lion tamarin, and the black-faced lion tamarin was described that same year by Lorini and Persson (1990) from the skin of a female adult from the island of Superagüi, close to the mainland in the north of the state. A large part of the island had been decreed a national park (21,400 ha) in 1989 as part of the Guaraqueçaba Área de Proteção Ambiental (APA—Environmental Protection Area), and the national and international interest that the black-faced lion tamarin attracted highlighted the serious problems that the park itself was facing.

The discovery of the black-faced lion tamarin was announced at the PVA Workshop on *Leontopithecus,* held in Belo Horizonte in June 1990. An IRMC was formed as a result and officially recognized in September 1992 (Chapter 3 this volume). Admiral Ibsen de Gusmão Câmara and Jeremy Mallinson cochaired the committee. Based on the little that was known at the time of the workshop, it was already possible to indicate that *L. caissara* (named after local inhabitants of the island, called *caiçaras*) was highly endangered. Preliminary surveys had found them to be rare and, besides on the island, occurring over a very small area on the coastal lowlands, east of the Serra do Mar (Lorini and Persson 1990).

An emergency action plan was drawn up and presented to IBAMA in June 1993. It summarized the data available on the species' distribution and population size, listed 15 specific factors identified as threats, and outlined research and conservation priorities. The plan especially considered the need to improve the status of the Superagüi National Park and to deal with the numerous threats to its integrity (Câmara 1993, 1994). Lorini and Persson carried out surveys and vegetation analyses throughout the range of the species in the state of Paraná (Lorini and Persson 1994a, 1994b). Rodrigues and Martuscelli surveyed the northern parts of its range and identified a number of new localities in the extreme southeast of the state of São Paulo, including the Jacupiranga State Park (150,000 ha, much of it deforested) (Martuscelli and Rodrigues 1992; Rodrigues et al. 1992). Rodrigues also attempted to set up a field study but suffered not only from the species' elusiveness but also from the hostility of those illegally extracting palm hearts (*palmito*) and carrying out other illicit activities in the region. There was some controversy concerning the reliability of some of the localities reported because they were derived in most cases from local interviews rather than actual sightings. Martuscelli and Rodrigues (1992) conservatively put the population in São Paulo state at about 200 individuals over 130 km^2, whereas Lorini and Persson (1994b) estimated a population of about 260 individuals (52 groups) throughout its known range of 173 km^2. Valladares-Padua et al. (2000a) were unable to find any evidence for the occurrence of *L. caissara* in the Jacupiranga State Park.

There was enthusiasm in the early days for the establishment of a captive colony of *L. caissara*. At the 1997 meeting of the IRMC, however, considerations of the difficulties involved and the impossibility of obtaining enough founders from the already severely reduced wild population sufficient for genetic health (Mansour and Ballou 1994) resulted in a decision against the development of a captive breeding program.

Protected Area: The Superagüi National Park

Lorini and Persson (1994b) estimated that one-third of the small geographic range of *L. caissara* lies within two protected areas, part of the Superagüi National Park (21,400 ha) and part of the Jacupiranga State Park in São Paulo. The remainder, however, falls within the two large APAs: Guaraqueçaba (313,400 ha, created in 1985) in Paraná and Cananéia-Iguape-Peruíbe in São Paulo (created in 1984, but its limits were changed in 1985). Both APAs are subject to environmental controls and restricted development (Amato et al. 2000; Ribas Lange 1997). Martuscelli and Rodrigues (1992) recorded *L. caissara* on the Rio Ipiranguinha and Rio Taquari in the Jacupiranga State Park, as well as a series of localities within the Cananéia-Iguape-Peruíbe Environmental Protection Area. They pointed to a series of threats to these areas, including deforestation for cattle ranching and charcoal production, palm heart (*Euterpe edulis*) extraction, and the construction of highways.

The Guaraqueçaba Environmental Protection Area was created to protect one of the last remaining areas of Atlantic Forest, besides serving as a buffer zone for the Guaraqueçaba Ecological Station, which contains an important area of mangroves. A wide range of projects is underway in the region, from faunal surveys to community education and the effective implementation of management to protect the natural ecosystems of the region (Margarido et al. 1997; Ribas Lange 1997). Serious problems were also identified by Câmara (1993) for the Superagüi National Park. They included urbanization and development for tourism along the coast of the island (not then included in the park), cattle farming, extraction of natural resources (especially for the palm heart), highway construction, illegal occupation of land, and the accompanying forest destruction. Guadalupe Vivekananda, the director of Superagüi National Park, reviewed the situation and problems at the time (Vivekananda 1994), and some have or are being resolved with an improved infrastructure and capacity for patrolling. In November 1997, the limits to the park were changed, increasing its size, and it now includes the coastal area formerly excluded and covers other areas where *L. caissara* still occurs (Capobianco 1998). Superagüi National Park was made a World Heritage Site in 1999. Vivekananda (2001) reviewed the current socioeconomic situation within the park and the impact of humans on conservation efforts.

Field Research and Environmental Education

Surveys of the black-faced lion tamarin populations and vegetation types have been carried out by two teams conducting censuses in the two states where *L. caissara* occurs. The surveys provide valuable information on the extent of its range and habitat requirements. Martuscelli and Rodrigues (1992; Rodrigues et al. 1992) carried out surveys in São Paulo, and Lorini and Persson (1994a, 1994b; Persson and Lorini 1991, 1993) obtained information on its habitat, ecology, and behavior in Paraná. In 1995, a field study was initiated on the island of Superagüi by Valladares-Padua and Prado (Prado 1999; Prado and Valladares-Padua 1997; Prado et al. 2000; Valladares-Padua and Prado 1996). Further surveys to delineate better the range of *L. caissara* were begun by Prado and Valladares-Padua in 1999 (Valladares-Padua et al. 2000a). These are urgently needed, not least because continental populations in many localities may well be disappearing already.

Environmental education programs in the region are continuing, with the SPVS and IPÊ being most active (Chapter 15 this volume).

ACKNOWLEDGMENTS

We gratefully acknowledge the help of many for jogging our memories and supplying many facts and details, especially Jonathan Ballou, James Dietz, Lou Ann Dietz, Gustavo Fonseca, Maria Cecília Kierulff, Suzana Padua, Luiz Paulo de Souza Pinto, Alcides Pissinatti, and Denise Rambaldi.

ANTHONY B. RYLANDS, MARIA CECÍLIA M. KIERULFF, AND LUIZ PAULO DE SOUZA PINTO

2
DISTRIBUTION AND STATUS OF LION TAMARINS

The four species of lion tamarins, *Leontopithecus* Lesson, 1840, are endemic to the lowlands of the Atlantic Forest in Brazil. They have small and disjunctive distributions in the south of the state of Bahia and extreme northeast of Minas Gerais (the golden-headed lion tamarin, *L. chrysomelas*), the lowland forest of the state of Rio de Janeiro (the golden lion tamarin, *L. rosalia*), the northeast tip of coastal Paraná and neighboring São Paulo (the black-faced lion tamarin, *L. caissara*), and the inland forests of the state of São Paulo (the black lion tamarin, *L. chrysopygus*) (Coimbra-Filho 1969; Coimbra-Filho and Mittermeier 1973, 1977; Hershkovitz 1977; Lorini and Persson 1990).

Three of the lion tamarins, *L. rosalia, L. chrysopygus,* and *L. caissara,* are classified as "critically endangered" in the *2000 IUCN Red List of Threatened Species* (Hilton-Taylor 2000), the internationally recognized listing of threatened species drawn up by the Species Survival Commission (SSC) of the World Conservation Union (IUCN), based in Gland, Switzerland. They were placed in the critically endangered category because of their small areas of occurrence and the severely fragmented populations (Criterion B1), continuing decline in the extent of occurrence (Criterion B2), and low numbers (fewer than 250 mature individuals, Criterion C) (IUCN 1994). The fourth, *L. chrysomelas,* is ranked as "endangered" due to its limited distribution of less than 5,000 km^2, the severely fragmented habitat (Criterion B1), a continuing decline in the extent of occurrence (Criterion B2), and mature individuals numbering less than 2,500, with populations declining and fragmented (Criterion C2a) (IUCN 1994). All four lion tamarins are included on the Brazilian Official List of Species Threatened with Extinction (*Lista Oficial de Espécies Brasileiras Ameaçadas de Extinção*, Edict No. 1.522/19 December 1989, see Bernardes et al. 1990; Fonseca et al. 1994) and likewise on the regional threatened species lists of the states of Paraná (Brazil, Paraná, SEMA 1995), Minas Gerais (Machado et al. 1998), São Paulo (Brazil, São Paulo, SMA 1998), and Rio de Janeiro (Bergallo et al. 2000). Here we describe the geographic distributions of the four lion tamarins, and summarize their conservation status.

THE ATLANTIC FOREST

The Atlantic Forest in Brazil, conservatively estimated to have covered 1.2 to 1.4 million km^2 in pre-Columbian times, has been reduced to less than 8 percent of its original size (Dean 1995; Fundação SOS Mata Atlântica/INPE 1998). With this enormous loss of forest, island biogeography theory predicts a much higher rate of vertebrate extinction than has been evidenced to date (Brooks and Balmford 1996; Brooks et al. 1999). However, based on the number of threatened bird species, Brooks and Balmford (1996) concluded that, without strong conservation measures, it is only a matter of time for the predicted extinctions to be realized. Endemic species are the most vulnerable when entire biomes are as severely threatened as the Atlantic Forest is. Of the 229 mammals occurring there, 32 percent are endemic (Costa et al. 2000; Fonseca et al. 1999). Endemism is especially marked in the primates: 21 of the 24 species and subspecies occur nowhere else. Eighteen of them are threatened (9 critically endangered and a further 3 endangered) (Rylands and Rodríguez-Luna 2000; Rylands et al. 1997).

The principal reason for the endangered status of the lion tamarins is the widespread devastation of their natural habitats (Coimbra-Filho and Câmara 1996; Dean 1995; Fonseca 1985). In 1995, the remaining forest cover in the principal four states where lion tamarins occur was estimated to be 9.3 percent in Paraná, 21.6 percent in Rio de Janeiro, 6.2 percent in Bahia, and 8.8 percent in São Paulo (Fundação SOS Mata Atlântica/INPE 1998). However, in both São Paulo and Rio de Janeiro, the majority of the remaining forests are montane; lion tamarins are restricted to lowland forests. Their continued survival rests on the capacity of their small populations to hang on in minute and isolated forest fragments. Such populations are likely not viable in the long term (Ballou et al. 1998; Seal et al. 1990) and will inevitably corroborate Brooks and Balmford's (1996) warning that "without immediate conservation action, the numerous Atlantic Forest birds (and untold numbers of other taxa) currently threatened will soon be extinct." The major conservation programs that have been developed for the lion tamarins since the early 1980s (see Chapter 1 this volume) have undoubtedly deferred their demise, even though the Atlantic Forest continues to shrink and suffer insidious depredation through such activities as clear cutting, logging, hunting, and unsustainable resource extraction.

DISTRIBUTION OF *LEONTOPITHECUS*

Leontopithecus rosalia (Linnaeus, 1766)

Type locality: Brazil, restricted to the coast between 22° and 23°S, from the Cabo de São Tomé to the municipality of Mangaratiba, by Wied-Neuwied (1826), further restricted to right bank, Rio São João, Rio de Janeiro, by Carvalho (1965) (see Coimbra-Filho 1969; Coimbra-Filho and Mittermeier 1973; Hershkovitz 1977).

The center of the range of *L. rosalia* is considered to be the basin of the Rio São João in the state of Rio de Janeiro. The original distribution was first clarified by Coimbra-Filho (1969, 1976a; Coimbra-Filho and Mittermeier 1973, 1977) (Figure 2.1). It covered the majority of the lowland coastal region of the state of Rio de Janeiro, below 300 m above sea level. The easternmost record for the species is Mangaratiba, on the coast in the southeast of the state. The original distribution included all or parts of the following municipalities: Mangaratiba, Itaguai, Nova Iguaçú, Nilópolis, São João do Meriti, Duque de Caxias, Rio de Janeiro, Magé, São Gonçalo, Niterói, Itaborai, Maricá, Araruama, Silva Jardim, Saquarema, Rio Bonito, Cachoeiras de Macucu, São Pedro da Aldeia, Cabo Frio, Casimiro de Abreu, Macaé, Conceição de Macabu, Campos, and São João da Barra.

Ruschi (1964) reported that *L. rosalia* could also be found to the north, in the state of Espírito Santo (see Hershkovitz 1977), in the municipalities of Iconha, Anchieta, and Itapemerim, although already extremely rare in these areas. Likewise, Von Ihering (1940) indicated its occurrence in Espírito Santo. Between 1815 and 1817, Prinz Maximilian von Wied-Neuwied (1826) traveled from Rio de Janeiro through the entire state of Espírito Santo, north into Bahia. He described

---

Figure 2.1. (**A**) The original distribution of *Leontopithecus rosalia* in the state of Rio de Janeiro, Brazil, and (**B**) enlarged area of current fragmented distribution of *L. rosalia* (based on Coimbra-Filho 1969; Kierulff 1993a; Kierulff and Procópio de Oliveira 1996). Localities of the current distribution are as follows: (1) Poço das Antas Biological Reserve, (2) vicinity of Poço das Antas, (3) hillsides of the Serra do Mar, (4) Campos Novos, (5) Centro Hípico, (6) Emerências, (7) gallery forest of the Rio São João and AGRISA, (8) Sobara, (9) Cabista, (10) Otacílio Melo, (11) Saquarema, (12) Fazenda (Ranch) of Luiza Brunet, (13) Angelim, (14) União Biological Reserve (a translocated population). Localities 6 to 13 contain isolated groups. Populations derived from the reintroduction program are represented by *R*. (Map by Stephen D. Nash/Conservation International)

BRAZIL
MINAS GERAIS
ESPIRITO SANTO
Vitória
Itapemirim
21°
Campos
MINAS GERAIS
22°
Casimiro de Abreu
Silva Jardim
RIO DE JANEIRO
ATLANTIC OCEAN
SÃO PAULO
Itaguaí
Cabo Frio
23°
Saquarema
Rio de Janeiro
0 50 100km
44°
43°
42°
41°

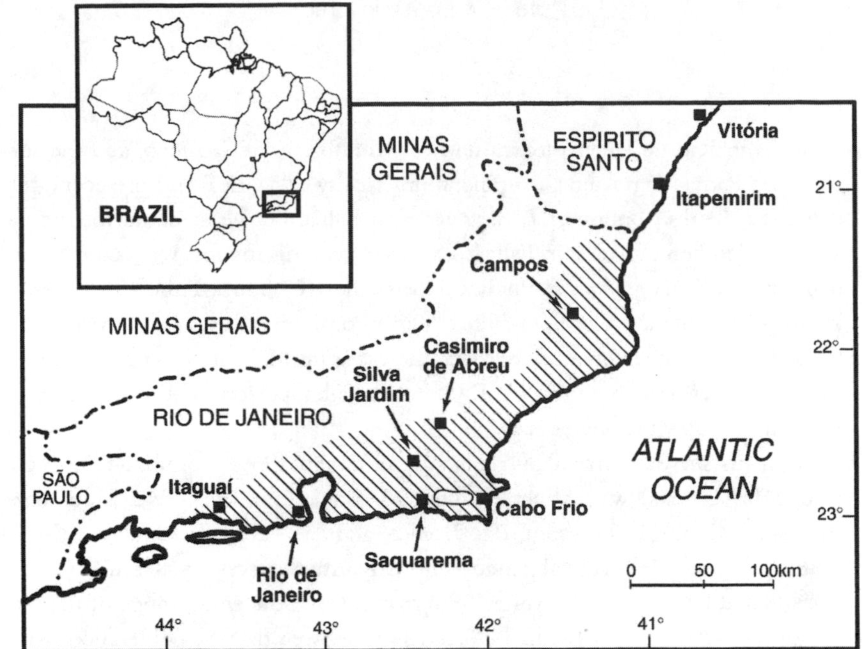

A

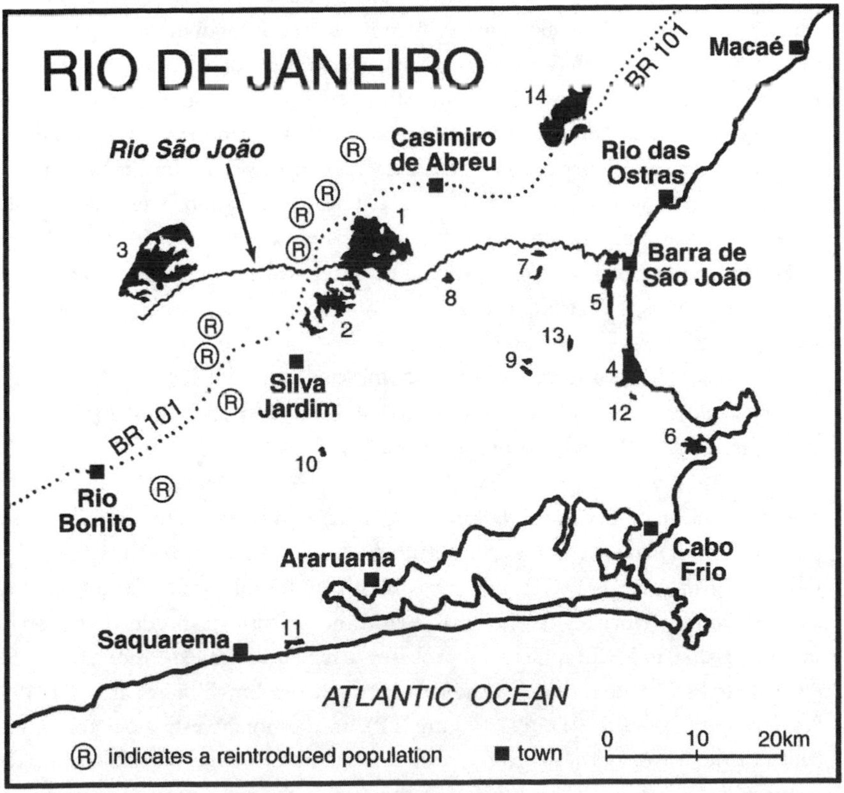

RIO DE JANEIRO
BR 101
Macaé
14
Rio São João
Casimiro de Abreu
Rio das Ostras
1
3
7
Barra de São João
8
5
2
13
9
4
Silva Jardim
12
BR 101
6
10
Rio Bonito
Cabo Frio
Araruama
11
Saquarema
ATLANTIC OCEAN
Ⓡ indicates a reintroduced population
■ town
0 10 20km

B

L. rosalia in Rio de Janeiro (Serra da Inoã, the forests of São João, and the surrounding Ponta Negra and Gurupina, municipality of Maricá) and also accurately delimited the distribution of *L. chrysomelas* in Bahia, but he made no mention at all of golden lion tamarins in Espírito Santo, where his interest was taken particularly by *Callithrix geoffroyi* in the Rio Doce valley (Coimbra-Filho 1969; Hershkovitz 1987). At first Coimbra-Filho (1969) considered *L. rosalia* extinct in Espírito Santo, but he later concluded that there was insufficient evidence to indicate that it had ever occurred there (A. F. Coimbra-Filho, personal communication, in Kierulff 1993a; Mittermeier et al. 1982).

From his surveys carried out between 1962 and 1969, Coimbra-Filho concluded that *L. rosalia* was extinct in all but 7 (Silva Jardim, Cabo Frio, Saquarema, Araruama, Casimiro de Abreu, Rio Bonito, and São Pedro da Aldeia) of the 24 municipalities of its original range. The exhaustive survey by Kierulff (1993a; Kierulff and Procópio de Oliveira 1996) that covered the entire range of the species during 18 months between 1990 and 1992 showed that the golden lion tamarins remain in only 104.5 km^2 of the forests in three regions: (1) near the coast (the Centro Hípico de Cabo Frio, with an estimated 29 individuals, and Campos Novos, with an estimated 36 individuals), (2) in the Poço das Antas Biological Reserve and adjacent forests (with an estimated 360 individuals), and (3) on the hillsides of the Serra do Mar (with an estimated 74 individuals), here largely restricted to lowland forest patches (Table 2.1). A further nine localities contained 12 isolated groups, totaling 60 individuals. These populations were registered in just four of the municipalities reported by Coimbra-Filho (1969): Silva Jardim, Cabo Frio, Saquarema, and Araruama. The latter two, however, maintained only a single group each (Kierulff 1993a).

Leontopithecus chrysomelas (Kuhl, 1820)

Type locality: Ribeirão das Minhocas, left bank of upper Rio dos Ilhéus, southern Bahia, Brazil (see Hershkovitz 1977).

The distribution of *L. chrysomelas* originally extended between the Rio de Contas (northern limit) and the Rio Pardo (southern limit) in southern Bahia (Coimbra-Filho and Mittermeier 1977), but it has also been found south of the Rio Pardo along its middle reaches to the Rio Jequitinhonha on the border between the states of Bahia and Minas Gerais—probably a recent range extension due to forest destruction and the silting of the Rio Pardo (Coimbra-Filho et al. 1991/1992; Rylands et al. 1988, 1991/1992; Figure 2.2). In the northwest, it occurs on both banks of the lower Rio Gongojí, a southern tributary of the Rio de Contas, but along its middle reaches it is limited to the east of the river and to the east of the

Table 2.1
Summary of the Status of the Lion Tamarins in Captivity and in the Wild

	GLT	GHLT	BLT	BFLT
Captive Populations				
Size	489 (stable)	598 (stabilizing)	112+ (growing)	None
Institutions	141	109	11	—
Founders	47 (3 alive)	130 alive	35 (6 alive)	—
Wild Population—*L. rosalia* (CR)				
Distribution	154 km^2 (actual occurrence)			
Total population size	±1,000			
Protected areas	Poço das Antas Biological Reserve (6,300 ha of which 3,500 ha is forested). Population ±230 individuals			
	União Biological Reserve (3,200 ha of which 2,400 ha is forested). Population ±120 individuals			
Reintroduced population	21 ranches (3,100 ha). Population ±360			
Other areas	12 areas. Population ±300			
Wild Population—*L. chrysomelas* (EN)				
Distribution	19,000 km^2 (geographic range)			
Total population size	6,000–15,000			
Protected area	Una Biological Reserve (7,059 ha). Population ±450			
Other areas	Numerous			
Wild Population—*L. chrysopygus* (CR)				
Distribution	±444 km^2 (actual occurrence)			
Total population size	±990			
Protected areas	Morro do Diabo State Park (34,441 ha of which 23,800 ha is forest). Population ±820			
	Caetetus State Ecological Station (2,178 ha). Population ±40			
Other areas	7 privately owned ranches (7,800 ha). Population ±130			
Wild Population—*L. caissara* (CR)				
Distribution	±300 km^2 (geographic range)			
Total population size	±260			
Protected area	Superagüi National Park (21,400 ha). Population ±160			
Other areas	Uncounted			

Sources: Captive populations, GLT—Ballou and Houle 2000 (data current 31 December 1999). GHLT—Leus and De Vleeschouwer 2001 (data current 31 December 1999). BLT—C. Valladares-Padua, studbook keeper, in litt. 5 June 2001 (data current 31 December 1999). *Wild populations,* GLT— Franklin and Dietz 2001; Kierulff 1993a, 1993b, 2000; Chapter 12 this volume. GHLT—Pinto 1994; Pinto and Rylands 1997; Pinto and Tavares 1994. BLT—Valladares-Padua and Cullen 1994; Chapter 14. BFLT—Lorini and Persson 1994a, 1994b; Prado 1999.

Note: GLT = *L. rosalia* (golden lion tamarins), GHLT = *L. chrysomelas* (golden-headed lion tamarins), BLT = *L. chrysopygus* (black lion tamarins), and BFLT = *L. caissara* (black-faced lion tamarins). EN = endangered, and CR = critically endangered (IUCN 1994).

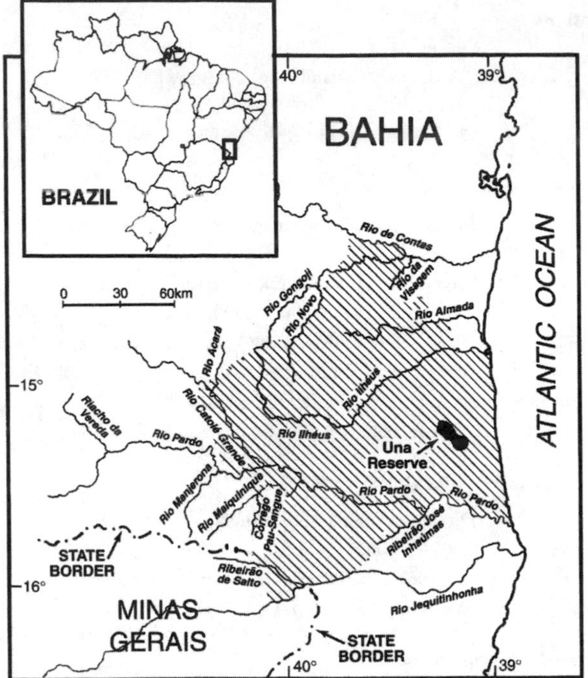

Figure 2.2. The distribution of *Leontopithecus chrysomelas* in southern Bahia, Brazil (based on Coimbra-Filho and Mittermeier 1977; Pinto 1994; Pinto and Rylands 1997; Rylands et al. 1991/1992). (Map by Stephen D. Nash/ Conservation International)

Rio Novo. It crosses the Rio Gongojí, westward, again at its headwaters, and occurs in the basin of the Rio Catolé Grande, a northern tributary of the Rio Pardo, which forms the westernmost extent of its range (Pinto and Rylands 1997). The western limits are defined by vegetational changes (mesophytic forest changing to liana forest in the west of its range) associated with an increase in altitude approaching the plateau of Vitoria da Conquista. The westernmost point is about 150 km from the coast. To the south, *L. chrysomelas* crosses the Rio Pardo and occurs in the lower basin of the Rio Maiquinique and east of the Córrego Pau-Sangue, south to the Rio Jequitinhonha in extreme northeast Minas Gerais (Pinto and Rylands 1997). The range of the golden-headed lion tamarin extends over approximately 19,000 km² (Table 2.1).

There are two lacunae in the range, one in the north near the coast, south of the lower Rio de Contas to the mouth of the Rio Ilhéus, and the other between the lower reaches of the Rio Pardo and the Rio Jequitinhonha (Pinto and Rylands 1997; Pinto and Tavares 1994). Coimbra-Filho and Mittermeier (1973, 1977) argued that the original range of *L. chrysomelas* extended south only to the Rio Pardo and that deforestation and the silting up of the river (see Coimbra-Filho and Câmara 1996) has resulted in lion tamarins crossing it along its middle reaches in recent times (the last century), thus explaining the golden-headed lion tamarin's presence there today and its absence between the Jequitinhonha and Pardo downriver, farther east. There is no obvious explanation for the absence of *L. chrysomelas* between the Rio de Contas and Rio Ilhéus to the northeast of its range.

Leontopithecus chrysopygus (Mikan, 1823)

Type locality: Ipanema (= Varnhagem or Bacaetava, near Sorocaba), São Paulo, Brazil (see Coimbra-Filho and Mittermeier 1973; Hershkovitz 1977).

This species formerly occurred along the north (right) margin of the Rio Paranapanema, west as far as the Rio Paraná, and between the upper Rio Paranapanema and Rio Tietê in the state of São Paulo (Coimbra-Filho 1976a, 1976b; Hershkovitz 1977; Figure 2.3). Today, it is known only from nine widely separated forest patches covering approximately 444 km^2. *Leontopithecus chrysopygus* occurs in two state-protected areas, the Morro do Diabo (Devil's Hill) State Park, municipality of Teodoro Sampaio, and the Caetetus State Ecological Station, municipality of Gália (Coimbra-Filho 1970a, 1970b, 1976a, 1976b; Coimbra-Filho and Mittermeier 1973, 1977; Valladares-Padua et al. 1994; Chapter 14 this volume). All surviving populations are in the central to western part of its former range and concentrated in the region called the Pontal de Paranapanema, except for Buri, municipality of Buri, where a tiny population was discovered in the eastern part, about 200 km east of the Caetetus State Ecological Station, by Valladares-Padua et al. (2000a).

Leontopithecus caissara Lorini and Persson, 1990

Type locality: Barra do Ararapira, Ilha de Superagüi, municipality of Guaraqueçaba, Paraná, Brazil, (25°18′ S, 48°11′ W) (Lorini and Persson 1990).

The black-faced lion tamarin occupies the southernmost limits of the distribution of the callitrichids (Figure 2.4). The type locality is on the northeastern part the island of Superagüi, on the coast of the state of Paraná. Other groups have been found elsewhere on the island, except in the extreme north and some higher ele-

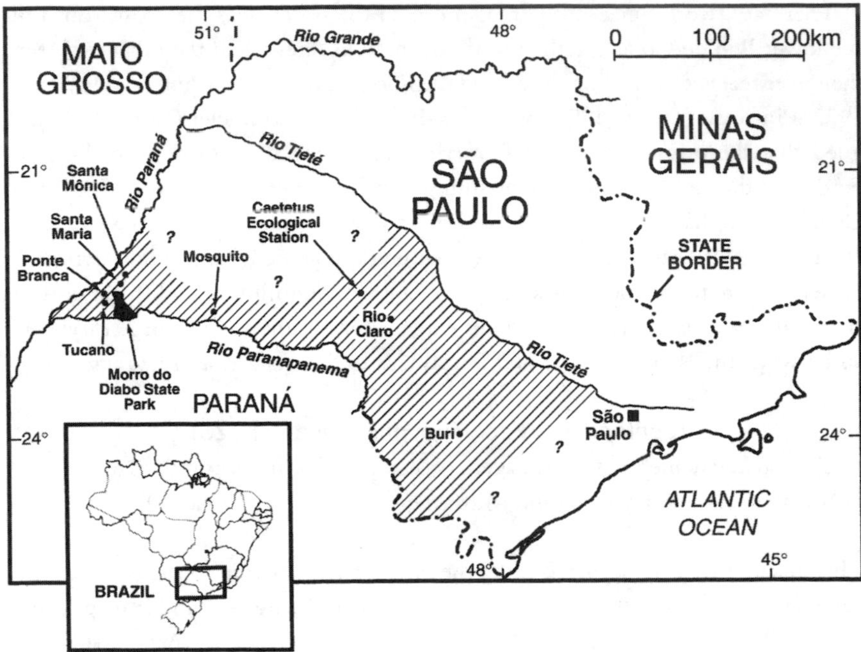

Figure 2.3. The original (shaded) and current (nine locations) distribution of *Leontopithecus chrysopygus* in the state of São Paulo, Brazil (based on Coimbra-Filho 1976b; Valladares-Padua and Cullen 1994; Valladares-Padua et al. 2000a). (Map by Stephen D. Nash/Conservation International)

vations in the southwest (Persson and Lorini 1991, 1993). Persson and Lorini found *L. caissara* on the mainland, in parts of the valleys of the Rio Sebuí and the Rio dos Patos, and limited in the north by the Rio Varadorzinho and in the west by the Serra da Utinga, Morro do Bico Torto, Morro do Poruquara, and Serra do Gigante. Persson and Lorini (Lorini and Persson 1994b; Persson and Lorini 1991, 1993) estimated that its entire range is less than 300 km^2. Four groups have been found to the north, also on the coast, in the municipality of Cananéia in the state of São Paulo (Persson and Lorini 1993). Martuscelli and Rodrigues (1992) reported four localities in the extreme southeast of São Paulo, two in the basin of the Rio do Turvo (Rio do Turvo and Morro do Teixeira, localities 1 and 3, map p. 922) and two farther north in the region of Itapitangui (localities 13 and 14, map p. 922), opposite the Ilha Cananéia. As a result of interviews of local people, Martuscelli and Rodrigues (1992) also indicated that *L. caissara* may occur farther

Figure 2.4. The current distribution of *Leontopithecus caissara* in the states of Paraná and São Paulo, Brazil (based on Lorini and Persson 1994a, 1994b; Martuscelli and Rodrigues 1992; Valladares-Padua et al. 2000a). (Map by Stephen D. Nash/ Conservation International)

inland, at two localities: the Rio Taquari (locality 11, map p. 922) and the Rio Ipiranguinha (locality 12, map p. 922). The latter may refer to Jacupiranga State Park (100,000 ha, although a large part of it is no longer forested), but neither of these localities has been confirmed. Field surveys by Valladares-Padua et al. (2000a) in the municipalities of Jacupiranga and Pariqueraçu failed to obtain any evidence of the existence of *L. caissara*. They were able, however, to confirm its presence between the villages of Ariri and Taquari, in the municipality of Cananéia, as had been reported by Martuscelli and Rodrigues (1992), and Valladares-Padua et al. (2000a) suggested that its range may extend only a short distance north.

STATUS IN THE WILD

Leontopithecus rosalia

Deforestation in the state of Rio de Janeiro began in the sixteenth century, with successive cycles of development supporting sugar cane plantations, coffee plantations, and, in the last century particularly, cattle breeding. Persistent logging, charcoal production, and clearing for urbanization have also contributed to deforestation (Dean 1995). The state is one of the most populous regions of Brazil, and today *L. rosalia* is limited to some few isolated forest patches. Approximately 20 percent of the original range of *L. rosalia* is still forested, but 60 percent of this total is made up of patches of 1,000 ha or less, 96 percent of which are less than 100 ha. The average size of the forest patches is 35 ha—smaller than the home range of a single lion tamarin group (Kierulff and Procópio de Oliveira 1996).

Early estimates of population size ranged from 200 to 600 (Coimbra-Filho 1969; Coimbra-Filho and Mittermeier 1973, 1977), but it was not until 1991 and 1992 that a full and thorough census was carried out by Kierulff (1993a, 1993b; Kierulff and Procópio de Oliveira 1996), using playback techniques developed by Kierulff and Kleiman (Kierulff et al. 1997). Not including the population in the Poço das Antas Biological Reserve, the estimated number of golden lion tamarins was 272 in 55 groups. They were divided among 14 forests—four main subpopulations with 6 or more groups each and 12 groups isolated in 10 forest fragments, each of 200 ha or less in area (Table 2.1). The total area of forest containing golden lion tamarins was 104.5 km^2. The majority of groups (29) were located in the municipality of Silva Jardim (53 percent), 24 groups were located in Cabo Frio (43 percent), one group each was found in the municipalities of Saquarema and Araruama. At the time, the Poço das Antas Biological Reserve was known to harbor about 290 lion tamarins (Kierulff 1993a), so there was a total population of 562 (or a range 470–631), close to the estimate of Coimbra-Filho in 1969.

During Kierulff's (1993a, 1993b) census, the population of reintroduced lion tamarins was 118, but by December 2000, the number had risen to 359 (Chapter 12 this volume). A translocation program, begun in 1994, established a new and currently thriving population in the União Biological Reserve, with six introduced groups resulting in a population of over 120 lion tamarins in the reserve by mid-2001 (Chapter 12 this volume). The forests targeted for reintroduction are now believed to be at carrying capacity, and Kierulff and Procópio de Oliveira (1996) estimated that the União Biological Reserve can hold no more than about 158 lion tamarins (33 groups). At 2,400 ha the União Biological Reserve has the second largest single block of lowland forest in the state, the first being the marginally larger 3,500 ha of forest in the 6,300 ha Poço das Antas Biological Reserve.

Beginning in 1997, golden lion tamarin numbers in the Poço das Antas Biological Reserve declined due to predation. An estimate of only 220 golden lion tamarins was recorded in this reserve in December 2000 (see Franklin and Dietz 2001). The current estimate of *L. rosalia* in the wild is now believed to have reached 1,000, mainly due to the growth of the reintroduced and translocated populations. However, there are very few forests available for further expansion of the population. Kierulff and Procópio de Oliveira (1996) identified a further four areas (two forest blocks, Rio Vermelho with 9 km^2 and the Morro de São João with 16 km^2, and two areas of fragmented forests, one of 16 km^2 along the BR101 at Gaviões and the other of 12 km^2 bordering the municipalities of Casimiro de Abreu and Silva Jardim) totaling 7,500 ha, which could hold a further 500 lion tamarins (about 100 groups). Occupying the entire habitat available, the golden lion tamarin population as a whole would still remain below the minimum viable of 2,000 recommended by Ballou et al. (1998) and Seal et al. (1990). Metapopulation management and reforestation are key strategies for the continued survival of this species in the wild (Kierulff 1993a; Kierulff and Procópio de Oliveira 1996). The Associação Mico-Leão-Dourado (AMLD—Golden Lion Tamarin Association), together with private landowners, Instituto Brasileiro do Meio Ambiente e dos Recursos Naturais Renováveis (IBAMA—Brazilian Institute of the Environment and Renewable Natural Resources) staff, and the Jardim Botânico do Rio de Janeiro (Rio de Janeiro Botanical Garden), is implementing strategies for habitat restoration and the development of corridors between fragments (Chapter 3 this volume).

Leontopithecus chrysomelas

There are records of over 100 localities where *L. chrysomelas* still occurs through the region bounded by the Rio de Contas in the north and the Rio Jequitinhonha in the south. More populations of *L. chrysomelas* remain than of all the other three species combined. The total wild population is estimated at 6,000 to 15,000 (Pinto 1994; Pinto and Rylands 1997) over approximately 19,000 km^2 (Table 2.1). The remaining forests are being destroyed, however, at an unprecedented rate for the region, and the surviving golden-headed lion tamarin populations are seriously depleted and fragmented (Figure 2.5). An important aspect that has contributed to the more favorable situation of the golden-headed lion tamarins is the traditional and fairly widespread use of the *cabruca* system for shading cacao trees. Some of the original canopy trees are left standing, and this allows for connectivity between forest patches. If well managed, this could be an important management tool for future conservation efforts, although the use of the *cabruca* system is declining.

The Una Biological Reserve, created to protect *L. chrysomelas,* has an estimated population of 450 animals, not large enough to be considered viable. A key strategy that has guided conservation efforts over the last decade has been to promote the preservation of the forests adjacent to the reserve (Alger and Araújo 1996; Alger et al. 1996; Blanes and Mallinson 1997; Santos and Blanes 1997, 1999). Threats to golden-headed lion tamarins come from socioeconomic transformations resulting from the expansion of alternative crops, notably African palm oil and coconuts, after the decline of the cocoa industry due to low prices and disease epidemics. The cocoa industry had dominated the region over the last 15 years (Alger and Caldas 1994). In the west of the range of *L. chrysomelas,* the forest has been increasingly destroyed and fragmented as a result of cattle ranching (Pinto 1994; Pinto and Rylands 1997).

Leontopithecus chrysopygus

Leontopithecus chrysopygus is now known to survive in nine localities in the state of São Paulo. In Morro do Diabo State Park, which contains 23,800 ha of forest, the population is estimated at approximately 820 (Valladares-Padua and Cullen 1994). The Caetetus State Ecological Station contains about 2,000 ha of forest and a *L. chrysopygus* population estimated at 40, but the other four major localities, consisting of fragments of between 400 and 800 ha, together harbor only about 130 individuals. The total population is estimated at about 1,000, spread through nine isolated forests, eight of which are certainly too small to be viable in the mid- to long term (Table 2.1). As with *L. rosalia,* the key threat for this species is the isolation and small size of the existing populations. This is being addressed by a metapopulation management program that includes the captive population. Current efforts focus on maintaining the genetic health of these populations through translocation, managed dispersal, and reintroduction (Chapter 14 this volume), environmental education, the preservation of remaining forest fragments with and without lion tamarins (Valladares-Padua et al. 1997), and the creation of corridors to link forest patches to establish larger areas of continuous forest (Padua and Valladares-Padua 1997; Valladares-Padua et al. 2000b)

Leontopithecus caissara

With a very restricted distribution and few individuals known to exist, this species is extremely endangered (Lorini and Persson 1990) and probably the rarest and most endangered of all the callitrichids, despite the fact that part of the island of Superagüi, along with the Ilha das Peças, was decreed a national park (without

Figure 2.5. Continued deforestation in the vicinity of the Una Biological Reserve is still a critical problem for the survival of the golden-headed lion tamarin. (Photo by Anthony B. Rylands)

knowledge of the existence of the lion tamarins) of 21,400 ha in 1989. Lorini and Persson (1994b) estimated a population not exceeding 260 animals (Table 2.1), divided into three subpopulations: that on the island of Superagüi and two on the adjacent mainland. The black-faced lion tamarins were found in the valleys of the Rio Patos and Rio Branco (estimated at 35 individuals) and the valleys of the Rio Varadouro and Rio Araçauba (estimated at 100 individuals). The northern limits to the range of *L. caissara* identified by Lorini and Persson (1990) were extended north into the state of São Paulo to the Serra do Cordeiro through the surveys of Martuscelli and Rodrigues (1992), but many localities were based on reports from interviews that have yet to be confirmed (Valladares-Padua et al. 2000a). The northernmost confirmed localities to date are those in the region Ariri, municipality of Cananéia (Rodrigues 1998), and Valladares-Padua et al. (2000a) indicated that under any circumstances the populations there are extremely scarce. Both Martuscelli and Rodrigues (1992) and Valladares-Padua et al. (2000a) had difficulty using interviews as a guide to surveying the region. People knew little or confused the species, felt intimidated, and in some cases were hostile. The threats to, and conservation strategies for, surviving *L. caissara* populations have been discussed by Câmara (1993, 1994) and Vivekananda (1994). The main threats come from forest destruction and degradation due to agriculture, squatters, hunting and other extraction of natural resources, especially for palm hearts, and, most seriously, burgeoning human occupation through land speculation and tourism.

STATUS IN CAPTIVITY

Three of the four lion tamarin species are maintained in captivity (Table 2.1; see Chapter 4 this volume). Captive populations of *L. rosalia* and *L. chrysomelas* are maintained at around 500, with *L. chrysomelas* having the genetically healthiest population in terms of the greatest number of founders. The reintroduction of captive-born golden lion tamarins has contributed significantly not only to the numbers living in the wild but also to the protection of 3,100 ha of forests within their range (Beck et al. 1986a, 1991, 1994; Beck and Martins 2001; Chapter 12 this volume). Although not currently contributing to the conservation of golden-headed lion tamarins in the wild, the captive population of *L. chrysomelas,* which arose due to illegal commerce (Konstant 1986; Mallinson 1984), is an important genetic reservoir: a guarantee of retaining a portion of the genetic variation in this species while the wild populations continue to decline with ongoing deforestation. The captive population of *L. chrysopygus* is growing and, despite having few founders, is now also contributing significantly to the metapopulation management program currently being executed by Valladares-Padua and his team (Medici 2001; Valladares-Padua and Ballou 1998; Valladares-Padua and Martins 1996; Valladares-Padua et al. 2000a; Chapter 14 this volume). The first translocation of a wild *L. chrysopygus* group was carried out in 1995, and the first experimental reintroduction was carried out in July 1999 (Valladares-Padua et al. 2001; Chapter 14 this volume).

OVERVIEW OF STATUS OF *LEONTOPITHECUS*

The distribution of *Leontopithecus* species is the most highly fragmented of any of the Atlantic Forest primates. This genus is restricted to lowland forests, and the possible explanation for the lack of lion tamarins in Espírito Santo may lie in their isolation in refugia during the Pleistocene epoch (Kinzey 1982; Rylands et al. 1996). Their predilection for lowland forests may have been the key to their isolation, in a scenario where 18,000 years ago large expanses were lost and much of the remaining forests may have been restricted to the upland areas of the Serra do Mar. Populations of *L. chrysopygus* and *L. caissara* were evidently isolated in the inland plateau of São Paulo and the small area of coastal lowland forest in southern São Paulo and northern Paraná. Natori (1989), studying dental and cranial characters, found *L. rosalia* and *L. chrysopygus* to be more closely related to each other than either is to *L. chrysomelas*. From this it could be argued that the Bahia refuge was separated first and that refuges in the western part of the state of São Paulo and in Rio de Janeiro were more recent (Natori 1989).

Massive and widespread forest destruction in their diminutive geographic ranges is the overriding cause for the endangered status of the four lion tamarins: illegal commerce and hunting, more serious and widespread in the past, are significant but secondary causes. The key problem being tackled today to guarantee the survival of the lion tamarins is the maintenance of viable populations in the fragments where they remain, particularly in the few protected areas where they occur (Table 2.1). Conservation measures for the majority of species and subspecies of Atlantic Forest primates are limited to parks and reserves, and these are the mainstay for the hoped for survival of the lion tamarins: the Una Biological Reserve (*L. chrysomelas*), the biological reserves of Poço das Antas and União (*L. rosalia*), the Morro do Diabo State Park and Caetetus State Ecological Station (*L. chrysopygus*), and the Superagüi National Park (*L. caissara*). Most of the remaining populations in the wild are on private land, and their survival is dependent on stopping the regional sociopolitical and economic forces promoting the destruction and degradation of their forests and the creation and management of private protected areas. These goals are becoming increasingly emphasized in the conservation programs for *L. chrysomelas, L. chrysopygus,* and *L. rosalia* (Ballou et al. 1998). For the latter two species, securing all surviving populations is vital for the metapopulation management strategies to guarantee their genetic health and survival in the long term. A key strategy involves promoting the establishment of Reservas Particulares de Patrimônio Natural (RPPNs—Private Natural Heritage Reserves), private lands that are registered with the federal authorities for protection in perpetuity. In Rio de Janeiro, 10 RPPNs have been created (as of October 2001) that cover 1,581 ha of forest to protect golden lion tamarins, some a direct result of the reintroduction and education programs (D. Rambaldi, personal communication). For *L. chrysomelas,* the challenge is to secure forests in the face of ongoing felling for timber and agriculture before the number of populations and their isolation approximate the more precarious situations of *L. rosalia* and *L. chrysopygus* (Santos and Blanes 1999). Besides the Una Biological Reserve, there are now nine RPPNs covering 972 ha already established and four more being created covering 851 ha, all within the geographic range of the golden-headed lion tamarin (G. dos Santos, personal communication). Possibilities still exist in southern Bahia for further expansion of this private reserve network

(Saatchi et al. 2001). Serious problems in the Superagüi National Park identified by Câmara (1993) include urbanization and tourism development along the coast of the island (not then included in the park), cattle farming, resource extraction activities (especially for the palm heart), highway construction, and illegal occupation of land and the accompanying forest destruction. In 1997, the limits to the park were changed, increasing its size, including the coastal area formerly excluded and covering other areas where *L. caissara* still occurs (Capobianco 1998), and over recent years considerable success has been achieved in upgrading and improving the Superagüi National Park. The key challenge for this, the most endangered of the four species, is currently to understand better the distribution and status of the populations on the mainland.

APPENDIX
UNVERIFIED GAZETTEERS FOR THE LION TAMARINS

LEONTOPITHECUS ROSALIA

Type locality: Brazil, restricted to the coast between 22° and 23° S, from the Cabo de São Tomé to the municipality of Mangaratiba by Wied-Neuwied (1826), further restricted to right bank, Rio São João, Rio de Janeiro, by Carvalho (1965) (see Coimbra-Filho and Mittermeier 1973; Hershkovitz 1977).

Aldeia Velha, Rio, southern part, Rio de Janeiro (22°47′ S, 42°55′ W). Sighting (Coimbra-Filho 1969, p. 39; Hershkovitz 1977, p. 941, locality 336).

Araruama, municipality of, Rio de Janeiro (22°53′ S, 42°20′ W). Formerly present (Coimbra-Filho 1969, p. 38; Hershkovitz 1977, p. 941, locality 336).

Barro Branco, municipality of Magé, Rio de Janeiro (22°23′ S, 44°30′ W). Formerly present. Near to Raiz da Serra, specimen in the Museu Nacional, Rio de Janeiro (Coimbra-Filho 1969, p. 38; Hershkovitz 1977, p. 941, locality 336).

Cabo Frio, municipality of, Rio de Janeiro (22°53′ S, 42°01′ W). Bank of Lago Araruama (Wied-Neuwied 1815–1817), H. Burmeister (recorded 1854). Sighting by Coimbra-Filho (1969, p. 38; Hershkovitz 1977, p. 942, locality 338a).

Cachoeiras de Macacú, municipality of, southern part, Rio de Janeiro (22°28′ S, 42°39′ W). Formerly present (Coimbra-Filho 1969, p. 38; Hershkovitz 1977, p. 941, locality 336).

Campos, municipality of, Rio de Janeiro (Coimbra-Filho 1965, p. 38).

Campos Novos, municipality of Cabo Frio, Rio de Janeiro, Brazil (22°40′–43′ S, 42°00′–01′ W) (Kierulff 1993a, 1993b).

Casimiro de Abreu, municipality of, Rio de Janeiro (22°29′ S, 42°12′ W). Formerly present (Coimbra-Filho 1969, p. 38; Hershkovitz 1977, p. 941, locality 336).

Centro Hípico de Cabo Frio, municipality of Cabo Frio, area bordered by the Rio São João and the Rio Gargoá, Rio de Janeiro, Brazil (22°34′–38′ S, 42°00′–01′ W) (Kierulff 1993a, 1993b).

Conceição de Macabú, municipality of, Rio de Janeiro (22°04′ S, 41°52′ W). Formerly present (Coimbra-Filho 1969, p. 38; Hershkovitz 1977, p. 941, locality 336).

Correntezas, Rio de Janeiro (22°30′ S, 42°31′ W). Sighting (Coimbra-Filho 1969, p. 39; Hershkovitz 1977, p. 941, locality 336).

Duque de Caxias, municipality of, Rio de Janeiro (22°47′ S, 43°18′ W). Formerly present (Coimbra-Filho 1969, p. 38; Hershkovitz 1977, p. 941, locality 336).

Emerências, Serra das, municipality of Cabo Frio, Rio de Janeiro, Brazil (22°48′ S, 41°53′ W). Three groups in 3.4 km^2 of forest (Kierulff 1993a, 1993b).

Fazenda Cabista, municipality of Cabo Frio, Rio de Janeiro, Brazil. Two groups in two forest patches of 0.81 and 0.57 km^2, separated by pasture (22°42′ S, 42°08′ W and 22°41′ S, 42°08′ W) (Kierulff 1993a, 1993b).

Fazenda de Luiza Brunet, municipality of Cabo Frio, Rio de Janeiro, Brazil. One group of four individuals in a forest of 0.36 km^2 (Kierulff 1993a, 1993b).

Fazenda Otacílio Melo, municipality of Araruama, Rio de Janeiro, Brazil. One group in a forest of 0.33 km^2 (22°43′ S, 42°23′ W) (Kierulff 1993a, 1993b).

Fazenda Sobara, municipality of Cabo Frio, Rio de Janeiro, Brazil. A group in a forest of 0.92 km^2 (22°42′ S, 42°07′ W) (Kierulff 1993a, 1993b).

Fazenda Vermelho, Rio Alto Bacaxá, municipality of Rio Bonito (22°44′ S, 42°21′ W). Sighting prior to 1968 (Coimbra-Filho 1969, p. 39; Hershkovitz 1977, p. 941, locality 336).

Fazenda Vinhático (AGRISA) and Fazenda Alfa, municipality of Cabo Frio, Rio de Janeiro, Brazil. Two groups and one individual in gallery forest along the Rio São João, one of four individuals in a forest of 0.61 km^2 (22°34′ S, 42°04′ W), and a group of six individuals in a forest of 0.48 km^2 nearby (22°35′ S, 42°05′ W). A further group on the Fazenda Vinhático was found in a forest of 1.3 km^2 (22°36′ S, 42°06′ W) (Kierulff 1993a, 1993b).

Gaviões, Rio de Janeiro (22°34′ S, 42°33′ W). Sighting (Coimbra-Filho 1969, p. 39; Hershkovitz 1977, p. 941, locality 336).

Guapí, south of Lago Juturnaíba, Rio de Janeiro (22°51′ S, 42°06′ W). Sighting in September 1967 by Coimbra-Filho (1969, p. 39; Hershkovitz 1977, p. 941, locality 336).

Gurupina, Fazenda, Lagoa Gurupina, Rio de Janeiro (22°55′ S, 42°42′ W) (Hershkovitz 1977, p. 942, locality 337).

Hillsides of the Serra do Mar, municipality of Silva Jardim, Rio de Janeiro, Brazil (22°29′–33′ S, 42°29′–34′ W) (Kierulff 1993a, 1993b).

Inoã (Serra de), Rio de Janeiro (22°53′ S, 42°55′ W). Wied-Neuwied (1815–1817). (Hershkovitz 1977, p. 940, locality 334).

Itaboraí, municipality of, Rio de Janeiro (22°45′ S, 42°52′ W). Formerly present (Coimbra-Filho 1969, p. 38; Hershkovitz 1977, p. 941, locality 336).

Itaguaí, municipality of, Rio de Janeiro (22°52′ S, 43°47′ W). Present at least till 1945, sighted on the right bank of the Rio Itaguaí in 1942 by Coimbra-Filho (1969, p. 38; Hershkovitz 1977, p. 941, locality 336).

Jacarepiá, municipality of Saquarema, Rio de Janeiro, Brazil. One group in a forest of 0.97 km^2 (22°56′ S, 42°27′ W) (Kierulff 1993a, 1993b).

Juturnaíba, municipality of, Rio de Janeiro (22°38′ S, 42°18′ W). Formerly abundant (Coimbra-Filho 1969, p. 38; Hershkovitz 1977, p. 941, locality 336).

Macaé, municipality of, Rio de Janeiro (22°23′ S, 41°47′ W). Formerly present (Coimbra-Filho 1969, p. 38; Hershkovitz 1977, p. 941, locality 336).

Macaé, Serra de, Rio de Janeiro (22°23′ S, 42°06′ W). H. Burmeister (recorded 1854). (Hershkovitz 1977, p. 941, locality 336).

Magé, municipality of, Rio de Janeiro (22°39′ S, 43°02′ W). Formerly present (Coimbra-Filho 1969, p. 38; Hershkovitz 1977, p. 941, locality 336).

Mangaratiba, municipality of, Rio de Janeiro (22°57′ S, 44°02′ W). Formerly present (Coimbra-Filho 1969, p. 38; Hershkovitz 1977, p. 941, locality 336).

Maricá, municipality of, Rio de Janeiro (22°55′ S, 42°49′ W). Skins in Museu Nacional, Rio de Janeiro, collected by the Rockefeller Foundation (Coimbra-Filho 1969, p. 38). Wied-Neuwied (1826) (Hershkovitz 1977, p. 941, locality 336).

Nilópolis, municipality of, Rio de Janeiro (22°49′ S, 42°35′ W). Formerly present (Coimbra-Filho 1969, p. 38; Hershkovitz 1977, p. 941, locality 336).

Niterói, municipality of, Rio de Janeiro (22°53′ S, 43°07′ W). Formerly present (Coimbra-Filho 1969, p. 38; Hershkovitz 1977, p. 941, locality 336).

Nova Friburgo [*sic*], Rio de Janeiro (22°16′ S, 42°32′ W). (Hershkovitz 1977, p. 940, locality 335).

Nova Iguaçu, municipality of, Rio de Janeiro (22°45′ S, 43°27′ W). Formerly present (Coimbra-Filho 1969, p. 38; Hershkovitz 1977, p. 941, locality 336).

Parahyba do Sul, Rio de Janeiro (21°37′ S, 41°04′ W). Lidth de Jeyde; Schneider 1876 (Hershkovitz 1977, p. 940, locality 326).

Petrópolis (22°31′ S, 43°10′ W), Rio de Janeiro. Lowlands on road to Rio de Janeiro (Hershkovitz 1977, p. 940, locality 331a).

Poço d'Anta, Rio Iguape, municipality of Silva Jardim, Rio de Janeiro (22°35′ S, 42°17′ W). Sighting (Coimbra-Filho 1969, p. 39; Hershkovitz 1977, p. 941, locality 336).

Poço das Antas Biological Reserve, vicinity of, municipality of Silva Jardim, Rio de Janeiro, Brazil (22°34′–38′ S, 42°18′–23′ W) (Kierulff 1993a, 1993b).

Ponta Negra, Rio de Janeiro (22°58′ S, 42°42′ W). Wied-Neuwied 1815–1817 (Hershkovitz 1977, p. 942, locality 338b).

Rio Bonito, municipality of, Rio de Janeiro (22°43′ S, 42°37′ W). Formerly present (Coimbra-Filho 1969, p. 38; Hershkovitz 1977, p. 941, locality 336).

São Gonçalo, municipality of, Rio de Janeiro (22°51′ S, 43°04′ W). Formerly present (Coimbra-Filho 1969, p. 38; Hershkovitz 1977, p. 941, locality 336).

São João (Rio), Rio de Janeiro (22°36′ S, 41°59′ W). Sighting (Coimbra-Filho 1969, p. 39; Hershkovitz 1977, p. 941, locality 336).

São João (Rio), Rio de Janeiro (22°36′ S, 42°00′ W) (Hershkovitz 1977, p. 940, locality 332).

São João da Barra, municipality of, Rio de Janeiro (21°38′ S, 41°03′ W). Wied-Neuwied (1815–1817) (Coimbra-Filho 1969, p. 38; Hershkovitz 1977, p. 942, locality 336).

São João de Meriti, municipality of, Rio de Janeiro (22°48′ S, 43°22′ W). Formerly present (Coimbra-Filho 1969, p. 38; Hershkovitz 1977, p. 941, locality 336).

São Pedro d'Aldeia, municipality of, Rio de Janeiro (22°51′ S, 42°06′ W). Formerly present (Coimbra-Filho 1969, p. 38; Hershkovitz 1977, p. 941, locality 336).

Sapitiba (or Sepetiba), Rio de Janeiro (22°58′ S, 43°42′ W). Mata de Piaí, Johann Natterer and Pohl, April–May (?), August 1818 (Hershkovitz 1977, p. 940, locality 331b; Pelzeln 1883).

Saquarema, municipality of, Rio de Janeiro (22°56′ S, 42°30′ W). Formerly present (Coimbra-Filho 1969, p. 38; Hershkovitz 1977, p. 941, locality 336).

Silva Jardim, municipality of, Rio de Janeiro (22°39′ S, 42°23′ W). Exterminated by 1964. Sighting (Coimbra-Filho 1969, p. 38; Hershkovitz 1977, p. 941, locality 336).

Tijuca, Parque Nacional da (Guanabara) (22°58′ S, 43°17′ W) (Hershkovitz 1977, p. 940, locality 332).

LEONTOPITHECUS CHRYSOMELAS

Type locality: Ribeirão das Minhocas, Rio Ilhéus, Bahia (15°12′ S, 39°57′ W). Skins collected by Wied-Neuwied (1815–1817) (cited by Coimbra-Filho 1970a; Hershkovitz 1977).

Água Doce, 17 km northeast of Ibicuí, Ibicuí, Bahia (14°46′ S, 39°50′ W). Interview (Pinto 1994; Pinto and Rylands 1997).

Banco Central, vicinity of, Aureliano Leal, Bahia (14°30′ S, 39°23′ W). Interview (Rylands et al. 1991/1992).

Barragem do Funil road, right bank Rio de Contas, Fazenda São Jorge and the region of the Cachorro d'Água, Ubaitaba, Bahia (14°15′ S, 39°28′ W). Interview (Pinto 1994; Pinto and Rylands 1997).

Barro Vermelho, Rio Maruím, southeast border of the Una Biological Reserve, Una, Bahia (15°10′ S, 9°03′ W). Sighting (Rylands et al. 1991/1992).

Canavieiras, 12 km north of, BA-001 highway, swamp forest, Canavieiras, Bahia (15°35′ S, 38°59′ W). In captivity, caught locally (Pinto 1994; Pinto and Rylands 1997).

Canavieiras, Estação Experimental (CEPLAC), 16 km southeast of Una, Rio São Pedro, Canavieiras, Bahia (15°23′ S, 39°12′ W). Sighting (Pinto 1994; Pinto and Rylands 1997).

Canavieiras–Santa Luzia road, between Ponta Nova e Nova Betânia, Canavieiras, Bahia (15°28′ S, 39°15′ W). Sighting (Rylands et al. 1991/1992).

Coaraci, vicinity of, Serra da Palha, Bahia (14°39′ S, 39°33′ W). Interview (Pinto 1994; Pinto and Rylands 1997).

Córrego Angelim and Córrego Salinada, vicinity of, 19 km southeast of Potiraguá, Potiraguá, Bahia (15°43′ S, 39°45′ W). Interview (Pinto 1994; Pinto and Rylands 1997).

Córrego Pau-Sangue, junction with the Rio Maiquinique, Itarantim, Bahia (15°33′ S, 40°08′ W). Interview (Pinto 1994; Pinto and Rylands 1997).

Córrego Ribeirão Grande, junction with the Rio Catolé Pequeno, northeast of Itapetinga, Itambé, Bahia (15°11′ S, 40°21′ W). Interview (Pinto 1994; Pinto and Rylands 1997).

Djalma Bahia, Estação Experimental (CEPLAC), Una, Bahia (15°17′ S, 39°03′ W). Sighting (Pinto 1994; Pinto and Rylands 1997).

Esquina do Rocha, near Nova Palma, Gongojí, Bahia (14°18′ S, 39°35′ W). Interview (Pinto 1994; Pinto and Rylands 1997).

Fazenda Alegre, region of Barro Branco, southeast of the Rio do Meio, Itororó, Bahia (15°09′ S, 39°56′ W). Interview (Pinto 1994; Pinto and Rylands 1997).

Fazenda Alsácea (Uberlândia), 17 km north of Itarantim, right bank Rio Maiquinique, Itarantim, Bahia (15°30′ S, 40°04′ W). Sighting (Pinto 1994; Pinto and Rylands 1997).

Fazenda Anamam and Fazenda Faraó, 8 km south of Una, BA-001 highway (Ilhéus-Canavieiras), banks of the Rio São Pedro, Una, Bahia (15°20′ S, 39°03′ W). Interview (Pinto 1994; Pinto and Rylands 1997).

Fazenda Aporá, 11 km on the road to Água Doce, north of Santa Cruz da Vitória, Santa Cruz da Vitória, Bahia (14°48′ S, 39°50′ W). Skin (Pinto 1994; Pinto and Rylands 1997).

Fazenda Ássica, 15 km along the Una–São José road, Una, Bahia (15°14′ S, 39°11′ W). Interview (Pinto 1994; Pinto and Rylands 1997).

Fazenda Boa Sorte, near Tapirama, Gongojí, Bahia (14°14′ S, 39°39′ W). Sighting (Pinto 1994; Pinto and Rylands 1997).

Fazenda Boa Vista, left bank of Rio Jequitinhonha, Salto da Divisa, Minas Gerais (15°52′ S, 40°05′ W). Interview (Rylands et al. 1991/1992).

Fazenda Bolandeira, 10 km south of Una, BA-001 highway (Ilhéus-Canavieiras), Una, Bahia (15°21′ S, 39°00′ W). Sighting (Pinto 1994; Pinto and Rylands 1997).

Fazenda Borrachudo, Itabuna, Bahia (14°48′ S, 39°16′ W). Skin collected by Serviço do Estudo e Pesquisa sobre a Febre Amarela (SEPSFA 1945), cited by Hershkovitz (1977); specimen in the Field Museum of Natural History (FMNH), Chicago.

Fazenda Buenos Aires, Ribeirão dos Índios, between Ibicuí and Água Doce, Ibicuí, Bahia (14°48′ S, 39°54′ W). Interview (Pinto 1994; Pinto and Rylands 1997).

Fazenda Cabana da Serra, southern border of the Una Biological Reserve, Una, Bahia (15°13′ S, 39°04′ W). Interview (Pinto 1994; Pinto and Rylands 1997).

Fazenda Café sem Troco, 11 km on the road Santa Cruz da Vitória–Itajú do Colônia, Santa Cruz da Vitória, Bahia (15°03′ S, 39°48′ W). Interview (Pinto 1994; Pinto and Rylands 1997).

Fazenda Camponesa, road to the Rio Pardo ferry, south of Itapetinga, Itapetinga, Bahia (15°24′ S, 40°12′ W). Sighting (Pinto 1994; Pinto and Rylands 1997).

Fazenda Canal Torto, road to Água Doce north of Santa Cruz da Vitória, Santa Cruz da Vitória, Bahia (14°53′ S, 39°50′ W). Interview (Pinto 1994; Pinto and Rylands 1997).

Fazenda Cedrinho, 13 km southeast of Potiraguá, region of the Córrego Cedrinho and the Córrego Cedro, Potiraguá, Bahia (15°42′ S, 39°49′ W). Interview (Pinto 1994; Pinto and Rylands 1997).

Fazenda Consolo da Nega, Riacho Sapucaia, near Itapetinga, Bahia (15°13′ S, 40°11′ W). Interview (Pinto 1994; Pinto and Rylands 1997).

Fazenda Cotovelo, 14 km north of Canavieiras, Bahia (15°33′ S, 38°58′ W). Sighting (Pinto 1994; Pinto and Rylands 1997).

Fazenda de José Deodato Araújo, 14 km west of Una on the Una-Arataca road, Una, Bahia (15°17′ S, 39°12′ W). Sighting (Pinto 1994; Pinto and Rylands 1997).

Fazenda Dendhevea, 20 km east of Una, Una-Arataca road, Una, Bahia (15°14′ S, 39°13′ W). Sighting (Pinto 1994; Pinto and Rylands 1997).

Fazenda de Paulo Sérgio, 8 km southwest of Una, right bank of the Rio Aliança, Una, Bahia (15°18′ S, 39°08′ W). Sighting (Pinto 1994; Pinto and Rylands 1997).

Fazenda Diamantina, 5 km from Tapirama, Gongojí, Bahia (14°11′ S, 39°40′ W). Captive animal, locally caught (Pinto 1994; Pinto and Rylands 1997).

Fazenda Esplanada, Rio Lagoa do Carmo, northwest of Canavieiras, Bahia (15°36′ S, 39°01′ W). Interview (Pinto 1994; Pinto and Rylands 1997).

Fazenda Ferkau, near Olivença, Ilhéus, Bahia (14°58′ S, 39°02′ W). Interview (Pinto 1994; Pinto and Rylands 1997).

Fazenda Gebara, Esquina do Gongojí, Gongojí, Bahia (14°20′ S, 39°31′ W). Interview (Pinto 1994; Pinto and Rylands 1997).

Fazenda Graciosa, 25 km to the north of Ibicuí, Ibicuí-Ibitupã highway, Ibicuí, Bahia (14°41′ S, 39°53′ W). Interview (Pinto 1994; Pinto and Rylands 1997).

Fazenda Havaí, Serra Palmitar, left bank Rio Jequitinhonha, Itapebi, Bahia (15°53′ S, 39°41′ W). Interview (Pinto 1994; Pinto and Rylands 1997).

Fazenda HIASSU, near to the northeast of Palmares, Itapetinga, Bahia (15°20′ S, 39°48′ W). Interview (Pinto 1994; Pinto and Rylands 1997).

Fazenda Ipiranga, region of Acará after the Córrego da Pedra Mimosa, Caatiba, Bahia (15°00′ S, 40°16′ W). Interview (Pinto 1994; Pinto and Rylands 1997).

Fazenda Iracema, 7 km on the road Nova Canaã–Vila Icaraí, Nova Canaã, Bahia (14°51′ S, 40°10′ W). Interview (Pinto 1994; Pinto and Rylands 1997).

Fazenda Itajubá, Rio Piabanha, 16 km north of Itapetinga, Itambé, Bahia (15°06′ S, 40°13′ W). Interview (Pinto 1994; Pinto and Rylands 1997).

Fazenda Itapetinga, Serra do Felícimo, south of Itarantim, Bahia (15°48′ S, 40°09′ W). Interview (Pinto 1994; Pinto and Rylands 1997).

Fazenda Jaqueiral, southeast border of the Una Biological Reserve, Una, Bahia (15°10′ S, 39°03′ W). Interview (Pinto 1994; Pinto and Rylands 1997).

Fazenda Jueirana, eastern border of the Una Biological Reserve, Una, Bahia (15°12′ S, 39°09′ W). Interview (Pinto 1994; Pinto and Rylands 1997).

Fazenda Limeira, Sapucaieira, region of Rio Aguípe, Ilhéus, Bahia (15°03′ S, 39°04′ W). Sighting (Pinto 1994; Pinto and Rylands 1997).

Fazenda Limoeiro, 10 km along the road Nova Canaã–Itajaí, Nova Canaã, Bahia (14°53′ S, 40°08′ W). Sighting (Pinto 1994; Pinto and Rylands 1997).

Fazenda Mirassol, 6 km from Santa Cruz da Vitória on the BA-415 highway, Santa Cruz da Vitória, Bahia (14°56′ S, 39°47′ W). Interview (Pinto 1994; Pinto and Rylands 1997).

Fazenda Nova Aurora, 3 km southeast of Gurupá-Mirim, Potiraguá, Bahia (15°43′ S, 39°36′ W). Interview (Pinto 1994; Pinto and Rylands 1997).

Fazenda Nova Guaiaquil, near the Rio do Meio, Itororó, Bahia (15°08′ S, 39°57′ W). Sighting (Pinto 1994; Pinto and Rylands 1997).

Fazenda Palestina, 16 km east of Dário Meira on the BA-030 highway, Ibicuí, Bahia (14°26′ S, 39°46′ W). Interview (Pinto 1994; Pinto and Rylands 1997).

Fazenda Palmeira, Serra das Guaribas, left bank Rio Jequitinhonha, Itapebi, Bahia (15°57′ S, 39°38′ W). Sighting (Rylands et al. 1991/1992).

Fazenda Papuan, 19 km east of Dário Meira on the BA-030 highway, Ibicuí, Bahia (14°25′ S, 39°45′ W). Interview (Pinto 1994; Pinto and Rylands 1997).

Fazenda Piedade, eastern border of the Una Biological Reserve, Una, Bahia (15°11′ S, 39°12′ W). Sighting (Alves 1990).

Fazenda Pindorama, 10 km southwest of Una, right bank of the Rio Aliança, Una, Bahia (15°19′ S, 39°10′ W). Sighting (Pinto 1994; Pinto and Rylands 1997).

Fazenda Pirataquissé, Ilhéus, Bahia (14°50′ S, 39°05′ W). Skin collected by the Serviço do Estudo e Pesquisa sobre a Febre Amarela (SEPSFA 1944), cited by Hershkovitz (1977).

Fazenda Quatro de Julho, northern boundary of the Una Biological Reserve, Una, Bahia (15°08′ S, 39°11′ W). Interview (Pinto 1994; Pinto and Rylands 1997).

Fazenda Quixadá, 3 km to the south of BA-262 (Ilhéus-Uruçuca highway) on the road to Rio do Braço, Ilhéus, Bahia (14°41′ S, 39°14′ W). Interview (Pinto 1994; Pinto and Rylands 1997).

Fazenda Riacho Filó, region of Piancó, left bank Rio Gongojí, Gongojí, Bahia (14°18′ S, 39°41′ W). Sighting (Pinto 1994; Pinto and Rylands 1997).

Fazenda Rio Pardo, 36 km along the BA-670 highway (BA-415-Potiraguá), left bank Rio Pardo, Itororó, Bahia (15°27′ S, 39°55′ W). Interview (Pinto 1994; Pinto and Rylands 1997).

Fazenda Santa Clara, 9 km along the BA-270 highway (Canavieiras-Santa Luzia), Canavieiras, Bahia (15°34′ S, 39°04′ W). Sighting (Pinto 1994; Pinto and Rylands 1997).

Fazenda Santa Fé, 8 km east of Itororó on the BA-415 highway (Itabuna-Vitória da Conquista), Itororó, Bahia (15°06′ S, 40°00′ W). Interview (Pinto 1994; Pinto and Rylands 1997).

Fazenda Santa Inês, Córrego Cotinguiba, sul de Gurupá-Mirim, Itapebi, Bahia (15°46′ S, 39°40′ W). Sighting (Rylands et al. 1991/1992).

Fazenda Santa Inês and Fazenda Guanabara, Rio da Visagem, near Laje do Branco, Aureliano Leal, Bahia (14°25′ S, 39°26′ W). Interview (Pinto 1994; Pinto and Rylands 1997).

Fazenda Santa Júlia, 13 km north of Itapetinga, Rio do Ouro, Itambé, Bahia (15°09′ S, 40°13′ W). Sighting (Pinto 1994; Pinto and Rylands 1997).

Fazenda Santa Paula, 5 km south of Mascote, right bank Rio Pardo, Mascote, Bahia (15°37′ S, 39°18′ W). Interview (Pinto 1994; Pinto and Rylands 1997).

Fazenda Santa Rosa, 10 km south of Olivença, BA-001 highway (Ilhéus-Canavieiras), Ilhéus, Bahia (15°03′ S, 39°00′ W). Interview (Pinto 1994; Pinto and Rylands 1997).

Fazenda Santa Terezinha, 8 km east of Caatiba, banks of Rio São Bento, Caatiba, Bahia (14°57′ S, 40°20′ W). Interview (Pinto 1994; Pinto and Rylands 1997).

Fazenda Santa Terezinha, region of Barro Branco, southeast of the Rio do Meio, Itororó, Bahia (15°08′ S, 39°58′ W). Interview (Pinto 1994; Pinto and Rylands 1997).

Fazenda Santo Antônio, 12 km north of Una on the BA-001 highway (Ilhéus-Canavieiras), Una, Bahia (15°12′ S, 39°02′ W). Interview (Pinto 1994; Pinto and Rylands 1997).

Fazenda Santo Antônio, 6 km from the bridge over the Rio de Contas, road between Gongojí and Ubaitaba, Gongojí, Bahia (14°16′ S, 39°29′ W). Interview (Pinto 1994; Pinto and Rylands 1997).

Fazenda São Jorge, Ribeirão da Onça, leste de Itapetinga, Bahia (15°12′ S, 40°09′ W). Interview (Pinto 1994; Pinto and Rylands 1997).

Fazenda São José, 5 km along the BA-670 highway (BA-415-Potiraguá), Itapetinga, Bahia (15°13′ S, 40°04′ W). Interview (Pinto 1994; Pinto and Rylands 1997).

Fazenda São José, Rio do Braço, Ilhéus, Bahia. Skin collected by the Serviço do Estudo e Pesquisa sobre a Febre Amarela (SEPSFA 1944/1945), cited by Coimbra-Filho and Mittermeier (1973). Female juvenile. MNRJ8519-21.

Fazenda São Luis, Rio do Braço, Ilhéus, Bahia. Skin collected by the Serviço do Estudo e Pesquisa sobre a Febre Amarela (SEPSFA 1944/1945), cited by Coimbra-Filho and Mittermeier (1973). Female juvenile. MNRJ8522.

Fazenda Segredinho, 3 km north of Una on the BA-001 highway (Ilhéus-Canavieiras), Una, Bahia (15°15′ S, 39°03′ W). Sighting (Pinto 1994; Pinto and Rylands 1997).

Fazenda Sol Nascente, 7 km north of Una, Una, Bahia (15°15′ S, 39°03′ W). Interview (Pinto 1994; Pinto and Rylands 1997).

Fazenda Teimoso, southeast of Jussari, Jussari, Bahia (15°12′ S, 39°29′ W). Sighting (Lima 1990).

Fazenda Texana, 11 km along the BA-130 highway (Itapetinga-Macarani), Itapetinga, Bahia (15°21′ S, 40°16′ W). Interview (Pinto 1994; Pinto and Rylands 1997).

Fazenda UNACAU, northwest of the Una Biological Reserve, Una, Bahia (15°08′ S, 39°17′ W). Sighting (Alves 1990).

Floresta Azul, vicinity of, Bahia (14°52′ S, 39°39′ W). Interview (Rylands et al. 1991/1992)

Ilhéus, 7 km south of, BA-001 (Ilhéus-Canavieiras highway), Ilhéus, Bahia (14°51′ S, 39°02′ W). Interview (Oliver and Santos 1991).

Ilhéus-Canavieiras highway (BA-001), in coastal forest (*restinga*), Una, Bahia (15°12′ S, 39°01′ W). Sighting (Rylands et al. 1991/1992).

Itaibé, vicinity of, 8 km southwest of Teixeira do Progresso, Potiraguá, Bahia (15°45′ S, 39°32′ W). Interview (Pinto 1994; Pinto and Rylands 1997).

Jordânia–Salto da Divisa road, left bank Rio Jequitinhonha, Salto da Divisa, Minas Gerais (15°55′ S, 40°10′ W). Interview (Rylands et al. 1991/1992).

Jussari, vicinity of, Bahia (15°11′ S, 39°30′ W). Interview (Rylands et al.1991/1992).

Km 15, Una-Arataca road, Una, Bahia (15°15′ S, 39°12′ W). Interview (Pinto 1994; Pinto and Rylands 1997).

Lagoa do Mabaço, 13 km northeast of Una, near to the Rio Maruím, Una, Bahia (15°12′ S, 39°01′ W). Sighting (Rylands et al. 1991/1992).

Lagoinha, Morro Grande, Jordânia-Salto da Divisa road, Jordânia, Minas Gerais (15°58′ S, 40°08′ W). Interview (Pinto 1994; Pinto and Rylands 1997).

Lemos Maia, Estação Experimental (CEPLAC), Una, Bahia (15°17′ S, 39°05′ W). Sighting (Pinto 1994; Pinto and Rylands 1997; Rylands 1989b; Rylands et al. 1991/1992).

Oiteiro, 3 km west of, Una, Bahia (15°24′ S, 39°02′ W). Interview (Pinto 1994; Pinto and Rylands 1997).

Olivença, vicinity of, Ilhéus, Bahia (15°02′ S, 39°00′ W). Interview (Oliver and Santos 1991).

Outeiro-Una road, 2 km from the BA-001 highway, Una, Bahia (15°21′ S, 39°02′ W). Sighting (Rylands et al. 1991/1992).

Palmares, vicinity of, Itapetinga, Bahia (15°24′ S, 39°50′ W). Interview (Pinto 1994; Pinto and Rylands 1997).

Pardo, Rio, near the Rio ferry, south of Itapetinga, Itapetinga, Bahia (15°25′ S, 40°11′ W). Interview (Pinto 1994; Pinto and Rylands 1997).

Pimenteira, vicinity of, Ilhéus, Bahia (14°34′ S, 39°26′ W). Interview (Rylands et al. 1991/1992).

Piruna Agro Ltda. (Fazenda Pirelli), north of Colônia de Una, Una, Bahia (15°14′ S, 39°08′ W). Interview (Pinto 1994; Pinto and Rylands 1997).

Ponto do Astério, junction of highways BA-130 and BA-415, northeast of Firmino Alves, Bahia (14°57′ S, 39°54′ W). Interview (Pinto 1994; Pinto and Rylands 1997).

Poxim do Sul, 5 km west of, road to Sarampo, region of the Riacho Ribeira, Canavieiras, Bahia (15°30′ S, 39°00′ W). Interview (Pinto 1994; Pinto and Rylands 1997).

Riacho do José Antônio, right bank Rio Gongojí south of Iguaí, Nova Canaã, Bahia (14°53′ S, 40°05′ W). Interview (Pinto 1994; Pinto and Rylands 1997).

Riacho Ribeira, Poxim do Sul, Una, Bahia (15°28′ S, 39°01′ W). Sighting (Rylands et al. 1991/1992).

Ribeirão da Alegria, Serra da Bananeira, southwest of Itajú do Colônia, Itajú do Colônia, Bahia (15°14′ S, 39°44′ W). Interview (Pinto 1994; Pinto and Rylands 1997).

Ribeirão da Fortuna, Estação da Mata do Cacau and km 5 on the BA-251 highway (Buerarema-Ilhéus), Buerarema, Bahia (14°57′ S, 39°19′ W). Skins collected by the

Serviço do Estudo e Pesquisa sobre a Febre Amarela (SEPSFA 1944/1945), cited by Coimbra-Filho (1970a) and Hershkovitz (1977); specimens in the Museu Nacional, Rio de Janeiro. MNRJ-8518 and MNRJ-8523.

Ribeirão das Flores, Serra Tomba Morro northeast of Iguaí, Iguaí, Bahia (14°45′ S, 40°02′ W). Interview (Pinto 1994; Pinto and Rylands 1997).

Ribeirão das Inhaúmas, vicinity of, 8 km south of Mascote, right bank Rio Pardo, Canaviciras, Bahia (15°39′ S, 39°18′ W). Interview (Pinto 1994; Pinto and Rylands 1997).

Ribeirão do Café, vicinity of right bank Rio Novo, 15 km to the south of Ibitupã, Ibicuí, Bahia (14°39′ S, 39°52′ W). Interview (Pinto 1994; Pinto and Rylands 1997).

Ribeirão Mutuns, 5 km northeast of Itabuna, Itabuna, Bahia (14°44′ S, 39°18′ W). Interview (Pinto 1994; Pinto and Rylands 1997).

Ribeirão Vermelho, west of Oiteiro, limits southwest of the Fazenda Bolandeira, Una, Bahia (15°25′ S, 39°05′ W). Interview (Pinto 1994; Pinto and Rylands 1997).

Rio do Meio, vicinity of, Itororó, Bahia (15°04′ S, 39°59′ W). Interview (Pinto 1994; Pinto and Rylands 1997).

Rio do Ouro, vicinity of, southeast of Ibitupã, Ibicuí, Bahia (14°33′ S, 39°44′ W). Interview (Pinto 1994; Pinto and Rylands 1997).

Rio Gongojí, Gongojí, Bahia (14°12′ S, 39°38′ W). Specimens collected by O. M. O. Pinto (1932), cited in Coimbra-Filho (1970a), Hershkovitz (1977), and Vieira (1955).

Salobrinho, vicinity of, Ilhéus, Bahia (14°47′ S, 39°12′ W). Interview (Oliver and Santos 1991).

Santa Cruz da Vitória, vicinity of, Bahia (14°58′ S, 39°49′ W). Interview (Pinto 1994; Pinto and Rylands 1997).

São Roque, 8 km east of Almadina, Coaraci, Bahia (14°43′ S, 39°34′ W). Interview (Pinto 1994; Pinto and Rylands 1997).

Sapucaieira, region of Rio Aguípe, Ilhéus, Bahia (15°02′ S, 39°04′ W). Interview (Pinto 1994; Pinto and Rylands 1997).

Serra Pelada, 7 km south of Caatiba, Bahia (15°02′ S, 40°24′ W). Interview (Pinto 1994; Pinto and Rylands 1997).

Siqueiro Grande, east of Itajuípe, Itajuípe, Bahia (14°41′ S, 39°26′ W). Interview (Pinto 1994; Pinto and Rylands 1997).

Tapirama, 7 km east of Barra do Rocha, right bank Rio de Contas, Gongojí, Bahia (14°13′S, 39°40′ W). Interview (Pinto 1994; Pinto and Rylands 1997).

Una Biological Reserve (Rio Maruím), Una, Bahia (15°11′ S, 39°03′ W). Sighting (Dietz et al. 1994c; Rylands et al. 1991/1992).

Una-Colônia de Una road, Una, Bahia (15°17′ S, 39°06′ W). Sighting (Rylands et al. 1991/1992).

Una–Ribeirão da Serra road, in the direction of the Una Biological Reserve, Una, Bahia (15°16′ S, 39°06′ W). Sighting (Rylands et al. 1991/1992).

Una–Santa Luzia road, between Ribeirão Navalha and Ribeirão Pimenta, near to the

Fazenda Pindorama, Una, Bahia (15°20′ S, 39°09′ W). Sighting (Pinto 1994; Pinto and Rylands 1997).

LEONTOPITHECUS CHRYSOPYGUS

Type locality: Ipanema (or Varnhagem), near Sorocaba, São Paulo (23°26′ S, 47°36′ W), 600 m above sea level. Johann Natterer, March–September 1822 (Hershkovitz 1977, p. 942, locality 347) (see Coimbra-Filho and Mittermeier 1973).

Bauru, São Paulo (22°19′ S, 40°04′ W). Museu de Zoologia, São Paulo. O. Hume 1905 (Coimbra-Filho and Mittermeier 1973; Hershkovitz 1977, p. 942, locality 343b). Skin, female. Department of Zoology, University of São Paulo. DZSP2063.

Buri, municipality of Buri, São Paulo (23°45′ S, 48°45′ W). A riparian forest of 100 ha. Two groups seen (Valladares-Padua et al. 2000a).

Caetetus State Ecological Station, municipality of Gália, São Paulo (22°24′ S, 49°42′ W). Formerly Fazenda Paraiso (Coimbra-Filho 1976b; Valladares-Padua and Cullen 1994).

Fazenda Kitayama, Presidente Wenceslau, São Paulo (21°52′ S, 51°50′ W). (Coimbra-Filho and Mittermeier 1973; Hershkovitz 1977, p. 942, locality 339b).

Fazenda Mosquito, municipality of Narandiba, São Paulo (22°40′ S, 51°30′ W). A forest of 2000 ha (Valladares-Padua and Cullen 1994).

Fazenda Ponte Branca, municipality of Euclides da Cunha Paulista, São Paulo (22°24′ S, 52°30′ W). A forest of 1,195 ha (Valladares-Padua and Cullen 1994).

Fazenda Rio Claro, municipality of Lençóis Paulista, São Paulo (22°55′ S, 49°46′ W). A forest of 1600 ha (Valladares-Padua and Cullen 1994).

Fazenda Santa Maria I, municipality of Teodoro Sampaio, São Paulo (22°14′S, 52°50′ W). A forest of 441 ha (Valladares-Padua and Cullen 1994).

Fazenda Santa Mônica Farm, municipality of Marabá Paulista, São Paulo (22°08′ S, 52°16′ W). A forest of 584 ha (Valladares-Padua and Cullen 1994).

Fazendas Tucano and Rosanella, municipality of Euclides da Cunha Paulista, São Paulo (22°30′ S, 52°30′ W). A forest of 1,989 ha (Valladares-Padua and Cullen 1994).

Morro do Diabo State Park, municipality of Teodoro Sampaio, São Paulo (22°30′ S, 52°20′ W). Coimbra-Filho (1970b 1976b; Hershkovitz 1977, p. 942, locality 339b; Valladares-Padua and Cullen 1994).

Paranapanema, Rio, north bank, São Paulo (23°00′ S, 49°47′ W) (Hershkovitz 1977, p. 942, locality 342). Note by Hershkovitz (1977): "Numbered at confluence of Rio Itararé with which it forms boundary between the states of São Paulo (north) and Paraná."

Presidente Epitácio, São Paulo (21°46′ S, 52°06′ W) (Hershkovitz 1977, p. 942, locality 339b).

Vargem Grande, São Paulo (23°39′ S, 47° W). Johann Natterer, 1822 (Hershkovitz 1977, p. 942, locality 349).

Vitória, municipality of Botucatu, São Paulo (22°47′ S, 48°24′ W). E. Garbe, July 1902 (Coimbra-Filho and Mittermeier 1973; Hershkovitz 1977, p. 942, locality 344). Skins, male and female. Department of Zoology, University of São Paulo. DZSP21450-41.

LEONTOPITHECUS CAISSARA

Type locality: Barra do Ararapira, Ilha de Superagüi, municipality of Guaraqueçaba, Paraná, Brazil (25°18′ S, 48°11′ W). Museu Nacional, Rio de Janeiro, MN28861. 25 February 1990 (Lorini and Persson 1990).

Araçupeva, municipality of Cananéia, São Paulo. Skin. Centro de Pesquisas Aplicadas em Recursos Naturais da Ilha de Cardoso (CEPARNIC), Cananéia, São Paulo. P. Martuscelli (Persson and Lorini 1993).

Ariri/Taquiri road, approximately 10 km north of Ariri, municipality of Cananéia, state of São Paulo, Brazil (25°10′74″ S, 48°06′05″ W) (Valladares-Padua et al. 2000a).

Ariri/Taquiri road, approximately 10 km north of Ariri, municipality of Cananéia, state of São Paulo, Brazil (25°10′54″ S, 48°05′28″ W) (Valladares-Padua et al. 2000a).

Ariri/Taquiri road, approximately 10 km north of Ariri, municipality of Cananéia, state of São Paulo, Brazil (25°21′25″ S, 48°08′42″ W) (Valladares-Padua et al. 2000a).

Cananéia–Pariquera-açú highway, SP-226, municipality of Cananéia, at sea level (24°54′ S, 47°58′ W). Three individuals seen, 15 October 1991 (Martuscelli and Rodrigues 1992, interpreted as locality 13, listed as SP 226, map p. 922, coordinates not supplied by authors).

Morro do Itapitangui, municipality of Cananéia, São Paulo (24°53′ S, 47°55′ W). Sighting of a single individual 26 May 1991. Two hundred meters above sea level (Martuscelli and Rodrigues 1992, interpreted as locality 14, listed as Rio Iririaia-açu, map p. 922, coordinates not supplied by authors).

Morro do Teixeira, municipality of Cananéia, São Paulo (25°10′ S, 48°04′ W). Sighting of a group of three adults, 26 May 1991, 86 m above sea level (Martuscelli and Rodrigues 1992, locality 1, map p. 922, coordinates not supplied by authors).

Praia Deserta, Rio Pereque, Ilha de Superagüi, municipality of Guaraqueçaba, Paraná, Brazil (25°18′ S, 48°11′ W). Skin, Museu de História Natural "Capão de Imbuia," Curitiba, MNHCI 1681. Considered a topotype (Persson and Lorini 1993).

Rio do Turvo, municipality of Cananéia, São Paulo (25°10′ S, 48°05′ W). Skin of individual obtained 26 October 1990, killed about 3 years earlier (Martuscelli and Rodrigues 1992, locality 3, map p. 922, coordinates not supplied by authors).

DENISE MARÇAL RAMBALDI, DEVRA G. KLEIMAN,
JEREMY J. C. MALLINSON, LOU ANN DIETZ,
AND SUZANA M. PADUA

3

THE ROLE OF NONGOVERNMENTAL ORGANIZATIONS AND THE INTERNATIONAL COMMITTEE FOR THE CONSERVATION AND MANAGEMENT OF *LEONTOPITHECUS* IN LION TAMARIN CONSERVATION

In recent years, both Brazilian and international nongovernmental organizations (NGOs) concerned with the conservation of endangered species and the environment have played an increasingly important role in the conservation of natural resources in Brazil and abroad. The growth of environmental and conservation-oriented NGOs has paralleled a number of global political and economic changes, including the expansion in the numbers of democratically elected governments worldwide, the greater privatization of formerly government-controlled economic sectors, a rise in globalization and multinational corporations, an enormous increase in computing power, and new technologies that have made global communications rapid and inexpensive. Responsibilities that were formerly of national governments have increasingly been assumed by nonprofit organizations with a specific focus, especially in areas having to do with culture, the environment, and social services.

Nongovernmental organizations are especially important today in a number of arenas where government action is insufficient or does not exist, and some have been decisive in their contribution to the development of biodiversity conservation in Brazil. Nongovernmental organizations have become important advisory bodies and an essential link between government and local communities for the development and execution of public policies regarding the use and conservation of natural resources (Fonseca and Pinto 1996). Additionally, the Brazilian government has established several international committees to advise it on the management and conservation of endangered species (comparable, in part, to Endangered Species Recovery Teams in the United States). These committees are a model mechanism for incorporating the best multidisciplinary technical information available in Brazil and internationally into the development of government policy as well as forging the interinstitutional (government and nongovernment, international, national, and local) partnerships needed to address the complex issues affecting the long-term survival of endangered species and their habitat.

When compared with government agencies, NGOs benefit from their relative independence because they choose the projects they take on and the means to execute them. Funding is achieved independently or through networks and partnerships, and some NGOs are even able to supplement their budgets by exploiting the consumer market. Since they control the costs of their programs, they can maximize the programs' benefits. Nongovernmental organizations have increasingly demonstrated their importance as a means for civilians to implement alternatives to the unsustainable development models ultimately responsible for the destruction of so many natural ecosystems. Not least amongst these systems is the Atlantic Forest, globally outstanding for its biodiversity and local endemism yet one of the most threatened tropical biomes in the world (Mallinson 2000b; Myers et al. 2000). Among the endemic species in danger are the lion tamarins: *Leontopithecus caissara, L. chrysomelas, L. chrysopygus,* and *L. rosalia.* No more than 2 percent of the forests remain within the lion tamarins' original geographic ranges.

The conservation of these species in their natural habitats demands action with regard to (1) the recovery, protection, and expansion of native habitat, (2) surveys to identify, protect, and monitor all remaining populations, (3) the scientific management of captive and wild populations, (4) environmental education to inform local communities of the importance of biodiversity conservation and the sustainable use of natural resources and to promote the communities' involvement in local conservation, (5) training to build a cadre of professionals with proficiency in the myriad of disciplines needed for conservation management and the ability to work together to implement complex strategies, (6) influencing the formulation of public policy, (7) fundraising by public and private institutions at the national and international levels, and (8) dissemination of information and training in sustainable land use and management, compatible with biodiversity conservation.

In addition to the growing international NGO and Brazilian government contributions, a number of technical, scientifically oriented Brazilian NGOs have evolved in the last 15 years to help reach these long-term and wide-reaching objectives within the geographic range of the lion tamarins. Also, the International Committee for the Conservation and Management (ICCM) of *Leontopithecus* has been created to act in an advisory capacity to the Instituto Brasileiro do Meio Ambiente e dos Recursos Naturais Renováveis (IBAMA—Brazilian Institute of the Environment and Renewable Natural Resources), the Brazilian federal environmental agency responsible for issues concerning research, management, and conservation for the four species. This committee performs the crucial role of linking the efforts of the diverse institutions involved in lion tamarin conservation and guiding their actions toward achieving the long-term goal of conserving viable populations of the four species in their natural habitat.

This chapter discusses the evolution and functioning of the *Leontopithecus* ICCM and reviews the missions and contributions of six Brazilian national and four international NGOs that have been particularly involved in lion tamarin conservation. The growing collaboration among these and other government and NGOs both within Brazil and around the world is maturing into true partnerships and serves as an example to other biodiversity conservation initiatives.

THE INTERNATIONAL COMMITTEE FOR THE CONSERVATION AND MANAGEMENT OF *LEONTOPITHECUS*

In 1981, an International Research and Management Committee was elected for the global management of *L. rosalia,* then in captivity (see Kleiman 1982; Chapter 4 this volume); the committee was set up by the Smithsonian National Zoological Park (SNZP) in Washington, D.C., under the leadership of Devra Kleiman, then international studbook keeper for the species. In 1984, as part of the effort to confiscate and repatriate the *L. chrysomelas* exported illegally from Brazil to Europe and Japan (see Mallinson 1984, 1987a; Chapter 1 this volume), IBAMA (then IBDF—Instituto Brasileiro de Desenvolvimento Florestal, or Brazilian Forestry Development Institute) asked Jeremy Mallinson, director of the Jersey Zoo, and Adelmar F. Coimbra-Filho, then director of the Centro de Primatologia do Rio de Janeiro (CPRJ—Rio de Janeiro Primate Center), to establish an International Recovery and Management Committee (IRMC) (Mallinson 1984, 1987a). The *L. chrysomelas* IRMC was modeled after the IRMC for the management of the captive *L. rosalia* outside Brazil. The *L. chrysomelas* IRMC was formed primarily to deal with the restoration of the illegally held captive *L. chrysomelas* and the development of a captive population. A similar IRMC, chaired by Faiçal Simon, of the São Paulo Zoo, and Devra Kleiman, was formed for *L. chrysopygus* in 1987 due to the impending capture of black lion tamarin groups in the inundation area of the Rosana hydroelectric dam on the Rio Paranapanema in São Paulo state, as well as the need for better coordination of conservation efforts among state and federal agencies with an interest in the species.

Meetings of the *L. chrysomelas* IRMC and the *L. chrysopygus* IRMC were held ad hoc, but by 1988, it was clear that all three committees would profit by much closer interaction, especially since in situ and ex situ efforts to conserve the three species had become much more integrated (e.g., a reintroduction program for

L. rosalia was begun in 1984). The chairs of the *L. chrysomelas* and *L. chrysopygus* committees were elected to the *L. rosalia* committee later that year, and in 1990, all three met together for the first time at a Population Viability Analysis (PVA) Workshop, which also involved consideration of the newly described black-faced lion tamarin, *L. caissara* (Rylands 1993/1994; Seal et al. 1990). The IRMCs were composed of international and national experts in various aspects of lion tamarin biology and conservation—zoo biologists, studbook keepers, curators, and administrators, as well as field biologists, protected area managers, educators, and representatives of NGOs active in coordinating conservation efforts for the species. After 1992, they met together annually to review the conservation status of each of the species and to recommend priority actions. In 1990, a decree made three of the four committees official advisers to the Brazilian government (the decree for the *L. caissara* committee was published in 1992) (Brazil, IBAMA 1990, 1992). The sphere of activity for the IRMCs included all conservation and research efforts for the species both in situ and ex situ.

The recognition by IBAMA of the commitment of the international community to the conservation of the genus was reaffirmed by the extraordinary 1991 agreement by zoos around the globe to return ownership of captive golden lion tamarins to the people of Brazil (see Kleiman and Mallinson 1998; Chapter 4 this volume). By 1993, the committees decided to have regular symposia on the progress in research, education, and conservation for the genus *Leontopithecus* because the amount of work being accomplished had increased to the extent that it was difficult to keep up-to-date. Chapter 1 summarizes the major workshops and conferences that have contributed to our knowledge of the biology and conservation of these unique primates, including the 1997 symposium that resulted in this volume. In 1999, IBAMA consolidated the committees further by merging them under the new title of the Comitê Internacional para a Conservação e Manejo dos Micos-Leões (ICCM of lion tamarins) (Brazil, IBAMA 1999).

The functions of the *Leontopithecus* ICCM are presented in Table 3.1 (see also Kleiman and Mallinson 1998). This single committee is responsible for reviewing proposals, assessing progress, and making recommendations to IBAMA (which has ultimate decision-making authority) regarding the research and conservation activities for captive and wild populations. Conservation and research proposals and ongoing projects are reviewed for scientific soundness, feasibility, and their contribution to the species' conservation. For *L. rosalia* and *L. chrysopygus,* long-term master plans that involve metapopulation management of in situ and ex situ fragmented populations have been drawn up and approved (Ballou et al. 1998; Chapters 4 and 14 this volume). The *Leontopithecus* ICCM encourages partnerships to enhance the potential for the conservation of the four lion tamarin spe-

Table 3.1
The Functions of the International Committee for the Conservation and Management (ICCM) of Lion Tamarins

1. To promote the lion tamarins as "flagship" species for the preservation of the Atlantic Forest.
2. To evaluate and make recommendations to IBAMA concerning proposals for research and conservation.
3. To develop new methods and models for the conservation of species, especially for the management of metapopulations (for example, reintroduction and translocation).
4. To promote collaboration in research and conservation of the four species in nature and captivity.
5. To make recommendations to IBAMA concerning the distribution and management (genetic and demographic) of the populations of the species in captivity.
6. To promote the development of environmental education programs and programs to improve the socioeconomic conditions in the region.

Note: IBAMA = Instituto Brasileiro do Meio Ambiente e dos Recursos Naturais Renováveis (Brazilian Institute of the Environment and Renewable Natural Resources).

cies, especially when it seeks support from multiple national and international NGOs, zoos, and universities.

Since the survival of the four species depends on the protection and augmentation of their critical habitat, the *Leontopithecus* ICCM is also active in promoting the lion tamarins as "flagship" species for the conservation of the Atlantic Forest and its unique fauna and flora (Dietz et al. 1994c). The ICCM has endorsed and promoted the acquisition of land for reserves, the relocation of squatters, the reduction of human impact on the habitat, and the restoration of degraded areas by reforestation. Much of this is accomplished through advocacy of goals to individuals and agencies of the state and federal governments in Brazil. Recognizing that the long-term conservation of lion tamarin populations requires the support of local communities and other stakeholders, the *Leontopithecus* ICCM also promotes community conservation education programs targeting not only school children but also landowners and decision makers who influence community development, land use, and socioeconomic issues (Dietz and Nagagata 1995; Padua 1994a, 1994b; Chapter 15 this volume). Ultimately, the *Leontopithecus* ICCM promotes an integrated conservation and development effort that includes local communities as a means to achieve its fundamental mission of restoring viable *Leontopithecus* populations in their native habitat.

Unlike recovery programs for endangered species in other countries, the *Leontopithecus* ICCM has no dedicated staff or budget. Under the leadership of the president of the *Leontopithecus* ICCM, Maria Iolita Bampi (IBAMA, Departamento de Vida Silvestre, Diretoria de Ecossistemas), IBAMA staff takes the initiative in organizing annual meetings and supporting the participation of Brazilian members. All members take part on an unpaid, voluntary basis. Although meetings occur annually, there is regular contact throughout the year (Kleiman and Mallinson 1998); the near universal use of e-mail by *Leontopithecus* ICCM members has facilitated this frequent contact.

Most recently, the *Leontopithecus* ICCM has been active in raising funds for the Lion Tamarins of Brazil Fund (LTBF), which provides financial support to field research and conservation efforts for the four species. The LTBF was formed in the early 1990s to target specifically the institutions holding lion tamarins in captivity, to mobilize financial support for ongoing field conservation and research efforts, and to launch critical new initiatives (Mallinson 1994a). In recent years, the Margot Marsh Biodiversity Foundation, managed by Conservation International, has matched funds raised by the LTBF (Mallinson 1997a, 2001a). A newsletter, *Tamarin Tales,* is edited and published annually by Jonathan Ballou, of the SNZP, Washington, D.C., to provide LTBF contributors and participating zoos with information on the field conservation and research efforts.

Although the *Leontopithecus* ICCM provides support and general guidance to conservation and research activities for *Leontopithecus* species, it does not micromanage. Independent teams develop the individual projects and raise the necessary funds for implementation. A variety of organizations, including NGOs, government agencies, and universities, have financed lion tamarin research and conservation, and the independence of the project teams encourages greater flexibility and risk taking, which ultimately leads to the development of creative new techniques and methodologies (Kleiman and Mallinson 1998). This is especially important since the threats facing each of the species are different and require that the organizations focus on different approaches (e.g., see Chapter 15 this volume).

The participatory decision-making structure and the lack of hierarchical control make up an unusual, but highly effective, combination in contrast with most government-funded endangered species recovery programs, especially in the United States. Although IBAMA has the ultimate responsibility and authority for conservation of these endangered species, it has sought out and been responsive to national and international experts. In turn, the deliberations and recommendations of the ICCM foster collaboration, creativity, and synergy among the participating implementation teams, resulting in improved effectiveness of their conservation actions.

The *Leontopithecus* ICCM is undoubtedly an important model for the organization and implementation of conservation programs internationally. Brazil's IBAMA already uses this approach for the management of other endangered species, such as Spix's macaw (*Cyanopsitta spixi*), Lear's macaw (*Anodorhynchus leari*), the buffy-headed capuchin (*Cebus xanthosternos*), and the pied tamarin (*Saguinus bicolor*).

Recommendations from the *Leontopithecus* ICCM have had a major impact on defining the strategies for and influencing actions concerning the conservation of the four species. Specific examples include the expansion and consolidation of the Superagüi National Park for *L. caissara* in Paraná, creation of the União Biological Reserve for *L. rosalia* in Rio de Janeiro, final removal of squatters from the Una Biological Reserve for *L. chrysomelas* in Bahia, cessation of efforts to manage centrally *L. chrysomelas* confiscated in Bahia, a decision not to reintroduce *L. chrysomelas* in the near future, and a recommendation not to develop a captive population of *L. caissara*.

INTERNATIONAL NONGOVERNMENTAL ORGANIZATIONS

The Worldwide Fund for Nature

The Worldwide Fund for Nature's (WWF's) Global 200 ecoregions—the most outstanding examples of the world's major habitat types where large-scale conservation action must be mobilized—include the Atlantic Forest because of its globally important biodiversity and endemism as well as its being one of the most threatened tropical forest ecosystems in the world. The WWF recognizes the endemic lion tamarins as flagship species: a means to mobilize professional and public support for the conservation of the entire ecoregion.

Since the early 1970s the WWF network has been directly involved in in situ conservation efforts for the genus *Leontopithecus*, beginning with the WWF-U.S. Primate Action Fund created by Russell Mittermeier (Konstant 1996/1997). The WWF-U.S. Primate Action Fund supported a number of field surveys of primates in the Atlantic Forest, as well as grants to improve reserve protection and to build public awareness of the importance of these species and their forests. The current field conservation programs for all four lion tamarin species originated from these action grants. From the late 1980s, WWF broadened its focus to support the development of projects targeting the conservation of each of the (then) three lion tamarin species and thereby mobilizing actions for the conservation of a large portion of the remaining lowland Atlantic Forest. The WWF provided equipment, infrastructure, and training to support greater protection of the government-protected areas harboring each of the four lion tamarin species. Over the period from 1988 to 2001, WWF matched government funds from IBAMA to provide compensation to over 100 squatters to relinquish their claims within the Una Biological Reserve. The WWF also mobilized a multinational, multi-institutional partnership to purchase land to increase the size of the Una Biological Reserve and supported field research there on golden-headed lion tamarins (Dietz et al. 1994c, 1996) for future conservation recommendations.

To build public support for forest protection, WWF mobilized the development of targeted community conservation education programs through a continuing program of grants, technical assistance, and training for project leaders. The WWF also helped several of the new NGOs advance toward self-reliance through grants for strategic planning, developing fundraising strategies, technical assistance, and staff training in organizational and management skills.

The establishment in 1996 of WWF-Brazil as an independent Brazilian national organization within the WWF international network made it possible for WWF to work with local Brazilian partner organizations on national policy in Brazil through advocacy and communications activities. Some of these initiatives have included promotion of the establishment of private reserves, state tax incentives to municipalities for establishing protected areas, a campaign to abolish illegal wildlife trade, efforts to transform the presidential decree prohibiting cutting of the Atlantic Forest into permanent law, and a national campaign to prevent continued attempts by the agricultural lobby in the Brazilian Congress to reduce the 1967 Forest Code requirements for the preservation of the Atlantic Forest on private land (current law requires maintenance of forest cover along streams and on slopes, as well as on 20 percent of private properties in the Atlantic Forest region).

Major support has been provided by WWF-Brazil to building the capacity of Brazilian conservation professionals, and it has assisted a significant number of Brazilian graduate students in field thesis research related to lion tamarin conservation. In partnership with the State University of New York and the IIEB (the International Institute for Brazilian Education), WWF has provided training for the majority of professionals involved in lion tamarin conservation through workshops, courses, support for participation in congresses, and exchanges in varied areas.

The WWF supported both the 1990 PVA and the 1996 Population and Habitat Viability Analysis (PHVA) workshops (Ballou et al. 1998; Seal et al. 1990) for lion tamarins and has used the results together with recommendations of the ICCM for lion tamarins to prioritize its conservation investment. The lion tamarins are currently the only Atlantic Forest species for which enough information is available to estimate habitat requirements for viable populations. The WWF looks to lion tamarin conservation initiatives to demonstrate to the broader conservation community the methodologies for planning and implementing landscape designs in development to ensure there are viable populations of each of the lion tamarin species.

Specific WWF contributions on behalf of the lion tamarins are detailed in Table 3.2 as an example of one NGO's conservation support. Contributions by other NGOs are presented in the following, but in less detail.

Table 3.2

A Detailed Example of One NGO's Contributions to Lion Tamarin Conservation and Research: Support of WWF (Worldwide Fund for Nature, Known as World Wildlife Fund in the United States and Canada) for Conservation, Research, and Management of Lion Tamarins

---

Primate Surveys in the Atlantic Forest, Captive Breeding Programs, and Workshops

Support for the establishment of the Tijuca Biological Bank for breeding lion tamarins and other endangered Atlantic Forest primates and, from 1979, the Rio de Janeiro Primate Center (Coimbra-Filho and Mittermeier 1977; Coimbra-Filho et al. 1986a)

A major survey of Atlantic Forest primates and protected areas in 1980/1981 (Mittermeier et al. 1981, 1982), with a second phase in 1983/1984 (Rylands et al. 1991/1992; Santos 1983, 1984; Santos et al. 1987)

Determination of status and distribution of endemic arboreal mammals of the Atlantic Forest of southern Bahia and Espírito Santo, Brazil, in 1985/1986 (Oliver and Santos 1991)

Support for the *Leontopithecus* PVA Workshop, Belo Horizonte, 1990 (Rylands 1993/1994; Seal et al. 1990)

Support for the *Leontopithecus* PHVA Workshop, Belo Horizonte, 1997 (Ballou et al. 1998)

Support for case study on partnerships of NGOs with protected areas—Poço das Antas Biological Reserve, Caetetus State Ecological Station, and Guaraqueçaba Ecological Station

Leontopithecus rosalia

A survey of golden lion tamarins and the vegetation types in the Poço das Antas Biological Reserve (Green 1980; Kleiman et al. 1988)

Support for Golden Lion Tamarin Conservation Program field studies component in the Poço das Antas Biological Reserve between 1983 and 1985. Ongoing support from 1985 for the administration of the field conservation projects, community environmental education, habitat restoration, and public policy

Technical assistance and financial support for institutional development of AMLD

Support for IBAMA in setting up a fire prevention system (including the construction of fire towers and communications equipment) and in combating fires in the Poço das Antas Biological Reserve

Establishment of a radio communication network between IBAMA and AMLD specifically with a view to combating fires, patrolling, and security in the Poço das Antas Biological Reserve

Support for Brazilian interns, academic exchange, and short-term training for conservation professionals offered by the Golden Lion Tamarin Conservation Program

Technical assistance and short-term training for IBAMA and AMLD staff in the development of ecotourism, reserve management, environmental education, organizational development, financial administration, and public policy

Support for eight Brazilian graduate students' research, including studies of marsupials, rodents, and bats in the Poço das Antas Biological Reserve, a survey of *L. rosalia* populations outside the reserve, translocation of threatened groups of *L. rosalia,* genetic and foraging ecology studies, improved database management for *L. rosalia,* and habitat restoration and reforestation research (e.g., Grativol et al. 2001; Kierulff 1993a, 2000; Kolb, 1993; Procópio de Oliveira 1993)

Press coverage through the production of "Wildlife Story of the Year, a Video News Release" for international television distribution and other feature film material used extensively in community education and media efforts

Continued on next page

Table 3.2 continued

Coordination of an international celebration of the one thousandth golden lion tamarin, "Mico Mil," to be born in the wild and a fundraising campaign for forest corridors (2001)

Support for the production of a documentary video for ranch owners concerning private reserves, with the participation of IBAMA

Emergency support for purchase of minimum infrastructure for the director of the new União Biological Reserve before a government budget was available (1998)

Advocacy at the national level in Brazil for the creation of the União Biological Reserve (1998), for increased government support for the União and Poço das Antas Biological Reserves (2001), and for the creation of the São João Basin Environmental Protection Area (2001)

Leontopithecus chrysomelas

A field study of the ecology and behavior of *L. chrysomelas* in the Lemos Maia Experimental Station, Una, Bahia, in 1980 (Rylands 1980, 1982, 1983)

Technical and financial support provided for the director of the Una Biological Reserve/IBAMA to compensate over 100 squatters who subsequently gave up their claims in the reserve (1986–2001)

Support for the purchase of land to consolidate the Una Biological Reserve (Coimbra-Filho et al. 1993)

Support for short-term training of the Una Biological Reserve director and IESB staff in developing community support for conservation of protected areas and in development of GIS

Support for an environmental education program in the vicinity of the Una Biological Reserve (Alves 1991, 1992; Santos 1995)

Support for an evaluation of a community-based conservation education program in the vicinity of the Una Biological Reserve (Nagagata 1994a, 1994b)

In the Una region, training and technical assistance for the systematic development and evaluation of three environmental education initiatives targeting landowners, school children, and residents of land settlements

Support for acquisition of satellite images and GIS studies of forest cover in southern Bahia to establish spatial priorities for protecting sufficient forest to support a minimum viable population of golden-headed lion tamarins (Allnutt 1997; Stith 1990)

Advocacy to promote IBAMA and international support for resolving the land tenure situation of the Una Biological Reserve as well as increasing financial support for protection and management of the reserve

Support for studies of the demography and ecology of *L. chrysomelas* in the Una Biological Reserve for a population viability analysis (Dietz 1993; Dietz et al. 1994c, 1996)

Support for ongoing doctoral thesis research of three students, the research on *cabruca* ecology and the responses of bats and other mammals to habitat fragmentation and degradation

Support for a socioeconomic survey (including land use mapping and analysis of economic viability of alternatives to cacao) of the region surrounding the Una Biological Reserve (Alger and Caldas 1994)

Leontopithecus chrysopygus

Support for a survey concerning the status and conservation of the black lion tamarin in São Paulo and research in 1985/1986 (Valladares-Padua et al. 1994)

Technical assistance, training, and financial support for the environmental education programs for the Morro do Diabo State Park and Caetetus State Ecological Station beginning in 1989 (Padua 1994a, 1994b)

Table 3.2 continued

Capacity building for the São Paulo Forestry Institute staff responsible for protected areas (1987–1999)
Support for vegetation studies in the Caetetus State Ecological Station (Kim and Passos 1994)
Support for ecological studies of *L. chrysopygus* in the Caetetus State Ecological Station (Passos 1991, 1992, 1994, 1997a, 1997b, 1999)
Support for two students' ongoing doctoral thesis research on the behavioral ecology of managed groups of *L. chrysopygus* and a conservation assessment for forest fragments in the Pontal do Paranapanema
Support for short-term professional development training of IPÊ staff
Technical and financial support for the organizational development of IPÊ

Leontopithecus caissara
Support for an environmental education program coordinated in Iguape, São Paulo
Support for the renovation of a historic building for IBDF (now IBAMA) offices, researcher dormitory, and environmental education space (1986)
Support for a survey of primates in Paraná by the Capão Imbuia Museum (1989)
Support for a survey of *L. caissara* in the state of São Paulo (Rodrigues et al. 1992)
Technical assistance and training for staff members of SPVS in the development of environmental education programs and ecotourism
Technical assistance and financial support for organizational development of SPVS

Note: AMLD = Associação Mico-Leão-Dourado (Golden Lion Tamarin Association), GIS = geographic information systems, IBAMA = Instituto Brasileiro do Meio Ambiente e dos Recursos Naturais Renováveis (Brazilian Institute of the Environment and Renewable Natural Resources), IBDF = Instituto Brasileiro de Desenvolvimento Florestal (Brazilian Forestry Development Institute), IESB = Instituto de Estudos Sócio-Ambientais do Sul da Bahia (Institute for Social and Environmental Studies of Southern Bahia), IPÊ = Instituto de Pesquisas Ecológicas (Institute for Ecological Research), PHVA = Population and Habitat Viability Analysis, PVA = Population Viability Analysis, and SPVS = Sociedade de Pesquisa em Vida Selvagem e Educação Ambiental (Society for Wildlife Research and Environmental Education).

Conservation International

The Atlantic Forest biome was identified by the international NGO Conservation International (CI) as a "Hotspot" due to the extent to which it has been destroyed, its remarkable biodiversity, and the extraordinarily high degree of endemism (Mittermeier et al. 1999b; Myers et al. 2000). The Brazilian regional program of CI, Conservation International do Brasil (CI-Brasil), has carried out two priority-setting workshops for biodiversity conservation in the Atlantic Forest. The northern Atlantic Forest (including southern Bahia) was the focus of one in Recife, Pernambuco, in 1993, organized with the Sociedade Nordestina de Ecologia (the Northeastern Society for Ecology) (Conservation International 1995; Fonseca et al. 1995), and the entire Atlantic Forest was reviewed in a workshop in 1999, in collaboration with the Brazilian government, in Atibaia, São Paulo (Conservation International do Brasil 2000). Both workshops involved the identification and mapping of key areas in terms of biological importance and the establishment of a database on current knowledge of the areas to provide orientation for research and conservation efforts. The regions where the four lion tamarins occur were all pinpointed as key areas for conservation action in these workshops.

Conservation International do Brasil currently has as a major partner, SOS Mata Atlântica, a Brazilian NGO dedicated to the protection of the Atlantic Forest. Also, CI and CI-Brasil are major partners with the Instituto de Estudos Sócio-Ambientais do Sul da Bahia (IESB—Institute for Social and Environmental Studies of Southern Bahia) (see below), providing technical and scientific support and searching together for funds to support their work in conservation. Conservation International has supported all the major recent workshops on conservation of the genus *Leontopithecus*. More important, CI was key to the creation of the Margot Marsh Biodiversity Foundation (MMBF) in 1996, which now provides annual support for lion tamarin conservation projects through matching the funds raised by the LTBF (comanaged by Jeremy Mallinson, Devra Kleiman, and Jonathan Ballou). The MMBF supported the Second *Leontopithecus* PHVA Workshop, golden lion tamarin translocation and genetics studies, and golden-headed, black, and black-faced lion tamarin field studies and conservation. Also CI has contributed substantially to the professional development of Brazilian primatologists and conservation biologists through a variety of grants.

The Wildlife Trusts

Beginning in the 1970s, the Durrell Wildlife Conservation Trust (DWCT) (formerly the Jersey Wildlife Preservation Trust—JWPT) and its New World offshoots the Wildlife Trust (formerly the Wildlife Preservation Trust International—WPTI) and the Wildlife Preservation Trust–Canada (WPT-Canada) have played major roles in the conservation of the lion tamarins. One of the DWCT's first contributions was through its training program; a number of key researchers in lion tamarin conservation, including Cristina Alves, Ilmar Santos, Inês Castro, and Cláudio Valladares-Padua, attended its Training Centre at the Jersey Zoo, Channel Islands.

Donations by the DWCT have been critical to the success of research on the translocation of threatened *L. rosalia* (Kierulff and Procópio de Oliveira 1994, 1996) and field research on *L. chrysomelas* (Oliver and Santos 1991; Pinto and Tavares 1994). Gerald Durrell, founder of the Jersey Zoo, established the LTBF in 1991 and distributed the first international appeal for funds for field research and conservation for *Leontopithecus* species (Mallinson 1994a). By 2002, more than $280,000 (U.S.) had been raised from zoos internationally, as well as matching funds from the MMBF, in aid of lion tamarin conservation work in Brazil.

Over the years, DWCT has provided considerable support to the original IRMCs (Mallinson 1986), as well as for workshops, symposia, and the IUCN/SSC (World Conservation Union/Species Survival Commission) Primate Specialist Group's newsletter *Neotropical Primates* and its supplement of the proceedings of the Second Symposium on *Leontopithecus* (Rylands and Rodríguez-Luna 1994). As mentioned above, then director of Jersey Zoo Jeremy Mallinson spearheaded the return of illegally exported *L. chrysomelas* and, as head of the IRMC for the species, set up the international breeding program and the first studbook (Mallinson 1984, 1986, 1987a, 1987b, 1989). The DWCT was also a major donor for the purchase of land for a corridor within the Una Biological Reserve; the corridor links important areas of the reserve and thus contributes to its future viability. As a partner with IESB, the DWCT has supported surveys of the distribution of *L. chrysomelas* and educational efforts with the landowners in the Una region to encourage better agroforestry practices (Blanes and Mallinson 1997; Santos and Blanes 1997, 1999).

For several years, the Wildlife Trust in the United States sponsored the Projeto Mico-Leão-Baiano (Bahian Lion Tamarin Project) (Alves 1991, 1992), an environmental education program in the Una region. More recently, it has developed a formal partnership with the Instituto de Pesquisas Ecológicas (IPÊ—Institute for Ecological Research) in promoting its research, conservation, education, and training programs.

The Smithsonian Institution, the Smithsonian National Zoological Park, and the Friends of the National Zoo

The Smithsonian National Zoological Park (SNZP), a semigovernmental organization, became involved with golden lion tamarin conservation in the early 1970s through its endangered species breeding programs and its support of the workshop "Saving the Lion Marmoset" (Bridgwater 1972a). The SNZP developed a research and management program for captive golden lion tamarins, including oversight of the international studbook and the dedication of considerable resources, which eventually led to the establishment of a self-sustaining captive population of this species (see Chapter 4 this volume). The SNZP then took the lead in establishing those partnerships in and outside of Brazil that resulted in the Golden Lion Tamarin Conservation Program (GLTCP). From the inception of this research and conservation effort in 1972 until the present time, the parent Smithsonian Institution (SI), SNZP, and the Friends of the National Zoo (FONZ) have provided significant resources for every sector of the GLTCP, including management of the global captive population; development of the infrastructure for the NGO Associação Mico-Leão-Dourado (AMLD) in Brazil; conservation education and media awareness programs; reintroduction, translocation, and ecological studies of lion tamarins and other mammalian species; and habitat analyses and forest rehabilitation and restoration. All of these activities required a functional field station (e.g., dormitories, kitchen, and laboratories) and vehicles—a physical infrastructure funded primarily by SI, SNZP, and FONZ. The extensive involvement of SNZP staff members Devra Kleiman, Benjamin Beck, and Jonathan Ballou and the research conducted by other SNZP staff, for example, Richard Montali and Mitchell Bush, as well as the SNZP support for both Brazilian and non-Brazilian graduate students and postdoctoral fellows conducting research on golden lion tamarins, resulted in significant in-kind contributions to lion tamarin conservation. From 1983 to 1992 (when the AMLD was established), SI and SNZP had complete responsibility for overseeing the GLTCP in situ and ex situ research and conservation activities, including its total budget.

The SNZP has also supported the captive breeding programs for three of the species and the two PHVA workshops (Ballou et al. 1998; Seal et al. 1990). Numerous Brazilians have been trained through the Wildlife Conservation and Management Training Program at the SNZP Conservation and Research Center in Front Royal, Virginia (initially a Primate Conservation Training Program), as well as through the partnership that the Wildlife Conservation and Management Training Program has with IPÊ. Additional Brazilian students were brought into the GLTCP through a partnership with the Federal University of Minas Gerais (UFMG) in the late 1980s/early 1990s and through many internships. Several staff members of the SNZP have comanaged the LTBF and are members of the *Leontopithecus* ICCM.

BRAZILIAN NONGOVERNMENTAL ORGANIZATIONS

Fundação Brasileira para a Conservação da Natureza

The Fundação Brasileira para a Conservação da Natureza (FBCN—Brazilian Foundation for the Conservation of Nature), the first nongovernmental Brazilian environmental foundation, was created in August 1958. Throughout the past 44 years, the foundation has contributed enormously to the conservation and management of natural areas and endangered species. In 1968, José Candido de Melo Carvalho (chairman of the FBCN), Alceo Magnanini, and Adelmar F. Coimbra-Filho were responsible for placing *L. chrysomelas, L. chrysopygus,* and *L. rosalia* on the Brazilian Threatened Species List. During the 1970s and 1980s, a partnership between the SNZP, Washington, D.C., and FBCN coordinated and administered research and campaigns to protect the golden lion tamarin. The FBCN also played a key role in the development of the original management plan for the Poço das Antas Biological Reserve (Brazil, MA/IBDF/FBCN 1981), as well as the initial action plan drawn up for the black-faced lion tamarin, *L. caissara* (Câmara 1993, 1994). The FBCN's support was crucial to the organization, development, and implementation of GLTCP activities from 1983 to 1992 (the role was later taken over by the AMLD).

Associação Mico-Leão-Dourado

The GLTCP, a loosely knit group of researchers and conservationists based mainly in the United States, along with the conservation programs for the other lion tamarin species, collaborated with the IUCN/SSC Conservation Breeding Specialist Group (CBSG) and the Fundação Biodiversitas in organizing the first major PVA Workshop in 1990 (Lacy 1993, 1993/1994; Rylands 1993/1994; Seal et al. 1990). A recommendation of this workshop was the need to establish a Brazilian base and infrastructure for overseeing and coordinating the conservation and research activities for the golden lion tamarin (Seal et al. 1990). This resulted in the creation of the NGO Associação Mico-Leão-Dourado in 1992, the Brazilian successor to the SNZP's GLTCP. The AMLD had the stated aim of protecting biodiversity in the Atlantic Forest of the lowland areas in northern Rio de Janeiro state, with the golden lion tamarin as the flagship species.

The mission of the AMLD is (1) to maximize the probability of survival of a naturally evolving population of golden lion tamarins, (2) to expand and apply leading edge science in conservation biology, (3) to increase public awareness and involvement in the conservation of golden lion tamarins and their habitat, (4) to enhance professional training in biology and conservation, with preference given to Brazilians, and (5) to integrate activities with other conservation programs with similar methods and goals.

In order to guarantee continuity and to increase its sphere of action, the AMLD is integrating the management of the golden lion tamarin populations and their habitats with the needs and expectations of the local communities to produce a technical/scientific model that combines applied research with organizational development and includes the following components.

1. Demographic and behavioral studies, underway since 1983, in the Poço das Antas Biological Reserve. Over 60 percent of the wild population has been marked, habituated, and regularly observed in the last 15 years.
2. A reintroduction program concentrating particularly on the development of techniques for preparing captive-born golden lion tamarins for the wild and releasing them and on establishing populations on private lands.

Initial resistance by the ranch owners, due to fear that they might lose the right to parts of their lands, has been transformed into pride and local and national status as a result of their contribution to protecting a critically endangered species. Twenty-one ranches were participants in the program as of December 2001 (Chapters 12 and 13 this volume).

3. The translocation, monitoring, and observation of isolated and threatened golden lion tamarin groups identified in the exhaustive census carried out by Maria Cecília Kierulff (1993a, 1993b; Kierulff and Procópio de Oliveira 1996; Chapter 12 this volume).
4. A habitat restoration program, which has experimented with a variety of alternatives for increasing the amount of habitat available for lion tamarins (Camargo 1995; Kleiman et al. 1988). Initial steps involved mapping the forest cover of the Poço das Antas Biological Reserve and studying the floristic composition of the four vegetation types identified. In 1996, two experimental forest corridors and islands were set up inside the reserve. The lessons learned have been applied to projects to establish three corridors, using agroforestry techniques, between forest patches with reintroduced lion tamarins on privately owned land (Rambaldi 2000). The landowners are providing not only their land but also their employees to work on these corridors.
5. A community environmental education program, which has been a forerunner for conservation education programs in Brazil (see Chapter 15 this volume). The program works in close collaboration with government agencies for education, agriculture, and the environment, as well as with the relevant community associations. The program provides informal and formal learning possibilities through its Education Center (the only NGO structure within a federal reserve) and training programs and has developed the techniques for working with private landowners to register their land with IBAMA as permanent private reserves, Reservas Particulares de Patrimônio Natural (RPPNs—Private Natural Heritage Reserves).

The AMLD also works to build local community capacity and to evaluate its education and outreach programs systematically, which has resulted in the modification of techniques used to target different audiences, be they community leaders, teachers, the media, ranch owners, or public authorities. For example, a Rural Participatory Diagnosis (DRP) completed by the AMLD in the municipalities of Silva Jardim and Casimiro de Abreu has helped to explain recent changes in the rural communities and the land uses surrounding the reserve.

An important action of the AMLD has been the active participation in the Lake's Region Consortium of 12 municipalities along the São João river basin. This forum plays an important role in land-use planning and management at the river basin level. Since 80 percent of *L. rosalia*'s present area of occurrence is located within the São João river basin, decisions on land use in that region are of major significance for the species' future.

Informing stakeholders and interested parties of the work carried out in and around the Poço das Antas Biological Reserve is an important component of AMLD activities. This is done through the organization of, and participation in, scientific meetings, the publication of scientific articles (more than 130 to date), support for the production of national and international filmed documentaries, talks in universities and other research and teaching institutions, the publication of popular articles in magazines of national and international circulation, participation in community events such as science fairs and agricultural shows, and contributions to events organized by local municipal councils. The AMLD has been recognized as being of "public utility" by one local municipality and in 1993 was awarded the Diploma "Ação Verde" ("Green Action") in recognition of its work. The GLTCP and its individual participants have received numerous awards for their activities, including the 1988 Conservation Achievement Award from the Society for Conservation Biology and the American Zoo and Aquarium Association's (AZA's) Annual Conservation Award (1994), the Muriqui Award given to Denise Rambaldi by the National Council of the Mata Atlântica Biosphere Reserve (1998), and the Legislative Merit Medal awarded to Denise Rambaldi by the Silva Jardim Town Council (2001).

The GLTCP is one of very few conservation programs internationally that has requested and received a formal peer evaluation of its activities. In 1997, a formal review panel, chaired by Ross Simons of SI, and including Russell Mittermeier, John Robinson, Kent Redford, Suzana Padua, and José Márcio Ayres, reviewed the GLTCP's research and conservation activities and developed a set of recommendations, which were later incorporated into the AMLD's first strategic planning meeting in 1998.

In a recent opinion survey conducted by the Brazilian Environment Ministry and the Instituto de Estudos da Religião (ISER—Institute for Religious Studies) on "what Brazilians think about the environment and sustainable consumption," the AMLD was the second environmental institution most cited (25 percent); the first was IBAMA (78 percent). This survey was carried out in the biggest Brazilian cities in October 2001. Two thousand people were interviewed. It demonstrated that the message of the AMLD is reaching its target public in the Atlantic Forest and throughout Brazil.

Instituto de Pesquisas Ecológicas

Instituto de Pesquisas Ecológicas (IPÊ—Institute for Ecological Research) was created in 1992 in Piracicaba, state of São Paulo, and now has its headquarters at Nazaré Paulista, where it has built an education and training center for conservation. The IPÊ's mission is "Research and education for the conservation of biodiversity."

The IPÊ is the result of the expansion of the Black Lion Tamarin Project, which began in 1983 because of the imminent demise of part of a population of *L. chrysopygus* in the Morro do Diabo State Park due to the construction of the Rosana hydroelectric dam (Valladares-Padua 1987, 1993; Chapter 1 this volume). The goals of IPÊ are

1. to carry out research on rare and threatened species and to implement conservation strategies that integrate the management of captive and wild populations;
2. to promote and implement environmental education programs for local communities;
3. to work with local landowners to encourage the conservation of forest fragments and the planting of forest corridors, buffer zones, and stepping stones;
4. to introduce sustainable development practices that can lead to social and environmental benefits to local people and reduce the pressures on natural areas; and
5. to promote education and training programs in several areas of conservation.

Initially, the Black Lion Tamarin Project was focused on the Morro do Diabo State Park, administered by the Instituto Florestal de São Paulo (IF/SP—São Paulo Forestry Institute) and the major stronghold for the species. Efforts were later extended to the Caetetus State Ecological Station and smaller black lion tamarin populations found in private patches of forests.

One of the most important components of the Black Lion Tamarin Project has been training. A large number of Brazilian students have participated in many projects that have expanded over time. From a small group of three people in the Black Lion Tamarin Project in 1988, IPÊ now has 35 staff members, 5 of whom are now finishing doctoral degrees and 10 of whom have completed master's degrees.

Education and outreach have been among IPÊ's major achievements (see Chapter 15 this volume), and in 1997, IPÊ established a permanent training facility, the Brazilian Center for Conservation Biology, to train professionals and provide continuity for the conservation projects underway. Altogether, IPÊ has been involved in educating more than 400 students or midcareer professionals in courses that range from wildlife management to mapmaking, statistics for conservation, and environmental education. The courses are run with the collaboration of SNZP, the Wildlife Trust, the Center for Environmental Research and Conservation (CERC) at Columbia University, New York, IF/SP, the Forestry Foundation, and numerous universities.

In 1995, IPÊ signed a formal agreement to collaborate with the Wildlife Trust. Both institutions have similar goals, and for the United States–based Wildlife Trust, which had provided financial support for a number of years previously, the agreement represented the opportunity to be more closely involved in a grassroots project in one of the countries with the highest biodiversity in the world. The Wildlife Trust is part of the CERC consortium, based at Columbia University, which also includes the New York Botanical Garden, the American Museum of Natural History, and the Wildlife Conservation Society. The IPÊ was chosen to represent the consortium in Brazil. In the past 5 years, IPÊ has received a number of important national and international awards: the 1998 Ford Conservation Award, the 1999 Environmental Contribution Soroptimist Internacional Society, the 1999 Whitley Continuation Award of the Royal Geographic Society, the 1999 Wildlife Trust Conservationists of the Year, and the 2000 Society for Conservation Biology Institution of the Year.

Although the Black Lion Tamarin Project is the core of IPÊ's projects, a number of other key research and conservation programs are also underway. The IPÊ is now working on a site-specific approach in five different regions, including where the black lion tamarin is found west of the state of São Paulo; the Superagüi National Park, habitat of the black-faced lion tamarin; Nazaré Paulista, where IPÊ has its headquarters and training center; São Francisco Xavier, São Paulo State, where the largest primate of the New World, the muriqui (*Brachyteles* spp.), serves as a symbol for broader conservation initiatives; and finally, near Manaus, where IPÊ has been working in the Anavilhanas Ecological Station to integrate conservation with sustainable alternatives for local development. In all sites IPÊ has been using the model developed for the black lion tamarin with its components of research and education, public participation, sustainable development alternatives, training, and the influencing of policies.

Instituto de Estudos Sócio-Ambientais do Sul da Bahia

Concerned for the future of the Una Biological Reserve and its flagship species, the golden-headed lion tamarin (*L. chrysomelas*), and considering the reserve's key role in protecting the biodiversity of the region, CI, through its Brazil Program based in Belo Horizonte, supported the establishment of a major research and conservation project in 1993 (IESB 1996; Conservation International 1997). As a result of this project the forests around the Una Biological Reserve have been mapped and evaluated and economic alternatives to reduce forest cutting have been promoted. Although dealing with the socioeconomy of the entire region, it is centered on the Una reserve and offers real hope for the reserve's viability and continued integrity.

The partner NGO of CI-Brasil is the Instituto de Estudos Sócio-Ambientais do Sul da Bahia (IESB—Institute for Social and Environmental Studies of Southern Bahia). The institute was founded in 1994 by a group of research professors at the State University of Santa Cruz in Bahia with the objective of promoting the conservation of ecosystems and natural resources, especially the fauna and flora, through research and programs that support both human and environmental needs (Alger and Araújo 1996; Alger and Caldas 1994; Alger et al. 1996; IESB 1996). The activities of IESB have contributed directly to the creation of the Serra do Conduru State Park of 9,000 ha and the establishment of numerous private reserves (RPPNs) in the state of Bahia, including the IESB-owned and -operated ecotourism attraction Ecoparque de Una (Una Ecopark) (Mourão 1996). The IESB focuses its activities on developing partnerships with the rural agricultural community, supporting conservation-friendly activities (IESB 2000; Moreau et al. 1996; Reid and Blanes 1996; Santos and Blanes 1997, 1999), encouraging the creation of RPPNs, developing public policy toward conservation, encouraging biodiversity research within the forests of the Una Biological Reserve, and monitoring land use within the region by using remote sensing and geographic information systems (GISs). Researchers for IESB have also taken the lead in the development of corridor-scale planning tools for southern Bahia and the north of the state of Espírito Santo, in collaboration with the Projeto de Conservação e Utilização Sustentável de Diversidade Biológica Brasileira (PROBIO—Project for the Conservation and Sustainable Use of Brazilian Biological Diversity), a World Bank research group, and with CI's Center for Applied Biodiversity Science (CABS). The IESB staff includes political scientists, sociologists, forestry engineers, agronomists, conservation biologists, ecologists, economists, and lawyers.

Sociedade de Pesquisa em Vida Selvagem e Educação Ambiental

The Sociedade de Pesquisa em Vida Selvagem e Educação Ambiental (SPVS—Society for Wildlife Research and Environmental Education) is based in Curitiba in the state of Paraná. It has focused on promoting wildlife research and environmental conservation in Paraná, with a strong emphasis on environmental education programs. Initial research by SPVS led to the discovery of the black-faced lion tamarin, *L. caissara,* on the island of Superagüi near Guaraqueçaba (Oliveira and Pereira 1990).

The SPVS supported research on the black-faced lion tamarin from the beginning, with the goal of promoting conservation strategies in the entire region of Guaraqueçaba (Amato et al. 2000; Margarido et al. 1997; Ribas Lange 1997).

Fundação Biodiversitas

The Fundação Biodiversitas (Biodiversity Foundation) was founded by a group of researchers from the Federal University of Minas Gerais (UFMG) in Belo Horizonte. Members of the team had participated in the World Wildlife Fund Primate Program surveys in the Atlantic Forest from 1980 to 1983 (Mittermeier et al. 1981, 1982; Rylands et al. 1991/1992; Santos 1983, 1984; Santos et al. 1987) and, in the late 1980s, in collaboration with the DWCT, contributed to another survey in coastal Bahia and Espírito Santo that provided important data on the distribution and status of *L. chrysomelas* (Oliver and Santos 1991). The Fundação Biodiversitas has played a key role in a number of initiatives for the conservation of lion tamarins. They include the organization and hosting of the 1990 PVA and 1997 PHVA workshops for the genus (Ballou et al. 1998; Seal et al. 1990), responsibility for handling the logistics and legalities of the purchase of land for the consolidation and expansion of the Una Biological Reserve (Coimbra-Filho et al. 1993), and support for the distribution surveys of *L. rosalia* carried out by Kierulff (1993a, 1993b) and of *L. chrysomelas* by Pinto (Pinto 1994; Pinto and Rylands 1997; Pinto and Tavares 1994). In 2000, Fundação Biodiversitas took over the coordination of the Brazilian committee of the IUCN, and it is currently organizing a full revision and expansion of the official Brazilian Threatened Species List in collaboration with IBAMA.

CONCLUSIONS

The ICCM for the genus *Leontopithecus* has been a model and a pioneering force in the Brazilian government's development of endangered species conservation and recovery programs (Kleiman and Mallinson 1998), and IBAMA has promoted and supported a multi-institutional, multinational, multidisciplinary approach to species preservation and a participatory approach to decision making about conservation activities, even when some Brazilian institutions and individuals encouraged a more exclusive and solely Brazilian approach.

Because the IRMCs (now the ICCM) focused on four species in different regions and with different basic threats to the species and their habitats, the conservation efforts for the four species of *Leontopithecus* were diverse and allowed colleagues to benefit from the "lessons learned" within each program. While there is still room for improvement, the separate programs of the ICCM are increasingly taking advantage of the experience of their colleagues, and thus avoiding reinventing the wheel. The annual meetings provide a venue for discussions of each program's successes (and problems). There have also been several collaborative workshops across programs, for example, concerning the methods of approaching and explaining to landowners about the fiscal and other benefits of private reserves (RPPNs) and regular exchange visits to the study sites. The ICCM for the genus *Leontopithecus,* with its central coordinating function, can rapidly provide each program with appropriate contacts and information when new issues, problems, or questions emerge (Kleiman and Mallinson 1998).

The four species programs and their associated NGOs have different origins and backgrounds, which have affected their functioning. For example, the GLTCP arose outside of Brazil, as a consortium of conservation biologists and educators that focused on the recovery of an endangered Brazilian primate. It pioneered the integration of in situ and ex situ species conservation in Brazil and was well funded from non-Brazilian sources. It also originally had the strongest ties to the international zoo community. This contrasts with the programs for the other three species, which arose mainly in Brazil and focused somewhat more on habitat protection than on species protection and recovery.

The original program differences are disappearing, however, as the Brazilian conservation community has expanded, taking on more of the ultimate responsibility for these conservation programs, and Brazilian NGOs have grown to become equal partners with international NGOs. Moreover, all four species programs now consider biodiversity conservation and habitat preservation to be the main objectives of their programs; thus, the focus of attention has shifted to Brazil, with the international zoo community taking on a supportive role for field studies and local conservation efforts.

The success of the activities of Brazilian NGOs has been linked to their capacity for long-term fundraising and their ability to deal with local governments and communities. After the United Nations Conference on Environment and Development (UNCED) in Rio de Janeiro in 1992, many NGOs disappeared with the same speed with which they were created. Nevertheless, the professionalism of the NGOs mentioned in this chapter has enabled the continuity of the conservation strategies for the lion tamarins. Despite the difficulties of raising long-term support, NGOs can be more effective than the government in many arenas due to their simplified decision-making processes, greater flexibility, and considerable levels of idealism and scientific skill of their participants.

The successful outcome of partnerships and collaborative agreements between the Brazilian government and international NGOs, such as SNZP, WWF, and CI, has demonstrated the vital importance of such links for the long-term effectiveness of measures for the protection of habitats and species. Each has clearly distinct attributes that are complementary. The international NGOs usually contribute technical expertise and financial support unavailable to government organizations. The frequent changes in personnel and policies of government organizations in Brazil have required that partnership contracts with international NGOs be clearly defined in terms of the contributions and expectations of each in order to guarantee continuity.

Brazilian NGOs, such as IPÊ, AMLD, Fundação Biodiversitas, and SPVS, have developed agreements with state and federal governments when their activities occur in government-protected areas. Nevertheless, such agreements do not usually come with financial support, and the NGOs have to find support for their institutional infrastructure on their own. The national NGOs have deeply influenced public policies and government action in Brazil by translating scientific information and providing it to legislators and governmental agencies. This process has gone beyond writing letters defending environmental causes. Some NGOs have set up offices at the state and federal capitals in order to keep track of pending legislation, to inform and mobilize public opinion through the media, to encourage governmental sectors to include environmental issues in their agendas, and to promote the connection among different public sectors or stakeholders with a common interest in managing and protecting the environment.

As a result of the growing activity and scope of the Brazilian NGOs and the need to include these groups in national policy debates, the Brazilian Congress approved a law (1998) that created the Public Interest Civil Society Organizations (OSCIP) and set up a formal partnership agreement between the public sector and private nonprofit organizations to encourage greater openness about and commitment on actions taken on public issues, such as how to protect the environment.

Prior to this legislation, the Brazilian government only recognized NGOs acting in the areas of health, education, and social assistance and used its discretionary judgment to give them tax exemption status. Nowadays, it recognizes civil institutions acting in all areas, including the environment, that involve civil rights, and these organizations may apply for OSCIP status and for public resources (even to support institutional structures) on a competitive basis.

For environmental organizations, this change means the recognition and empowerment of activities supporting biodiversity conservation, protected areas preservation and management, and endangered species recovery and greater legitimacy in their mandate. This legal advance has opened a new institutional path to third-sector organizations, in line with the country's strategic needs.

ACKNOWLEDGMENTS

The following institutions have provided major support for the GLTCP and AMLD over the past 15 years: Brazilian Ministry of the Environment; IBAMA; Centro de Primatologia do Rio de Janeiro/Fundação Estadual de Engenharia do Meio Ambiente (CPRJ/FEEMA—Rio de Janeiro Primate Center/State Foundation for Environmental Engineering); SI/SNZP/FONZ; WWF-U.S. and WWF-Brasil; Jardim Botânico do Rio de Janeiro (JBRJ—Rio de Janeiro Botanical Garden); Conselho Nacional de Desenvolvimento Científico e Tecnológico (CNPq—Brazilian National Science Council); U.S. Fish and Wildlife Service (USFWS); U.S. National Science Foundation (NSF); Frankfurt Zoological Society (FZS); Durrell Wildlife Conservation Trust/Wildlife Trust (DWCT/WT); Whitley Wildlife Trust; Program of Development of the Brazilian Tropical Forests (PD/A); IUCN Netherlands Committee; TransBrasil Airlines; British Embassy; Canadian Embassy; National Geographic Society; LTBF; Calgary Zoo; Lincoln Park Zoological Society, Chicago; Philadelphia Zoological Society; Chicago Zoological Park; Fundação O Boticário de Proteção à Natureza (Boticário Foundation for Nature Protection); and municipalities of Casimiro of Abreu and of Silva Jardim.

The IPÊ and Black Lion Tamarin Project have received support from a large number of institutions, among them: Apenheul Zoo; Canadian Embassy in Brazil; CERC; Duratex S.A.; Fanwood Foundation; Fauna and Flora International; Fundação O Boticário de Proteção à Natureza; Fundo Nacional do Meio Ambiente (FNMA—National Environment Fund); IBAMA; ICCM of the Black Lion Tamarin; IF/SP; DWCT; Lincoln Park Zoo, Chicago; LTBF; The John D. and Catherine T. MacArthur Foundation; Town Council of Euclides da Cunha Paulista; SI; State University of New York (SUNY)–ADC Training Program (USAID); Wildlife Conservation Society; WT; Whitley Animal Protection Trust; and WWF-U.S..

The financing for IESB, in addition to CI, has come from the U.S. Agency for International Development (USAID), WWF-U.S., Ford Foundation, Banco Real S.A., Ford Motor Company, and Jurzykowski Foundation. Other national and local organizations that are important collaborators include IBAMA, Bahian Lion Tamarin Project (Alves 1991, 1992), and the NGO Jupará.

JONATHAN D. BALLOU, DEVRA G. KLEIMAN,
JEREMY J. C. MALLINSON, ANTHONY B. RYLANDS,
CLÁUDIO B. VALLADARES-PADUA, AND KRISTIN LEUS

4

HISTORY, MANAGEMENT, AND CONSERVATION ROLE OF THE CAPTIVE LION TAMARIN POPULATIONS

Captive breeding programs serve a wide variety of functions, ranging from last-ditch efforts to save species almost extinct in the wild (e.g., California condors, black-footed ferrets, Hawaiian geese) to maintaining common (i.e., nonthreatened) species simply for education, research, or recreational purposes (e.g., fallow deer; Frankham et al. 1986). The existence of captive populations is also often justified as an insurance policy against possible extinction of the species in the wild. Alternatively, the function of a captive population may be to learn more about the basic biology of the species.

The relationship between the function of the captive population and the species' status in the wild is clearly seen in the case of the lion tamarins. Three of the four lion tamarin species—golden lion tamarins (*Leontopithecus rosalia*), golden-headed lion tamarins (*L. chrysomelas*), and black lion tamarins (*L. chrysopygus*)—currently have international captive breeding programs. Despite the taxonomic, biological, and geographic closeness of these species, their captive breeding programs differ substantially in terms of their objectives and contributions to species conservation for two reasons: the history of the captive population and the status of each in wild.

Golden lion tamarins have been maintained in captivity since the sixteenth century (Mallinson 1996). They were popular as pets for the European aristocracy, and during the nineteenth and twentieth centuries many continued to be exported for the pet trade as well as for zoos and research laboratories. By the early 1970s, only a small captive population existed for the species, but like most captive populations at that time it was not self-sufficient (Kleiman 1972, 1977a). With a ban on the export of lion tamarins from Brazil beginning in 1968, zoos could no longer maintain the viability of their captive populations by supplementing with wild-caught imports. The current captive population was founded mainly by individuals imported before 1968, with only a few added after the population was being managed on a global scale.

Less is known about the history in captivity of the golden-headed (*L. chrysomelas*) and black lion tamarins (*L. chrysopygus*). There are records of golden-headed lion tamarins being held in captivity at the London Zoo in 1869 and in Rio de Janeiro in 1961. However, the only recorded captive colonies maintained for these species were those established by Adelmar F. Coimbra-Filho in 1972 (for golden-headed lion tamarins) and 1973 (black lion tamarins) at the Tijuca Biological Bank in Rio de Janeiro (later moved to the Centro de Primatologia do Rio de Janeiro [CPRJ—Rio de Janeiro Primate Center]) (Coimbra-Filho et al. 1986a). In both cases, the number of founders was small, and for many years the colonies were small and not self-sufficient. The black-faced lion tamarin (*L. caissara*), discovered in January 1990, has never legally been held in captivity.

In 1983 through 1984, a large number of golden-headed lion tamarins were illegally exported from Brazil to Europe and Japan (Konstant 1986; Mallinson 1984). Most were confiscated by the authorities and subsequently used to establish a well-founded global captive breeding program (Ballou 1989; Mallinson 1989). In 1986, the construction of a hydroelectric dam, which flooded part of the remaining habitat of *L. chrysopygus,* stimulated a minor expansion of the captive breeding program for black lion tamarins. However, this population remained small, with few genetic founders, and was limited to only two Brazilian institutions.

In 1990, and again in 1997, Population and Habitat Viability Analysis (PHVA) workshops were held to review the status of the genus and establish long-term action plans for its conservation (Ballou et al. 1998; Seal et al. 1990). By this time the major conservation focus had changed from captive propagation to the management of wild populations and protected habitat in Brazil. However, the captive populations were also considered an important resource in achieving these goals. By 1997, that for the golden lion tamarin was stable at fewer than 500 animals. Efforts were also underway to reduce and stabilize the captive population of golden-headed lion tamarins, which had grown to over 600. Additionally, there were discussions concerning the best options for managing the captive black lion tamarin population, which had very few founders and low reproduction and was consequently still small.

The three captive lion tamarin populations, with their varied histories, distributions, and genetic and demographic characteristics, exemplify the diversity of roles that captive breeding programs can play in conservation programs for endangered and threatened species. This chapter summarizes the histories and status of these three related captive breeding programs and discusses how the role and objectives of each have been molded by its past, by the status of existing wild populations, and by its in situ conservation program.

THE GOLDEN LION TAMARIN CAPTIVE BREEDING PROGRAM

Historical Review

A few marmosets and tamarins existed in zoos in the nineteenth and early twentieth centuries. The first birth of a golden lion tamarin was recorded at the London Zoo on 13 November 1872 (Zuckerman 1931). However, it was generally considered that the species was delicate in captivity and seldom successfully reproduced. A detailed account of the history of the golden lion tamarin in captivity outside Brazil is presented by Mallinson (1996). In 1962, Alceo Magnanini and Coimbra-Filho first attempted breeding *L. rosalia* in the hopes of establishing a reintroduction program; lion tamarins were already extinct throughout a large part of their original range (Coimbra-Filho and Magnanini 1968). In 1963, the project was moved to the Rio de Janeiro Zoo (Magnanini and Coimbra-Filho 1972; Magnanini et al. 1975), where Coimbra-Filho was the head of the research department (Coimbra-Filho 1965). Later, in 1972 through 1974, Coimbra-Filho and Magnanini established the Tijuca Biological Bank, sponsored by the International Union for Conservation of Nature and Natural Resources (IUCN), the World Wildlife Fund (WWF), and the Brazilian government, specifically for the conservation and captive breeding of lion tamarins (Coimbra-Filho and Mittermeier 1977; Coimbra-Filho et al. 1986b).

By 1965, the international zoo community was becoming aware of the endangered status of the golden lion tamarin. Through the efforts of Clyde Hill of the San Diego Zoo, the American Association of Zoological Parks and Aquariums (AAZPA; renamed the American Zoo and Aquarium Association, or AZA) formally recognized the golden lion tamarin as an endangered species in 1966 and agreed to support a ban on its importation to the United States. The AAZPA also recommended that the species be included in the IUCN "Red Data Book" for rare and endangered species. In 1967, the International Union of Directors of Zoological Gardens and Aquariums (IUDZG; in 2000 renamed the World Zoo Organization, or WZO) also pledged that its members would not import the species and furthermore would help to publicize the animal's endangered status (Hill 1970).

In 1966, the Wild Animal Propagation Trust (WAPT), one of the earliest zoo-based organizations established to conserve endangered species, created a Golden Marmoset Committee. In 1969, under the direction of Donald Bridgwater, the committee began to monitor the status of the captive golden lion tamarin population and to develop management policies, encourage loan agreements, and act as a clearinghouse for the exchange of individuals (Bridgwater 1972b).

In 1972, WAPT sponsored the groundbreaking conference "Saving the Lion Marmoset," at the Smithsonian National Zoological Park (SNZP), which brought together 28 European, American, and Brazilian biologists who reviewed and analyzed all available data on the lion tamarins and other callitrichids. Long-term recommendations for husbandry were developed for research and conservation activities, including support for the breeding program in Brazil, studies of breeding biology, protocols for captive husbandry and management, medical programs, hand-rearing guidelines, interinstitutional cooperation, and the establishment of a studbook and a data bank to record all aspects of the lion tamarins' captive propagation (DuMond 1972; Kleiman 1972).

Following this important meeting, SNZP made a major commitment to the captive propagation and conservation of the golden lion tamarin and launched a long-term investigation into the reproduction, social behavior, and husbandry of the species in captivity (Kleiman 1972). At the time, little was known about the zoo biology of the genus *Leontopithecus.* A lack of knowledge about the lion tamarins' mating and social system led to uncertainty about whether they were best kept as monogamous or polygynous breeding groups (Kleiman 1978b; Kleiman and Mack 1977). Research on behavior and husbandry overseen by Devra Kleiman began in earnest in the mid-1970s. The first international studbook on *L. rosalia* was published by Marvin Jones (1973) and subsequently maintained by Kleiman. Jonathan Ballou, who developed one of the first computerized database systems for managing studbooks, took it over in 1983 (Ballou 1983–1996).

Until 1975, the reproductive trends of the golden lion tamarin population were disheartening. Population increase was minimal, and the survivorship of both adults and young remained poor (Figure 4.1). During the period 1964 to 1974, there was a 1968 peak in numbers of *L. rosalia* in captivity outside of Brazil, but this was believed to be due to a flurry of last-minute exports before trade was banned. The captive population declined from then on (Coimbra-Filho and Mittermeier 1977). In an analysis of breeding performance and survival outside of Brazil through the mid-1970s, Kleiman (1977a, 1977b) concluded that "the breeding programme to date has had conspicuously poor success." It was clear that without a dramatic reversal over the next 2 to 3 years it would be unlikely to achieve a self-sustaining captive population of golden lion tamarins (Kleiman and Jones 1978).

At the end of 1975, there were 83 *L. rosalia* in 16 institutions outside of Brazil, and the Tijuca Biological Bank, Rio de Janeiro, held a further 39 specimens (Kleiman 1977b). These populations increased little over the following 3 years, but by

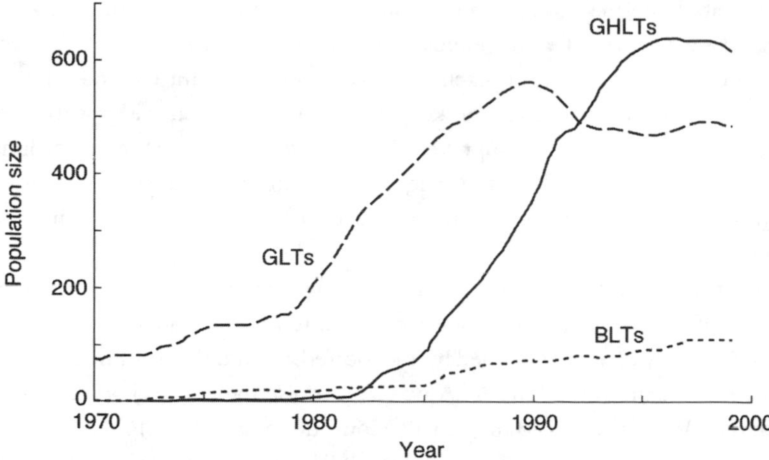

Figure 4.1. Growth of the captive lion tamarin populations, *Leontopithecus* species. GHLTs = golden-headed lion tamarins, GLTs = golden lion tamarins, and BLTs = black lion tamarins.

applying the results of captive research, and by using the studbook data to avoid inbreeding through the careful pairing of young animals, the numbers increased dramatically during 1979 and 1980.

By 1980, when the captive population was expanding exponentially, Kleiman organized a program where zoos owning the specimens began seeding new zoos with breeding groups on loan. In an unusual decision for the time, the zoos owning the majority of the captive population agreed to refrain from selling their stock to prevent the possibility of golden lion tamarins entering the animal trade. Then, in 1981, the zoos both owning and holding golden lion tamarins on loan agreed to formalize this arrangement and become part of an International Research and Management Committee, with all decisions concerning the management of the captive population being in the hands of an elected subgroup. The International Research and Management Committee consisted of representatives from the six institutions owning all the captive *L. rosalia* at the time, as well as elected representatives from zoos holding breeding pairs that were on loan.

The first task of the committee was to develop the Cooperative Research and Management Agreement (CRMA), which established the basis for the international captive breeding program. The CRMA stated, among other things, that (1) all signatories of the agreement would pool their specimens to form a founding stock that would be managed as a single unit, (2) no specimen would be sold, traded, or otherwise used in a commercial transaction, (3) all signatories would agree to abide by the management recommendations of the committee, and no animals would be transferred to another institution or bred without the specific consent of the committee, (4) each institution would submit data on an annual basis to the international studbook keeper; (5) any institution wishing to receive golden lion tamarins must be approved through a formal application procedure to the International Research and Management Committee and sign the CRMA, and (6) the agreement would remain in effect for the lifetime of the animals and their progeny.

All institutions holding *L. rosalia* at that time signed the agreement. The CRMA is still in effect and was the first such agreement to give management responsibility and authority of animals owned by and located at multiple institutions to a central oversight committee. The AZA was later to use the golden lion tamarin International Research and Management Committee and the CRMA as models for its Species Survival Plan (SSP©). Since 1981, the International Research and Management Committee has undergone several significant changes. On 28 September 1990, the Instituto Brasileiro do Meio Ambiente e dos Recursos Naturais Renováveis (IBAMA—Brazilian Institute of the Environment and Renewable Natural Resources) formally recognized the management committee as the official advisory body to the Brazilian government on both captive and wild conservation issues for *L. rosalia* (Kleiman and Mallinson 1998; Chapter 3 this volume). On 31 December 1991, six of the seven institutions that owned the majority of the captive *L. rosalia* population agreed to transfer ownership of their animals to the Brazilian government. Monkey Jungle in Miami, Florida, retained ownership of its animals for financial reasons but still participates fully in the program and remains a signatory of the CRMA. As a result, the Brazilian government is the owner of all but a handful of the 489 existing specimens of captive golden lion tamarins. This may be the first example of a species in which the ownership (but not possession) of all but a very few captive animals by multiple international zoos was returned to the jurisdiction of the native country. With this transfer of ownership and the new responsibilities of the committee, the membership has changed and now consists of members officially appointed by IBAMA, advisors, and elected representatives with expertise in both captive breeding and the conservation of lion tamarins in the wild.

The zoo population of golden lion tamarins has been managed globally since 1973, and in 1981 *L. rosalia* was one of the first species to be designated as part of the AZA's SSP©. The population is currently managed intensively by the International Committee for the Conservation and Management (ICCM) of *Leontopithecus* under the guidance of the studbook keeper.

Research on *L. rosalia* in captivity prior to 1972 was reviewed in Bridgwater

1972a, which includes a full bibliography to that date. Snyder (1974) also reviewed what was known of the species' behavior. Major publications focusing on the development of the captive population's reproduction and management since 1972 include Ballou 1983–1996; Ballou and Houle 1999–2000; Ballou and Mickelberg 2001; Ballou and Sherr 1997; Coimbra-Filho and Maia 1979b; Coimbra-Filho and Mittermeier 1977; Coimbra-Filho et al. 1981; French et al. 1989, 1996b; Hoage 1977, 1978, 1982; Inglett et al. 1989, 1990; Kleiman 1977c, 1978b, 1984a; Kleiman and Evans 1980–1982; Kleiman and Jones 1978; Kleiman et al. 1982, 1986; and Mallinson 1996.

Genetic and Demographic Management

The research conducted in the 1970s led to significant improvements in captive breeding success and a rapid increase in the global population. By the late 1980s the population reached over 500 individuals (Figure 4.1). At this time, the management goal was for rapid population growth. Little attention was paid to genetics beyond avoiding close inbreeding. As a result, there were substantial differences in reproductive success among breeding pairs (e.g., one pair, number 123 and number 195 at SNZP, had been used extensively for behavioral research and produced 55 offspring) (Figure 4.2). Because of the rarity of transferring individual golden lion tamarins between zoos at the time, the genetic contribution from many potential founders was not realized. The studbook shows that of 243 wild-caught individuals brought into the captive population internationally since 1960 (Ballou and Mickelberg 2001), many either failed or were not given the opportunity to reproduce, and consequently their lineages died out. By 1980, the captive population was descended from only 41 true founders (i.e., wild-caught animals that have left descendants in the 1980 population).

Although management objectives for the captive golden lion tamarins were still undefined, it was recognized that the population needed to be expanded, that inbreeding should be avoided and that the population should be managed as a single genetic unit. The overall intent was to secure a genetic and demographic reservoir for, but independent of, the wild population. The International Research and Management Committee was able to meet these objectives as the population expanded globally.

By 1982, the importance of maintaining genetic diversity was widely recognized, and from then until 1989, management protocols attempted to equalize the genetic contribution of founders, in addition to avoiding inbreeding (Ballou and Foose 1996). Genetic management had to compensate for the differential reproduction among early breeding pairs since genes from founders 123 and 195 (mentioned earlier) accounted for almost two-thirds of the diversity in the 1982 gene

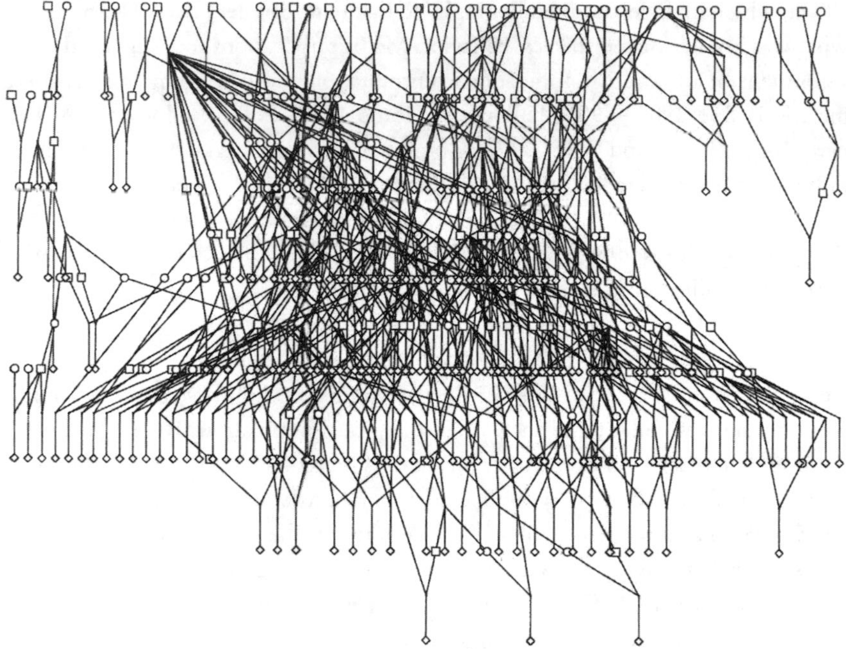

Figure 4.2. Pedigree showing increase and then decrease in the descendants of one pair of golden lion tamarins over time, before and after measures to equalize founder contribution.

pool. The challenge to the management group was to identify (and preferentially breed) descendents from underrepresented founders. A detailed pedigree analysis of founder contributions was the basis for the first Global Masterplan in 1982, and a specific recommendation to breed, transfer, or place in nonbreeding groups was made for all of the 328 individuals in the population.

The 1984 Workshop on Genetic Management of Captive Populations, held at SNZP's Conservation and Research Center in Front Royal, Virginia, provided the basis for formalizing the genetic objectives for this and other captive breeding programs. Soulé et al. (1986) recommended that captive populations be of sufficient size to maintain 90 percent of the wild species genetic diversity for a period of 200 years. (The time frame was later shortened to 100 years.) Given data on species biology and number of founders, it was possible to calculate directly the population sizes needed to reach this objective (Ballou 1987). For golden lion tamarins, this resulted in a target population of about 500, and demographic objectives were developed to manage the population at this size.

As the population approached 500 individuals, reproduction was controlled through selective contraception and single-sex groupings (Ballou 1996). Currently, breeding is accomplished solely to maintain demographic and genetic stability and to provide animals for reintroduction. Presently zero population growth requires breeding only 40 pairs (16 percent of the population) per year so a majority of the population is always held in nonbreeding situations. Careful consideration, therefore, has to be given as to which animals breed, how often, and with whom.

Until 1998, the studbook database was managed through a customized software system developed by Ballou in 1983 and 1984 in dBase (Ballou 1983–1996). The system included demographic and pedigree analyses routines as well as features specifically adapted to the golden lion tamarin population management (e.g., handling of social group information, incorporation of pending, and recommended transfers). The system was converted to SPARKS in 1998 (Scobie 1994).

The task of identifying the genetically most important individuals to breed can be formidable when pedigrees are complex. As that for *L. rosalia* grew more intricate, so too did the process of identifying the genetically most valuable individuals to breed. It was no longer clear that scanning and comparing the hundreds of tables of founder contributions (i.e., the percentage of genes in each individual descending from each founder) was sufficient to identify those animals from the underrepresented founders. Moreover, Lacy (1989) determined that equalizing founder contribution was not the best way to maintain genetic diversity in the managed populations. Soon after, Ballou and Lacy (1995) developed the concept of "mean kinship." Mean kinship is a numeric value calculated for each living animal as the average relationship between that individual and all the living animals in a population. Individuals with few relatives in the population have low mean kinships and are considered more genetically valuable than individuals with many relatives (and high mean kinship). By calculating the mean kinship of every individual, then sorting the mean kinship values lowest to highest by sex, it was easy to identify the most valuable genetic pairings. Mean kinship strategy is now routinely used to manage captive golden lion tamarins (and most other captive populations), resulting in a significant increase in genetic diversity in the population (Figure 4.3).

The golden lion tamarin population rose to 550 individuals in the 1980s and then declined as demographic management took effect, leveling the population off at about 490 animals. The population has achieved its demographic objective of zero population growth for the last 10 years. The population size in 2001 was 489 individuals at 143 zoos in 29 countries (Table 4.1). Despite its global distribution, the population is still managed as a single genetic and demographic unit, with regional coordinators in Australia, Europe, and South America. All management recommendations are based on analyses of the international studbook.

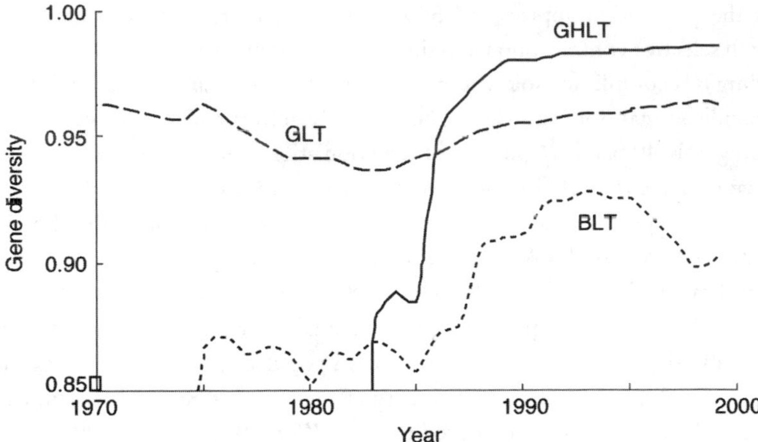

Figure 4.3. Genetic diversity (percentage of gene diversity of the wild population) in three captive lion tamarin populations over time. Without management, genetic diversity is predicted to decrease at a rate of $1/0.6\ N$ per generation (assuming effective size is 30 percent of actual size). In golden lion tamarins, genetic diversity has steadily increased since management for genetic diversity began in the mid-1980s. For black lion tamarins, genetic diversity increased but then decreased, primarily as a result of overreproduction of a few successful breeding pairs in recent years. For golden-headed lion tamarins, genetic diversity vastly increased in the mid-1980s due to the addition of new founders and coordinated management of breeding in the population. Continued management of the captive population and a regular input of new founders (as a result of forest clearance and confiscations) make it possible to maintain this high level of genetic variation. GHLT = golden-headed lion tamarin, GLT = golden lion tamarin, and BLT = black lion tamarin.

The initiation of the reintroduction program in 1984, and the discovery of a number of other fragmented populations in the wild in 1990 (Kierulff 2000; Kierulff and Procópio de Oliveira 1996), changed the role of the captive population in the context of the conservation of the species. The wild "metapopulation" of *L. rosalia* currently consists of about 1,000 individuals (Chapters 2 and 12 this volume) spread among 17 geographically isolated populations that range in size from just 1 family group to less than 300 in the Poço das Antas Biological Reserve (Franklin and Dietz 2001). This includes about 120 individuals in the population established through translocations into the União Biological Reserve and about 360 golden lion tamarins in the multiple sites of the reintroduction program (Chapters 12 and 13 this volume). Although there were more individuals than

Table 4.1
Status and Roles of Captive Lion Tamarin Populations

Species	Status in the Wild	Status in Captivity	Role of Captive Population
Golden lion tamarin (GLT)	$N = 640$ (wild) $N = 360$ (reintroduced) Many fragments, two larger ($N > 200$) protected populations Metapopulation management underway Reintroduction into unoccupied habitats Translocations from severely threatened to protected habitats	$N = 489$, stable since 1990 Globally managed, 143 institutions Levels of inbreeding and gene diversity adequate but not high Gene diversity 0.963 Average inbreeding 0.023	Genetic and demographic reservoir for wild populations Genetic diversity maintained at 90+ percent for 100 years Captive population actively involved in species metapopulation management "Source" population in GLT metapopulation. Provides animals for reintroduction Educational/exhibit/research value
Golden-headed lion tamarin (GHLT)	$N = 6{,}000–15{,}000$ Still relatively widely distributed Increasing deforestation projected to create small isolated populations and one larger population in Una Biological Reserve within 5–10 years Reintroduction not appropriate	$N = 598$, beginning to decline Extensive founder base Gene diversity 0.986 Average inbreeding 0.028	An independent and self-sufficient genetic and demographic reservoir for the species Current goal of 90 percent genetic diversity maintained for 100 years No reintroduction projected Education/exhibit/research value
Black lion tamarin (BLT)	$N = 990$ Highly fragmented One large protected population Eight small isolated populations Experimental translocations, recolonization of unoccupied habitat underway	$N = 112$, small, growing slowly Low genetic diversity (gene diversity 0.902) Average inbreeding 0.043 Until recently only at two Brazilian institutions Since 1995, small but expanding collections in Europe, the United States, and Australia	Nucleus population Expansion of $150 < N < 200$ Wild genetic diversity maintained at 95 percent Regular gene flow between other (wild) populations

thought prior to Kierulff's survey (Kierulff and Procópio de Oliveira 1996), the PHVA held in 1997 reinforced the impression that the species was still not viable over the long term without continued management interventions (Ballou et al. 1998). The conservation of *L. rosalia* was now seen as a metapopulation management challenge balancing the contributions (and costs) of each population, including the captive population, with the goal of the preservation of the species.

The captive population's role in this metapopulation took on an added dimension. Although still managed as a secure genetic and demographic reservoir for the species, it is no longer managed independently of the wild population. It now acts as a "source" within the metapopulation of sufficient genetic and demographic potential to supply animals, as needed, for reintroduction, for colonization, or as a supplement to other populations without jeopardizing its own genetic and demographic viability.

However, the captive population's role as a source to supply animals for the reintroduction program may be limited. The reintroduced population (reintroduced animals and their descendants) stands at over 360 individuals and is growing rapidly at about 25 percent per year (Chapter 13 this volume) due solely to reproduction. The PHVA projection models indicate that it has the capacity to grow steadily without additional reintroduced animals (Ballou et al. 1998). Furthermore, the growing population is rapidly expanding to fill the remaining available forest. Decisions will soon have to be made about the utility of continuing to reintroduce individuals or leaving the reintroduced population to expand on its own. Pedigree analysis indicates that the genetic diversity of the reintroduced population may still be enhanced through selectively reintroducing captive-born individuals that are unrelated to previously reintroduced animals (Ballou 1992).

In a more general sense, the role of the captive population as just a "source" of animals goes beyond the immediate need of reintroducing animals. The primary objective of the Golden Lion Tamarin Conservation Program (GLTCP) has been to preserve habitat using golden lion tamarins as a flagship species (Dietz et al. 1994b). Viability of the wild populations varies due to differences in abundance and vulnerability to threats (Kierulff and Procópio de Oliveira 1996). Future extinctions of some fragments are likely, and these areas should be recolonized through translocation of golden lion tamarins from other viable (i.e., source) wild populations (e.g., the reintroduced population), through reintroduction of captive animals, or through both. Vacant but suitable habitat should also be considered part of the *L. rosalia* metapopulation as a currently "extinct" population in need of recolonization. For example, the União Biological Reserve was uninhabited by golden lion tamarins but recolonized through translocations from other more threatened populations (Chapter 12 this volume).

The captive population serves other objectives as well. Probably the most important is the link it provides between the in situ programs and the global zoo and conservation community. Information and educational materials about golden lion tamarin conservation specifically, and conservation issues more generally, can be distributed to and exhibited at zoos with golden lion tamarins throughout the world, reaching tens of millions of visitors each year. Since 1997, the *Tamarin Tales* newsletter (Ballou 1997–2001) has been sent to all major donors and zoos involved with the breeding and conservation of lion tamarins. This newsletter provides updates on all research and conservation activities for all species of lion tamarins. In addition to the public education value, this venue also has tremendous potential for fundraising. For example, the newsletter is used to solicit contributions from zoos participating in the captive breeding program. These donations go to the Lion Tamarins of Brazil Fund (LTBF), which has raised over $280,000 (U.S.) for field research and conservation of lion tamarins (see Chapter 3 this volume).

THE GOLDEN-HEADED LION TAMARIN CAPTIVE BREEDING PROGRAM

Historical Review

Coimbra-Filho (1970a) published some behavioral observations of *L. chrysomelas* in captivity, based on three individuals kept at the Rio de Janeiro Zoo in 1961. He was the first to set up a small colony, at the Tijuca Biological Bank; it was later moved to CPRJ in the early 1980s.

The captive population exploded, however, in the mid-1980s as a result of a massive illegal export of golden-headed lion tamarins from Brazil. Between 1983 and 1984 large numbers found their way into the hands of animal dealers and private exotic animal collections, primarily in Belgium and Japan (Ballou 1989; Konstant 1986; Mallinson 1984, 1987b; Chapter 1 this volume). In March 1985, the Brazilian government environmental agency IBAMA invited Jeremy Mallinson to form and chair an International Recovery and Management Committee (IRMC) for the species. The initial objective was to work with IBAMA to repatriate the contraband animals, and indeed, the committee was successful in having some of the illegally exported lion tamarins returned to Brazil (Ballou 1989; Mallinson 1987a, 1987b). As a result of these illegal exports, zoos in Europe, Japan, the United States, and Brazil established additional breeding colonies. A preliminary international studbook was published by Mallinson (then at the Jersey Wildlife Preservation Trust, or JWPT) in 1987 (Mallinson 1987a), and for the next 5 years the official studbook was maintained by Georgina Mace (on behalf of Jersey Zoo), subsequently transferred to Helga de Bois in 1993, and finally to Kristin Leus (the last two of the Royal Zoological Society of Antwerp) in 1998 (Leus 1993–1999). Although most captive golden-headed lion tamarins are owned by Brazil, a number are still privately owned by animal dealers and breeders, especially in Europe, Japan, and Brazil. The IRMC and the studbook keeper manage the captive population following the strategies developed for the golden lion tamarins. Institutions wishing to participate in the captive breeding program must sign an official management agreement and have their application approved by the IRMC. The IRMC was formally recognized as an advisory committee by IBAMA during the 1990 PHVA, and like the pre-existing IRMC for *L. rosalia,* the mission was expanded to include consideration of the conservation, research, and management of wild populations (Mallinson 1986, 1989).

Confiscated animals, however, continued to be added to the population. Golden-headed lion tamarins were frequently kept as pets in Brazil, and the administration of the Una Biological Reserve in Bahia confiscated and held such animals in cages within the reserve. Due to potential negative effects on the wild population (e.g., disease transmission and the disruption of social behavior of wild golden-headed lion tamarins), the IRMC recommended to IBAMA that the colony be removed. In 1992 Cristina Alves organized the transfer of 25 golden-headed lion tamarins held captive in Una to a number of Brazilian zoos, and the cages within the Una Biological Reserve were demolished. Alves established a second facility for holding confiscated pet golden-headed lion tamarins in 1994, but it too was closed down in 1997 (due to lack of funding and the need to halt further population growth, as well as to discourage further confiscations of pet animals), and again the animals were sent to Brazilian zoos. At the end of 1999 the tenth international studbook listed the ex situ scientifically managed population of *L. chrysomelas* worldwide as 598 individuals (Leus and De Vleeschouwer 2001).

Research on the management and reproduction of captive golden-headed lion tamarins includes a comparative analysis of breeding performance of the three lion tamarin species at CPRJ (French et al. 1996b), observations on infant carrying (De Vleeschouwer 2000; Van Elsacker et al. 1992; Oliveira et al. 1999), studies on the use of contraceptive hormone implants (De Vleeschouwer et al. 2000b, 2000c; Van Elsacker et al. 1994), and social and reproductive behavior studies (Alonso et al. 1997a, 1997b; De Vleeschouwer 2000; De Vleeschouwer et al. 2000a, 2001; Chapters 6 and 9 this volume). The Durrell Wildlife Conservation Trust (DWCT) has undertaken a number of comparative studies including cross-generic food sharing between lion tamarins and other tamarins (Feistner and Price 1999; Price and Feistner 1993; Chapter 9 this volume), behavioral adaptation of captive-born *L. chrysomelas,* and hormonal studies in a free-ranging environment (Chaoui and Hasler-Gallusser 1999; Moore 1997).

Genetic and Demographic Management

The golden-headed lion tamarins placed in the captive breeding program through the confiscation of illegal exports in 1984 and 1985 provided the most substantial founder base for any captive population recorded: 96 contributing founders. The considerable knowledge of the zoo biology of lion tamarins, gained from research on *L. rosalia,* was applied to breeding the golden-headed lion tamarins. When the confiscated animals were placed in institutions with experience in breeding callitrichids, the population exploded. By the end of 1999, there were 611 animals in 99 institutions (Ballou et al. 1998; De Bois 1994; Leus 1998, 1999; Leus and De Vleeschouwer 2001; Mace and Mallinson 1992; Van Elsacker et al. 1994; Figure 4.1).

The importance of capturing the genetic representation of founder individuals was recognized at the outset, and as a result, low levels of inbreeding and very high levels of genetic diversity have been retained in this population. Probably the most significant complications were the rapid and continuing growth rate of the population (Figure 4.1) and the relatively large number of animals held by private breeders in Brazil, Japan, and Europe. While the population in North America has been controlled at 100 individuals over the last 10 years, and the one in Europe at 220 individuals during the last 6 years, it has continued to rise in Brazil, partly because of the continued confiscation of wild-caught animals.

At the start of the captive breeding program, very little was known about the golden-headed lion tamarins' status in the wild (Seal et al. 1990). The rapid destruction of the forest in this region argued for its threatened status and the coordination of the captive breeding program (Coimbra-Filho 1970a; Coimbra-Filho and Mittermeier 1972, 1973, 1977). A survey carried out from 1991 to 1993 found that the wild populations were not as fragmented as those of golden lion tamarins or black lion tamarins, and wild populations were estimated to be about 6,000 to 15,000 individuals (Pinto and Rylands 1997; Chapter 2 this volume). However, as forests continue to be cleared, in all likelihood the golden-headed lion tamarins' distribution will come to resemble those of golden lion tamarins and black lion tamarins: a few isolated and small populations in whatever forests are remaining with (probably) one larger protected population, numbering only in the hundreds at best, in the Una Biological Reserve (Ballou et al. 1998; Chapter 2 this volume). The captive population should therefore continue to be managed to ensure that it remains a genetic and demographic reservoir for the species. A goal of maintaining 90 percent of the species' genetic diversity for 100 years is an appropriate and realistic objective (Table 4.1). The current size certainly meets that objective (and is well above what is needed). Reintroduction is not considered necessary in the near future. As with the golden lion tamarin population, the captive golden-headed lion tamarin also serves as an educational ambassador representing the conservation efforts needed for the preservation of the Atlantic Forest of Brazil.

While additional founders will continue to be available as forests are cleared and animals confiscated, they are not currently needed for the genetic health of the population. In fact, confiscated animals pose more of a problem than a benefit because they are often in poor health, rarely breed, and take up valuable cage spaces when placed in zoos. From a demographic perspective, the captive population should be reduced. Maintaining 90 percent genetic diversity for 100 years requires only 350 animals. However, controlling reproduction for this species is more problematic than for golden lion tamarins. Melengestrol acetate contraceptives successfully used for golden lion tamarins appear to have low reversibility in golden-headed lion tamarins (De Vleeschouwer et al. 2000b, 2000c; Wood et al., submitted).

THE BLACK LION TAMARIN CAPTIVE BREEDING PROGRAM

Historical Review

At the time of the WAPT conference in 1972, black lion tamarins had just been rediscovered and had never been formally kept in captivity (Coimbra-Filho and Mittermeier 1972). The following year, Coimbra-Filho captured seven from Morro do Diabo State Park, the start of the first breeding colony for the species at the Tijuca Biological Bank (Coimbra-Filho 1976a, 1976b). However, the population grew very slowly, possibly due to the small number of founders and consequent inbreeding. In 1984 there were 26 *L. chrysopygus* at the CPRJ (Valladares-Padua 1987).

The inundation area of the Rosana hydroelectric dam under construction in 1985 included 3,000 ha in the Morro do Diabo Park, and a rescue operation was set up for the eight black lion tamarin groups located there (Sério 1986; Valle and Rylands 1986). One group of six was sent to CPRJ in November 1985. A further 31 lion tamarins were captured and maintained in cages, with the objective of translocating them to a nearby forest (Carvalho and Carvalho 1989). Deterioration in the health of these animals, however, resulted in six males and eight females surviving, these being transferred to the São Paulo Zoo in November 1986 (Simon 1988).

In 1987, an International Committee for the Preservation and Management of the Black Lion Tamarin was established to contribute to decision making regarding the rescue operation, field research, and the expanded captive population. Faiçal Simon, general curator of the São Paulo Zoo, chaired the committee, with Devra Kleiman as cochair, and developed the first studbook (Simon 1988, 1989; Valladares-Padua and Simon 1990). On 18 July 1990 IBAMA officially recognized the IRMC for *L. chrysopygus*.

The first international studbook was published in 1988 (Simon 1988). The São Paulo Zoo and CPRJ were the only institutions breeding *L. chrysopygus* until August 1990, when six animals were transferred to the Jersey Wildlife Preservation Trust (now the DWCT). Breeding at the Jersey Zoo was very successful, and as a consequence, the population began to grow more rapidly. Since then, animals have been sent to additional zoos in Europe, North America, and Australia. The 1999 studbook (31 December 1999) registered a captive population of 112 *L. chrysopygus* in 11 institutions.

Less has been published on captive black lion tamarins than the other lion tamarins (Coimbra-Filho 1976a, 1976b; Coimbra-Filho and Maia 1977, 1979b; Coimbra-Filho and Mittermeier 1977; Coimbra-Filho et al. 1981; Pissinatti et al. 1992; Price 1997, 1998; Rosenberger and Coimbra-Filho 1984; Wormell in press). French et al. (1996b) presented a detailed comparative analysis of captive breeding of *L. rosalia, L. chrysomelas,* and *L. chrysopygus* based on his work at CPRJ.

Genetic and Demographic Management

The black lion tamarin population was founded with fewer animals than the other two species, and it remained small because of poor reproduction and survival, as well as delays in decisions on procedures for its expansion. Furthermore, black lion tamarins have been considered more difficult to breed. The first captive Masterplan providing recommendations for breeding priorities and exchanges between the São Paulo Zoo and CPRJ was produced in 1991, but few of the recommendations were implemented.

In 1992, the IRMC felt that the lack of genetic diversity and the small size of the captive population seriously impaired any significant role it could have in the conservation of the species. Revitalization with new founders and population expansion were considered essential for the captive population to serve as a genetic reservoir in its own right (Mansour and Ballou 1994), but they were not feasible at the time. However, recognizing the security that a captive population could provide, the IRMC was reluctant to abandon the program. The solution was provided in 1996, when Valladares-Padua and Ballou (1998) proposed an overall metapopulation management program, with the captive animals playing the role of a "nucleus" population (Foose et al. 1986).

Like the wild golden lion tamarin population, black lion tamarins in the wild occur in one large protected (Morro do Diabo State Park) and many small isolated populations (Chapters 1, 2, and 14 this volume). Valladares-Padua and the conservation nongovernmental organization Instituto de Pesquisas Ecológicas (IPÊ—Institute for Ecological Research) initiated the metapopulation management project by translocating animals between fragments to recolonize patches of uninhabited forest (Valladares-Padua and Cullen 1994). In 1998, Valladares-Padua and Ballou proposed that the captive population be limited to 150 to 200 individuals in a small number of dedicated zoos and that genetic diversity be continuously maintained at 95 percent of that of the wild population. This is higher than the genetic objectives set for most captive populations and would require periodic supplementation of the captive population with new founders as well as reintroduction of captive-born animals back into the wild (Figure 4.4). This small but highly diverse population would then serve as a nucleus or seed from which a fully expanded, genetically healthy captive population could be developed, if needed.

Achieving these levels of genetic diversity would require an initial supplement of 14 founders from the wild in the next one or two generations, followed by pe-

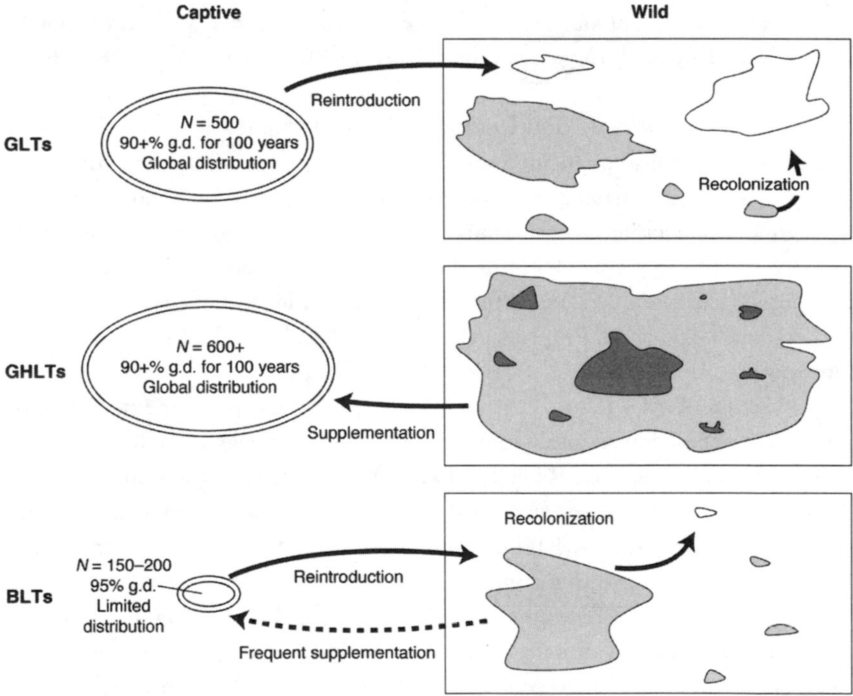

Figure 4.4. The relationship between the captive populations (circles) and the wild populations (squares) is shown for each of the lion tamarin species with captive populations. Solid arrows show most important current links and the dotted line represents a possible future link between populations. Light gray areas show current distributions. For golden-headed lion tamarins, the dark gray areas show the expected future distribution of the wild population. GLTs = golden lion tamarins, GHLTs = golden-headed lion tamarins, BLTs = black lion tamarins, and g.d. = percentage of gene diversity to preserve.

riodic supplements of 3 or 4 founders every generation. The viability of the wild populations will probably not be jeopardized by the acquisition of new founders because the areas containing them are believed to be at capacity density and the groups/individuals removed would be naturally replaced with subadult and adult dispersers from other groups. The IPÊ has begun research into the effects of these kinds of group and individual removals and reintroductions on the dynamics and ecology of the populations (Chapter 14 this volume). The metapopulation management program of the black lion tamarin has just begun and involves what Valladares-Padua et al. (2001; Chapter 14 this volume) referred to as "conservation shifts." One new founder was added to the captive population (1999), two groups were translocated (1995 and 1999), two mixed groups (captive-born and wild-born individuals) were reintroduced (1999 and 2000), and two males underwent a "managed dispersal" within the Morro do Diabo State Park (1999) (Chapter 14 this volume).

The major challenge for this captive population is to ensure that the demographic and genetic expansions of the population are achieved through reproduction of new founders rather than older, overrepresented lineages already in captivity. Overreproduction of the common gene lines has already caused a decrease in genetic diversity in recent years (Figure 4.3).

CONCLUSIONS

The roles of the captive populations in the conservation programs of the lion tamarin species differ not only due to their history but also due to their differing situations in the wild (Figure 4.4). The global captive breeding program for the golden lion tamarin began with individuals already existing in a few zoos in the 1970s—in some cases already two or three generations removed from the original wild-caught "founders." The population grew rapidly in the 1980s and expanded into an international captive breeding program, though still centrally managed. Because of the careful genetic management of the population, genetic diversity has been preserved and inbreeding avoided. As understanding of the status of the wild population grew, and in situ conservation programs developed, so too did the role of the captive population. It now serves as one of multiple populations in the species' metapopulation and enhances the wild population's viability by acting as a "source" for reintroduced animals. However, the captive population is still managed for self-sufficiency (i.e., at a size sufficient to retain high levels of genetic diversity and demographic security, even though there have been no further imports of new founders). Thus, it continues to serve as a genetic and demographic reservoir, providing a source of animals in the event of the loss of the wild population.

The captive golden-headed lion tamarin population was extremely well founded and has grown rapidly. It has developed into a large, self-sufficient, and genetically and demographically healthy population. In contrast to the management of the captive golden lion tamarins, the current management objectives for captive golden-headed lion tamarins are that they serve as an independent (rather than interactive) genetic and demographic reservoir for the species. Use of the captive population for reintroductions is not envisioned in the near future. In direct contrast to the objectives for the captive populations of the other two species, the objectives for the captive black lion tamarin population are that it remains relatively small but maintains higher levels of gene flow with the wild (Figure 4.4). Rather than serving as a genetic and demographic reservoir for the species, this captive population exists as a nucleus from which a larger population can be derived.

ACKNOWLEDGMENTS

The authors wish to thank all the zoos that participate in the captive breeding programs for the lion tamarin species. Their involvement and contributions are critical to the continued success of these conservation efforts. Kristel de Vleeschouwer, Center for Research and Conservation, Royal Zoological Society of Antwerp, kindly reviewed the manuscript and assisted with the international studbook for the golden-headed lion tamarin. The authors also wish to thank Maria Iolita Bampi and Rosemary Mamede of IBAMA. Their leadership and vision have been the foundation of the development of lion tamarin conservation in Brazil. Kristin Leus would like to thank Jeremy Mallinson of the DWCT for helpful comments on the manuscript.

Part Two

THE BIOLOGY OF LION TAMARINS

HÉCTOR N. SEUÁNEZ, ANTHONY DI FIORE,
MIGUEL ÂNGELO M. MOREIRA,
CARLOS ALBERTO DA S. ALMEIDA,
AND FLÁVIO C. CANAVEZ

5
GENETICS AND EVOLUTION OF LION TAMARINS

Lion tamarins (genus *Leontopithecus*) can be divided into four morphologically distinct types that are dispersed among a number of forest fragments in the Atlantic Forest of Brazil (Chapter 2 this volume). Presumably, these types have recently derived from a common ancestor that had a wider geographic distribution throughout the Atlantic Forest region and whose original range has been drastically reduced in recent years. All of the extant lion tamarins are presently considered endangered or critically endangered by the World Conservation Union (IUCN), and all are characterized by low population sizes and densities in the few places where they persist in the wild. Small effective population size, together with increasing fragmentation of their natural habitats, is likely to have important consequences for the structuring and maintenance of genetic diversity in these taxa. Indeed, it is very likely that each of the lion tamarin species is currently undergoing a significant population bottleneck in the wild, resulting in loss of natural genetic variability and increased levels of inbreeding.

In this chapter, we review the varied molecular studies that have addressed the systematics, evolution, and population-level genetic architecture of lion tamarins. To date, such studies have focused on four major issues: What is the phylogenetic status of the lion tamarins in relation to other platyrrhine primates and specifically to other callitrichids? What are the phylogenetic relationships among the four currently recognized lion tamarin morphotypes? How genetically variable are lion tamarins, especially in relation to other primates, and how is that variation partitioned among the different lion tamarin morphotypes and among different populations of each morphotype? How is genetic variation structured at the population level, and what can molecular studies reveal about the social systems and breeding patterns of these taxa?

One additional issue that we do not discuss at length but is nonetheless important concerns conflicting ideas over how many valid species of lion tamarins should be recognized. In his classic monograph on New World primates, Hershkovitz (1977) considered the genus *Leontopithecus* monotypic and classified the three then-known morphotypes of lion tamarins (*rosalia, chrysomelas,* and *chrysopygus*) as subspecies of *L. rosalia,* the golden lion tamarin. Not long thereafter, however, this classification was revised, and each of these subspecies was elevated to species status (Groves 1993). The recently described fourth lion tamarin morphotype *L. caissara* has also been accorded species status (Lorini and Persson 1990).

In this chapter we have opted to consider the four *Leontopithecus* morphotypes as valid species, in view of their allopatric distribution and ecological adaptations. Nonetheless, it is important to note that the degree of genetic differentiation present between these taxa at a number of genetic loci is less than that typically found between individuals of the same species or even the same subspecies in some mammalian taxa, and a number of the studies reviewed below suggest that the divergence of the various lion tamarins from one another has been recent. Although debate over the taxonomic status of these morphotypes might appear trivial, it is actually quite important to settle this point because taxonomic designation plays a central role in establishing and prioritizing conservation policy (O'Brien and Mayr 1991).

CALLITRICHID PHYLOGENETICS

Early Ideas

Of the four major issues concerning the evolution and genetics of *Leontopithecus* outlined above, the question of the phylogenetic position of the genus within the Callitrichidae has received the most attention. The earliest molecular data addressing this question come from the seminal immunological and protein electrophoretic studies of Cronin and Sarich (1975, 1978; see also review in Sarich and Cronin 1980). Cronin and Sarich (1975, 1978) used immunological cross-reactivity comparisons involving the serum albumins and transferrins to reconstruct generic-level phylogenetic relationships of the New World monkeys, and they further examined relationships among callitrichid genera and species using plasma protein electrophoresis. Cronin and Sarich concluded that lion tamarins were the earliest group to split off from the ancestral callitrichid stock. The remaining callitrichids then split into two lineages, one containing the Amazonian tamarins (the genus

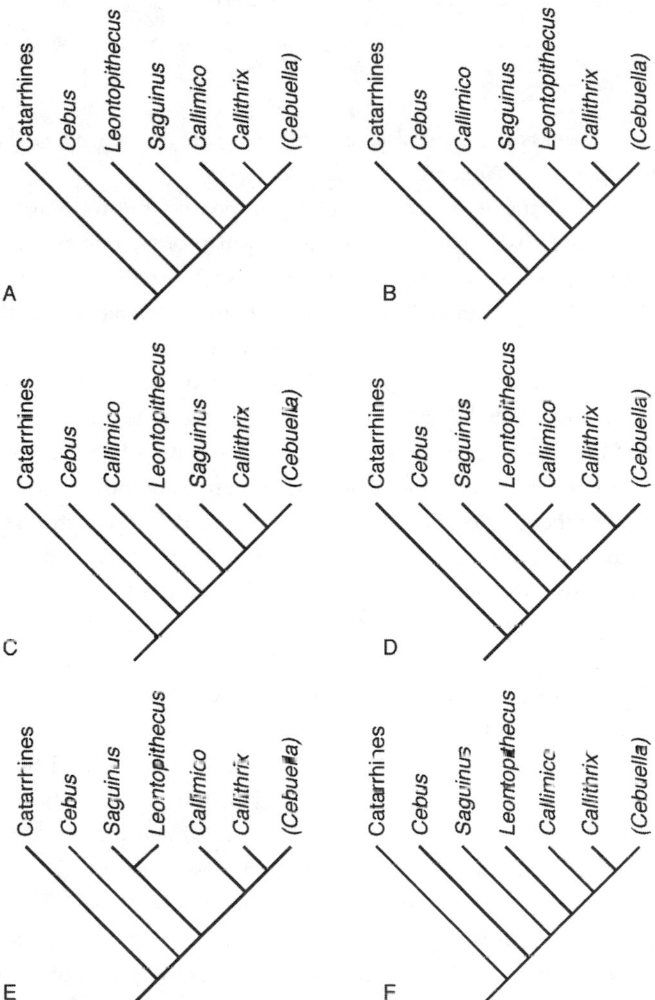

Figure 5.1. Alternative phylogenetic relationships among callitrichid primates. Early assessments of callitrichid phylogeny were made using (**A**) immunological cross-reactivity comparisons and plasma protein electrophoretic data (modified from Cronin and Sarich 1978), (**B**) cranial and postcranial morphological character data (modified from Ford 1986; Rosenberger 1981), and (**C**) dental character data (modified from Kay 1990). More recent phylogenetic arrangements have been reconstructed using nuclear DNA sequence data from (**D**) intron 1 of the interphotoreceptor retinol-binding protein (IRBP) gene (Barroso et al. 1997; Harada et al. 1995; Schneider et al. 1993, 1996), (**E**) coding and noncoding regions of the ε-globin gene (Harada et al. 1995; Porter et al. 1995, 1997a, 1997b; Schneider et al. 1993, 1996), and (**F**) noncoding introns of the glucose-6-phosphate dehydrogenase (G6PD) gene, both coding and noncoding regions of the β2-microglobulin gene (Canavez et al. 1999b; von Dornum and Ruvolo 1999), and a noncoding intron and short flanking region of the von Willebrand factor (vWF) gene (Chaves et al. 1999).

Saguinus) and one that included Goeldi's monkeys (the genus *Callimico*) and marmosets (the genera *Callithrix* and *Cebuella*) (Figure 5.1A).

During the following decade, however, this proposed set of relationships among the callitrichids was challenged. Using cladistic, parsimony-based analyses of morphological characters, Rosenberger (1981; see also Rosenberger and Coimbra-Filho 1984), Ford (1986), and Kay (1990) all relegated *Callimico* rather than *Leontopithecus* to the position of basal sister taxon within the Callitrichidae, while preserving the idea of a close sister relationship between *Callithrix* and *Cebuella*. Where their proposed phylogenies differed from one another was in their relative placement of the lion tamarins and Amazonian tamarins between the marmosets and Goeldi's monkeys. Using large sets of both cranial and postcranial characteristics, both Rosenberger (1981) and Ford (1986) placed *Leontopithecus* as the most closely related sister taxon to the marmosets, while Kay (1990), using only dental characters, envisioned *Saguinus* in this role, with *Leontopithecus* branching off earlier on the tree (Figure 5.1B and 5.1C). More recently, Snowdon (1993) used acoustic features of long-call structure to construct a "vocal taxonomy" of the callitrichids that coincided perfectly with the morphology-based phylogenies proposed by Rosenberger (1981) and Ford (1986).

To many, the view of *Callimico* as the earliest taxon to branch off from the callitrichid ancestral stock made logical sense: all of the callitrichids but Goeldi's monkeys seem to share important derived anatomical and physiological characteristics not seen in other platyrrhines, such as the loss of the third molar and the unusual habit of frequent twinning. However, as more molecular data have become available in recent years, the consensus seems to be swinging back toward the taxonomy proposed by Cronin and Sarich. In fact, nearly all of the currently available molecular data favors placing the both the lion tamarins and the Amazonian tamarins basal to a clade consisting of Goeldi's monkeys and marmosets. The various types of molecular data supporting this view, along with those that address the issue of intrageneric relationships among the lion tamarins, are reviewed below.

Chromosome Studies

Karyotypic comparison—the comparative study of chromosome number and form—can be a useful tool in phylogenetic analysis because it is often possible to reconstruct the evolutionary history of chromosomal rearrangements that have accompanied the radiation and diversification of taxa by using the same parsimony-based, cladistic procedures employed by morphologists and geneticists. Héctor Seuánez and colleagues (Canavez et al. 1996; Seuánez et al. 1988, 1989) have made extensive use of karyotype comparisons to address the question of phylogenetic relationships among the callitrichids. Their studies reveal that callitrichid primates comprise a tightly related group of taxa within which karyotypic evolution has been fairly conservative. Moreover, using *Cebus apella* as an outgroup, Seuánez and colleagues concluded that the chromosomal complement of *Leontopithecus* is basal with respect to that of *Callimico,* Amazonian and coastal *Callithrix* species, and pygmy marmosets. Although representatives of the Amazonian tamarins (genus *Saguinus*) were not included in their study, the recovered karyotype-based phylogeny (Figure 5.2) is consistent with that proposed earlier by Cronin and Sarich based on immunological cross-reactivity and protein electrophoretic data.

Unfortunately, karyotype comparisons have offered little insight into the subgeneric phylogeny of lion tamarins. For example, in a comparative chromosomal analysis of *L. rosalia, L. chrysopygus,* and *L. chrysomelas,* Seuánez et al. (1988) found that all three of these lion tamarins shared a diploid number of 46 and had identical G-band karyotypes. Because karyotypic identity does not necessarily imply a lack of underlying genomic diversity, captive management should still strive to preserve the reproductive isolation of the different lion tamarin morphotypes seen in the wild. Rhesus monkeys and baboons, which share identical G-C-R and Ag-NOR band karyotypes (Turleau et al. 1978), provide a clear case of how two species exhibiting differences in morphology, behavior, and ecology drastic enough to warrant placing them in separate genera can nonetheless manifest morphologically similar chromosomal complements.

Mitochondrial DNA Studies

Additional perspectives regarding the phylogenetic placement of *Leontopithecus* within the Callitrichidae come from DNA sequence data for several loci in the mitochondrial genome. The mitochondrial genome in mammals contains some 16 to 18 kilobase pairs and codes for 13 proteins, two rRNAs (ribosomal-RNAs), and 22 tRNAs (transfer-RNAs). Mitochondrial DNA has a much faster nucleotide substitution rate than nuclear DNA, which makes it ideal for studying phylogenetic relationships of very recently diverged species and populations (W. M. Brown 1985). Moreover, because the mammalian mitochondrial genome is inherited solely through the maternal line and seems to be subject to little to no recombination, mitochondrial haplotypes can become fixed much more rapidly in a population than nuclear genes and coalesce more quickly.

Among the 13 protein-coding genes in the mitochondrial genome, cytochrome *b* is frequently analyzed in phylogenetic studies because it contains evolutionarily conserved regions (Kocher et al. 1989). Moreira and colleagues (Moreira and Seuánez 1999; Moreira et al. 1996) have used PCR to amplify and sequence a 301–base pair segment of the cytochrome *b* region of single individuals of *L. rosalia, L. chrysomelas,* and *L. chrysopygus,* as well as representatives of *Cebuella pygmaea, Saguinus oedipus, Callimico goeldii, Cebus apella,* and two species of *Callithrix,*

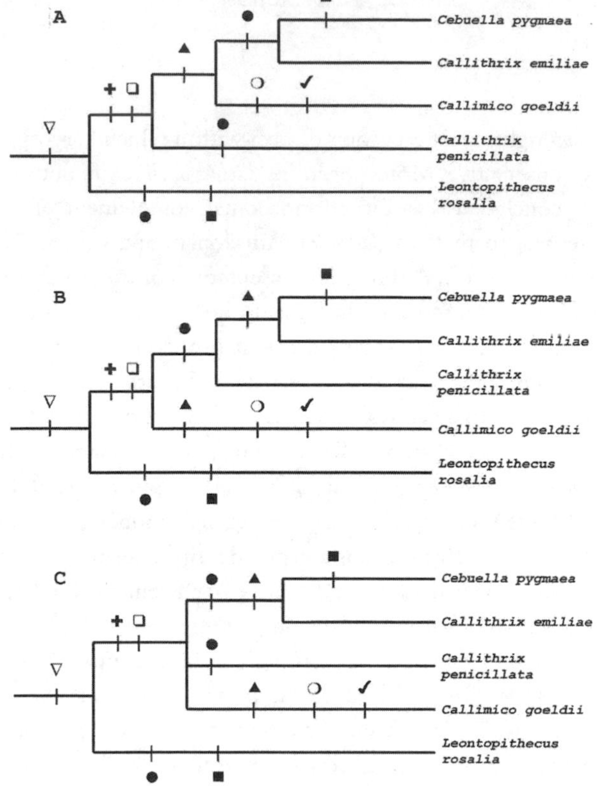

Figure 5.2. (**A** and **B**) Most parsimonious phylogenetic topologies based on karyotype data obtained by analyses of directional rearrangements between species (bidirectional rearrangements excluded). A consensus topology (**C**) of the two most parsimonious trees shows an association of *Cebuella pygmaea* and *Callithrix emiliae* forming a sister group with respect to *Callithrix penicillata* and *Callimico goeldii* (a trichotomy) and shows *Leontopithecus* as an outgroup. Note that this topology is somewhat less parsimonious than either (**A**) or (**B**), requiring 12 rather than 11 rearrangements.

Symbols represent specific rearrangements:

- ✓ Y-autosome translocation in *Callimico goeldii.*
- + Paracentric inversion in a chromosome arm similar to LRO 11q.
- ● Centric fusion involving a chromosome similar to CGO 14 and an unidentified chromosome.
- ■ Centric fusion involving a chromosome similar to CGO 16 and an unidentified chromosome.
- ○ Chromosome fission in *Callimico goeldii* resulting in CGO 13 and CGO 19.
- ❑ Pericentric inversion in a chromosome similar to LRO 14.
- ▲ Centric fusion of chromosomes similar to CJP 17 and CJP 14.
- ▽ Centric fusion involving a chromosome similar to CAP 14 and an unidentified chromosome.

in order to address the issue of phylogenetic relationships among the Callitrichidae and within lion tamarins. Parsimony- and distance-based phylogenetic analyses of both Moreira and Seuánez's (1999) original data set, and of a larger data set incorporating two additional species of *Saguinus,* squirrel monkeys (*Saimiri sciureus*), and additional hominoid outgroup taxa (orangutans and gibbons), yield yet another assessment of relationships among the callitrichids. In all of these analyses, the lion tamarins and Goeldi's monkeys together form a clade that is the sister group to the marmosets (*Callithrix* and *Cebuella*). Additionally, the parsimony analysis tentatively identifies the Amazonian tamarins (*Saguinus*) as the first genus to branch off within the callitrichids in that five of seven equally parsimonious shortest trees share this topology (Figure 5.3A). In the distance-based analysis, however, the Amazonian tamarin clade branches off from the platyrrhine ancestral stock basal to a clade comprising the remaining callitrichids plus *Cebus* and *Saimiri* (Figure 5.3B), although the inclusion of *Cebus* and *Saimiri* in this clade received relatively weak bootstrap support (60 percent). Together, these reconstructions of callitrichid phylogeny differ most markedly from those discussed previously, first, with regard to the placement of Goeldi's monkeys as the sister taxon to the lion tamarins and, second, in weakly implicating *Saguinus* as the first genus to split off from the ancestral callitrichid stock.

Horovitz and Meyer (1995) have also investigated the evolutionary relationships among New World monkeys using mitochondrial DNA sequence data, specifically a 542–base pair fragment of the gene coding for 16S rRNA. In contrast to the cytochrome *b* data, their results support the grouping of Goeldi's monkeys with the marmosets and are ambivalent as to whether *Leontopithecus* or *Saguinus* is the most closely related sister taxon for this clade. The former is supported in a parsimony-based, cladistic analysis weighted according to a transition:transversion ratio of at least 3:1, whereas the latter is supported if successive weighting using the rescaled consistency index (RC) is applied to the set of nine equal length trees that result from an initial parsimony analysis in which characters are weighted equally.

More recently, von Dornum (1997) used sequence data from the cytochrome C oxidase subunit II gene of the mitochondrion to investigate phylogenetic relationships among a large set of New World taxa, including the callitrichids. Analysis of her data set under maximum parsimony, based on a transition:transversion ratio of at least 5:1, also groups *Callithrix* and *Callimico* as sister taxa (as suggested by the 16S rRNA sequences) and, interestingly, joins *Leontopithecus* with *Saguinus* into a basal "tamarin" clade, an arrangement often assumed but supported by only limited molecular evidence. However, if transitions and transversions are given equal weighting in her analysis, the relationships among the callitrichid genera remain unresolved. Pastorini et al. (1998) also addressed the issue of callitrichid phy-

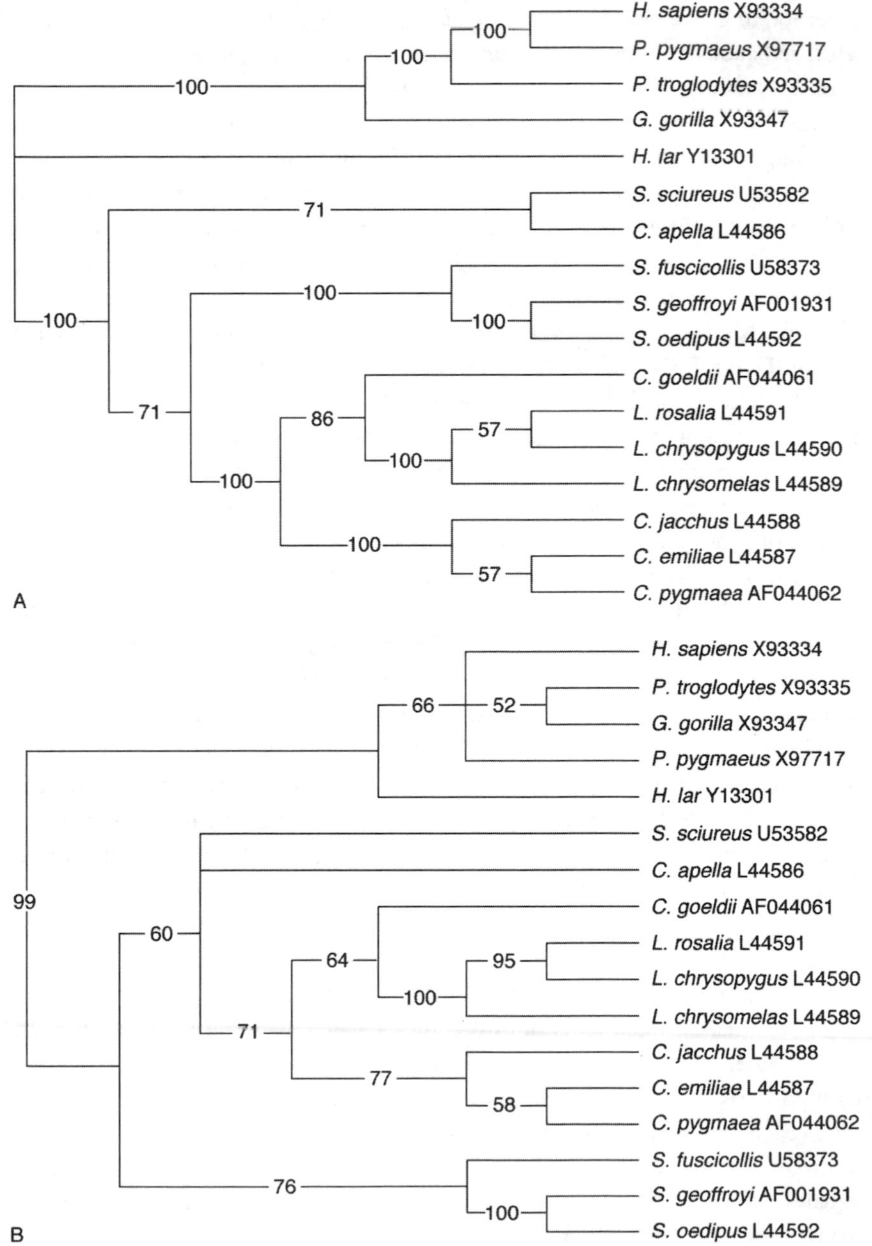
H. sapiens X93334
P. pygmaeus X97717
P. troglodytes X93335
G. gorilla X93347
H. lar Y13301
S. sciureus U53582
C. apella L44586
S. fuscicollis U58373
S. geoffroyi AF001931
S. oedipus L44592
C. goeldii AF044061
L. rosalia L44591
L. chrysopygus L44590
L. chrysomelas L44589
C. jacchus L44588
C. emiliae L44587
C. pygmaea AF044062
100
100
100
71
100
100
100
71
86
57
100
100
100
57
A
H. sapiens X93334
P. troglodytes X93335
G. gorilla X93347
P. pygmaeus X97717
H. lar Y13301
S. sciureus U53582
C. apella L44586
C. goeldii AF044061
L. rosalia L44591
L. chrysopygus L44590
L. chrysomelas L44589
C. jacchus L44588
C. emiliae L44587
C. pygmaea AF044062
S. fuscicollis U58373
S. geoffroyi AF001931
S. oedipus L44592
66
52
99
60
64
95
100
71
77
58
76
100
B

logenetics using sequence data from yet a different region of the mitochondrion—that coding for the ND4 protein and three transfer RNAs. Like von Dornum (1997), they found strong support for a clade comprising the marmosets (including *Cebuella*) and Goeldi's monkeys, to the exclusion of both Amazonian and lion tamarins, and found weak support for a basal callitrichid clade consisting of the two tamarin genera.

With respect to the intrageneric relationships among the lion tamarins, Moreira and colleagues' cytochrome *b* sequence data reveal that *L. rosalia* and *L. chrysopygus* share an identical sequence, one that differs from that of *L. chrysomelas* by five substitutions—four silent ones (consisting of T-C transitions) and one replacement substitution (an A-G transition that results in the substitution of alanine for threoinine in the protein product). However, these observed substitutions do not permit resolution of the phylogenetic relationships among *Leontopithecus* mitochondrial haplotypes in a conjoint parsimony analysis with other platyrrhine genera because the only differences between the morphotypes are due to the autapomorphies characterizing *L. chrysomelas*. If a 10 percent divergence rate per million years is assumed for these silent substitutions (Irwin et al. 1991), then the two identified *Leontopithecus* haplotypes may have been diverging for as much as

---

Figure 5.3. Phylogenetic arrangements for 10 callitrichids, squirrel monkeys and capuchins, and five hominoids reconstructed from cytochrome *b* nucleotide sequences corresponding to nucleotide positions 14,841 to 15,149 of the human mitochondrial genome (Anderson et al. 1981) (**A**) by maximum parsimony (numbers at nodes reflect the proportion of seven equally parsimonious trees containing the indicated clade) and (**B**) by neighbor-joining with Kimura's 2-parameter distance estimates (numbers at nodes are percentage bootstrap values [Felsenstein 1985] based on 10,000 replicates). All analyses were performed using PAUP* 4.0 beta 6 (Swofford 2000). Hominoids (*Hylobates lar, Pongo pygmaeus, Gorilla gorilla, Pan troglodytes,* and *Homo sapiens*) were used as outgroups in all analyses. Accession numbers L44586 to L44592 correspond to *Cebus apella, Callithrix emiliae, Callithrix jacchus, Leontopithecus chrysomelas, L. chrysopygus, L. rosalia, Saguinus oedipus,* respectively, and accession numbers AF044061 to AF044062 correspond to *Callimico goeldii* and *Cebuella pygmaea* (Moreira and Seuánez 1999). Sequences for two additional species of Amazonian tamarins (*S. fuscicollis* and *S. geoffroyi*) and for squirrel monkeys (*Saimiri sciureus*) have accession numbers U58373, AF001931, and U53582, respectively (Evans et al. 1998; Peres et al. 1997; Yoder et al. 1996). Accession numbers Y13301, X97717, X93347, X93335, and X93334 correspond to *Hylobates lar, Pongo pygmaeus, G. gorilla, Pan troglodytes,* and *Homo sapiens* (Arnason et al. 1996; Hall et al. 1998; Xu and Arnason 1996a, 1996b). **A** = 50 percent majority-rule consensus of seven trees, and **B** = bootstrap 50 percent majority=rule consensus tree.

133,000 years. However, this value should not be used as an estimate of the time of divergence of *L. chrysomelas* from the common lion tamarin stock because only single specimens were analyzed in this study and because one haplotype was shared by two allopatric morphotypes (*L. rosalia* and *L. chrysopygus*). Consequently, the existence of other common haplotypes shared by these two species and *L. chrysomelas* cannot be ruled out.

Nuclear DNA Studies

A number of additional studies of New World monkey phylogenetic relationships have directly or indirectly addressed the phylogenetic placement of *Leontopithecus* using sequence data from nuclear loci. For example, Schneider and colleagues (Barroso et al. 1997; Harada et al. 1995; Porter et al. 1995, 1997a, 1997b; Schneider et al. 1993, 1996) have sequenced two regions of the nuclear genome for a number of platyrrhine primates and found somewhat contradictory results concerning the placement of lion tamarins. Sequence data from intron 1 of the interphotoreceptor retinol-binding protein (IRBP) gene link Goeldi's monkeys and lion tamarins into a clade that is the sister group to the marmosets, and the data identify *Saguinus* as the basal sister taxon to all other callitrichids (Barroso et al. 1997; Harada et al. 1995; Schneider et al. 1993, 1996; Figure 5.1D), as found by Moreira and Seuánez (1999) for cytochrome *b*. Sequences from both coding and noncoding regions of the ε-globin gene, on the other hand, favor *Callimico* as the most closely related sister group to the marmosets and join *Leontopithecus* and *Saguinus* into their own distinct basal clade within the Callitrichidae (Harada et al. 1995; Porter et al. 1995, 1997a, 1997b; Schneider et al. 1993, 1996; Figure 5.1E). Interestingly, these ε-globin sequences, along with mitochondrial COII sequences (von Dornum 1997) and ND4 and tRNA sequences (Pastorini 1998), are the only data—molecular or morphological—that can be used to argue in favor of a monophyletic "tamarin" clade rather than the "tamarins" as a paraphyletic group.

Yet a third arrangement has more recently been suggested based on three additional nuclear gene regions. Like ε-globin nuclear sequences and 16S rRNA mitochondrial sequences, noncoding intron sequences from the glucose-6-phosphate dehydrogenase (G6PD) gene (von Dornum and Ruvolo 1999), β2-microglobulin gene sequences covering both coding and noncoding regions (Canavez et al. 1999a, 1999b), and intron sequences of the von Willebrand factor (vWF) gene (Chaves et al. 1999) all link marmosets and Goeldi's monkeys as derived callitrichids. However, none of these genes supports a distinct "tamarin" clade of *Saguinus* and *Leontopithecus,* and both genes identify lion tamarins as the sister group to the derived clade of marmosets and Goeldi's monkeys while implicating *Saguinus* as the first genus to branch off from the callitrichid ancestor (Figure 5.1F).

Even fewer molecular data from studies of nuclear DNA address the intrageneric relationships of the various lion tamarin species. Although β2-microglobulin sequences have been collected for several individuals of three of the four *Leontopithecus* species (Canavez et al. 1998), comparative analysis showed neither interspecific nor intraspecific differences, thus precluding resolution of the relationships among lion tamarins using these data. However, using sequence data from intron 1 of the IRBP gene, Mundy and Kelly (2001) argued in favor of a clear sister taxon relationship between golden lion tamarins (*L. rosalia*) and black lion tamarins (*L. chrysopygus*).

Ribosomal DNA Studies

Additional data that address the issue of evolutionary relationships among the lion tamarins come from studies of ribosomal DNA (rDNA) sequences. Ribosomal DNA sequences comprise hundreds to thousands of tandem repeats of a basic motif located within the nucleolar organizer regions (NORs) of chromosomes. This multigenic region of the genome is responsible for coding ribosomal RNA (rRNA), which is crucial for translation of messenger-RNA during protein synthesis. An important characteristic of rDNA repeats is their striking similarity within a given individual and between individuals of the same species (Hillis and Dixon 1991), a phenomenon termed *concerted evolution* (Dover 1982). This might be due to several different mechanisms that can account for sequence homogenization, such as unequal crossing over during meiosis, gene conversion, and slippage during replication.

Each rDNA repeat unit contains regions coding for 18S, 5.8S, and 28S rRNA subunits and a nontranscribed segment called the intergenic spacer (IGS). Different rDNA regions have evolved at different rates, resulting in either highly conserved regions (18S, 5.8S, and 28S) or very variable ones (IGS). The former are useful for phylogenetic studies of species that diverged a long time ago (some 80 to 500 millions years), and the latter are suitable for analyzing recently diverged species or populations.

A common approach in rDNA phylogenetic studies utilizes endonuclease restriction mapping (Hillis et al. 1996) to assess the presence (1) or absence (0) of particular restriction sites in IGS regions of rDNA, thus providing binary character data that can be analyzed using traditional phylogenetic methods. Seuánez and colleagues (unpublished data) used this procedure to conduct a phylogenetic analysis of three *Leontopithecus* and five *Callithrix* species using one representative of each taxon. Five restriction enzymes were used, and a total of 89 sites were mapped. Among lion tamarin species, four different fixed sites were found between *L. rosalia* and *L. chrysomelas* and between *L. rosalia* and *L. chrysopygus,* and two different fixed sites were found between *L. chrysomelas* and *L. chrysopygus*.

Seuánez and colleagues conducted a phylogenetic analysis of these data under Dollo parsimony, which is based on the assumption that informative characters—in this case restriction sites—are more frequently lost than gained because nucleotide substitutions at these sites alter the base pair sequences recognized by restriction enzymes. Their analysis grouped lion tamarin species into a distinct clade within which a closer relationship between *L. chrysomelas* and *L. chrysopygus* was observed, sustained by a shared gain of one *Hind*III and one *Eco*RI site. The closer relationship of black lion tamarins and golden-headed lion tamarins is in agreement with the distance-based allozyme data of Lisa Forman and her colleagues (1986) (discussed in more detail below) but contradictory to both the distance-based reconstruction derived from cytochrome *b* sequence data on the same individuals and to the recent results of Mundy and Kelly (2001) based on nuclear IRBP sequences.

Synthesis of Molecular and Genetic Studies

The major conclusion emerging from the growing number of molecular studies of callitrichid phylogeny, using many different types of molecular data, is that both lion tamarins and Amazonian tamarins branched off early from the callitrichid common ancestor, whereas Goeldi's monkeys, marmosets, and pygmy marmosets constitute a more derived clade. As noted above, this is in contrast to several earlier phylogenies based on cladistic analyses of morphological characters that all placed Goeldi's monkeys as the first taxon to branch off from the ancestral callitrichid lineage (Ford 1986; Kay 1990; Rosenberger 1981; Figures 5.1B and 5.1C).

However, two fundamental questions still remain unanswered with respect to the phylogenetics of lion tamarins. First, which tamarin genus diverged earlier from the ancestral callitrichid stock, *Saguinus* or *Leontopithecus,* or do these two genera themselves constitute a "tamarin" clade? Whereas the preponderance of the genetic data points to paraphyly for the "tamarins," and whereas more of the data seem to suggest an earlier divergence for Amazonian rather than lion tamarins—a conclusion also reached by von Dornum (1997)—this issue is far from resolved. Second, what are the phylogenetic relationships among the four recognized lion tamarin morphotypes? With respect to both of these questions, additional sequence data from multiple nuclear genes would be invaluable.

GENETIC VARIABILITY BETWEEN LION TAMARIN MORPHOTYPES

The earliest attempt to assess variability within and between *Leontopithecus* species and populations at the molecular level involved electrophoretic assays of allozyme loci. Forman and colleagues (1986) surveyed 47 standard allozyme loci for 140 golden lion tamarins (*L. rosalia*), eight golden-headed lion tamarins (*L. chrysomelas*), and 16 black lion tamarins (*L. chrysopygus*). Their study revealed extremely low genetic variability both within and between these morphotypes: 45 of the 47 loci they screened were monomorphic, with only one allele segregating in the total, combined sample of individuals from these three species. Both variable loci were characterized by only two alleles, and one common allele was found in all three lion tamarin morphotypes at each of these loci.

Forman et al. (1986) then calculated estimates of genetic distance (*D*) between the three morphotypes represented in their study by using their allele frequency data and found *L. chrysomelas* and *L. chrysopygus* to be genetically more similar to one another than either was to *L. rosalia.* However, on the whole, genetic differentiation between the three taxa was quite low relative to other primate populations, which suggests very recent divergence of these among the lion tamarins (Forman et al. 1986). Low genetic variability between the different morphotypes of *Leontopithecus* was also indicated by the small-scale study of cytochrome *b* sequences by Seuánez and colleagues discussed above. Genetic distance based on those sequences indicated a closer relationship between *L. rosalia* and *L. chrysopygus* than between any other alternative pair of lion tamarin morphotypes, in agreement with the proposition of Della Serra (1951) based on dental morphology but differing from the distance estimates of Forman et al. (1986).

GENETIC VARIABILITY WITHIN MORPHOTYPES

The allozyme survey by Forman et al. (1986) revealed extremely low genetic variability within each surveyed lion tamarin morphotype. As noted above, only 2 of 47 allozyme loci (4.3 percent) assayed in their study were variable within lion tamarins as a whole. Both of these loci were polymorphic in *L. rosalia,* but only one was variable in either *L. chrysomelas* or *L. chrysopygus.* Average individual heterozygosities across loci within each morphotype were similarly low (for each population, mean $H = 0.01$). Moreover, this near absence of genetic variation was seen in both wild and captive populations for the *L. rosalia* sample, indicating that low genetic variability is not merely the result of descent from a small captive founder population but also characterizes the species in its natural habitat. In terms of both the proportion of screened loci that were polymorphic (*P*) and the proportion of the genome estimated to be heterozygous (mean *H*), these values are among the lowest measures yet of genetic variability reported for nonhuman primates. The results found by Forman et al. (1986) are echoed in Valladares-Padua's (1987) survey of allozyme variation in *L. chrysopygus,* which was found to be monomorphic at all 25 loci examined.

More recently, Adriana Grativol and collaborators (2001) examined genetic variability and population genetic structure within wild populations of *L. rosalia* by assaying microsatellite loci rather than allozyme markers. Microsatellites are regions of the genome that contain tandem repeats of short nucleotide base pair motifs, usually 2 to 6 bases in length—for example, a microsatellite locus might contain of a stretch of DNA containing the 2–base pair sequence (CA) repeated a variable number of times. The different alleles segregating in the population at such a locus reflect different numbers of repeats of the motif. Microsatellites are particularly good genetic markers to use in studies of population structure because these loci are typically neutral. Moreover, the alleles at a microsatellite locus are codominant and can be scored directly using polyacrylamide gel electrophoresis because different alleles will migrate through a gel at different speeds since they are, essentially, DNA fragments of different lengths.

Grativol et al. (2001) identified five microsatellite loci for *L. rosalia,* of which four were found to be polymorphic. They then genotyped 57 individual tamarins from four geographically separated populations—one large population containing 420 animals and three smaller populations made up of ≤8 individuals each—at each of these polymorphic loci. The number of alleles per locus, calculated across all populations, ranged from 4 to 6, with a mean of 5.25 alleles per locus. However, no single sampled population—not even the largest—contained all of the alleles found in the entire sample at even a single locus, and both the mean number of alleles per locus and the mean heterozygosity were lower in the smaller populations (Grativol et al. 2001), although differences in heterozygosity between populations were not significant. Nonetheless, with only one exception, all of the populations sampled in this study were polymorphic at all loci, indicating that, as yet, even small isolated wild populations of lion tamarins do retain some degree of genetic variability. Grativol et al. (2001) also found considerable genetic divergence between their four study populations (mean Rho = 31 percent), and they suggested that isolation of populations due to forest fragmentation, combined with the extremely limited potential for dispersal between populations, may be responsible for this marked genetic substructuring.

MATING SYSTEMS AND PATERNITY

To date, no detailed intrapopulation studies of lion tamarin social systems using molecular techniques have been completed. However, James Dietz and colleagues (in preparation) have recently initiated a genetic investigation of the mating system of *L. rosalia* using 10 years' worth of blood and hair samples collected from the Poço das Antas Biological Reserve population, the largest remaining wild population of this species. Observational data on mating behavior within this population suggest that the breeding system of golden lion tamarins is largely monogamous, even for groups containing more than one adult male. Polyandrous groups make up about 40 percent of the population, and in these groups both males are seen to copulate with the breeding female; however, Baker et al. (1993; Chapter 8 this volume) found that dominant males were largely able to monopolize matings around the likely period of conception through aggressive mate defense and direct competition. These behavioral observations notwithstanding, the degree to which actual paternity is shared between males in polyandrous golden lion tamarin groups remains an important unresolved issue. Additionally, many recent molecular studies have found that extrapair paternity is not uncommon in putatively "monogamous" avian taxa, prompting Dietz and colleagues to begin addressing these questions for *L. rosalia* using molecular methods.

Although this study is ongoing, preliminary results suggest that some interesting things may be going on in this population with respect to breeding patterns. In at least 5 out of 152 cases (3 percent) in which genotype data for all resident males, resident females, and offspring are available for up to four variable microsatellite loci, it appears that intragroup males could not have sired the offspring; this conclusion is based on the presence of alleles in offspring that are not present in either the mother or any of the putative within-group fathers. If confirmed, this degree of extragroup paternity was not hinted at in previous behavioral studies. Moreover, in another four cases, offspring did not share any alleles with their putative mother—the sole resident adult female in the group—perhaps suggesting adoption of offspring by unrelated adult females.

CONSERVATION IMPLICATIONS

While the very low levels of genetic variability seen in lion tamarins may seem alarming from a conservation perspective, Pope (1996) noted that low levels of variability seem to characterize all callitrichid genera that have been studied to date, and she suggested that lion tamarins and other callitrichids "may in fact be adapted to a homozygous genetic background." For example, in three species of Amazonian tamarins and five species of marmosets screened for variability at 20 allozyme loci, the percentage of loci found to be polymorphic ranged from only 5 percent to 25 percent, and the mean homozygosity per locus ranged between 94 percent and 99 percent (Meireles et al. 1992; Melo et al. 1992; Pope 1996). These values, while indicating slightly greater genetic variability than seen in lion tamarins, are nonetheless still far lower than those reported for many other species of primates (see review in Forman et al. 1986). Pope (1996) suggested that the overall low levels of genetic diversity found in callitrichids are likely due to characteristic features of the social systems of these taxa—specifically to the reduction in effective population size and increased likelihood of inbreeding that are the result of reproductive suppression within groups and of the limited opportunities for individuals in highly saturated environments to disperse into new breeding positions.

On the other hand, lion tamarins show somewhat lower levels of variability than other callitrichids at these allozyme loci (Pope 1996), and even at species-specific microsatellite loci, golden lion tamarins (*L. rosalia*) seem to show lower levels of variability than at least one other callitrichid for whom species-specific microsatellite markers have been identified: the common marmoset (*Callithrix jacchus*). Nievergelt and colleagues (1998) identified and screened 16 novel markers for common marmosets and found 9 to be polymorphic; only 3 of these loci were variable in golden lion tamarins. Moreover, in their sample of 98 captive marmosets living in 19 family groups, they found an average of 5.9 alleles per locus (range = 3 to 12) and mean heterozygosities per locus ranging from 0.38 to 0.94, reflecting considerably greater genetic variation than that revealed in a comparable study of golden lion tamarins (Grativol et al. 2001). A wild population of 40 common marmosets screened at these same loci was less variable, containing an average of only 3.2 alleles per locus with mean heterozygosities per locus ranging from 0.08 to 0.75 (Nievergelt et al. 2000). Although these values are comparable to those seen in some populations of wild golden lion tamarins (Grativol et al. 2001), at present it seems most prudent to conclude that genetic diversity in lion tamarins may be quite low—perhaps even in comparison with other callitrichids—and to bear that conclusion in mind when developing and implementing conservation strategies for maintaining this diversity.

ADDENDUM

Since the time this chapter was submitted, two relevant publications have supplied valuable insights to the question of callitrichid phylogeny. Schneider et al. (2001) analyzed four different DNA sequence datasets, representing all extant neotropical primate genera, which comprised the nuclear genes interphotoreceptor retinol-binding protein (IRBP), ε-globin, glucose-6-phosphate dehydrogenase (G6PD), and β-2-microglobulin (β-2M). Datasets were aligned in tandem, totaling 6,763 base pairs with 2,086 variable characters and 674 informative sites and analyzed by maximum parsimony, maximum likelihood, and neighbor-joining. Within callitrichines, *Cebuella* merged with *Callithrix*, *Callimico* appeared as a sister group of *Callithrix/Cebuella*, *Leontopithecus* appeared as a sister group of the previous clade, and *Saguinus* was the earliest callitrichine offshoot.

Furthermore, Neusser et al. (2001) carried out an an extensive cytotaxonomic analysis of five callitrichine species (*Saguinus oedipus, Cebuella pygmaea, Callithrix argentata, Callithrix jacchus, and Callimico goeldii*) by establishing comparative genome maps by multicolor in situ hybridizations (FISH) with human, *S. oedipus*, and *L. lagothricha* chromosome-specific probes. These studies supported previous evidence in favor of the inclusion of *Callimico goeldii* in the marmoset clade, as a sister branch of *Callithrix* and *Cebuella*, and in proposing *S. oedipus* as the most basal callitrichine offshoot. However, phylogenetic relationships between tamarins could not be addressed because *Leontopithecus* was not included.

JEFFREY A. FRENCH, KRISTEL DE VLEESCHOUWER,
KAREN BALES, AND MICHAEL HEISTERMANN

6
LION TAMARIN REPRODUCTIVE BIOLOGY

Knowledge of reproductive biology is critical, both for the effective management of a captive breeding program and for the assessment of the viability and future status of wild populations. This information is essential for (1) understanding the basic biology of reproduction under normal conditions, (2) designing husbandry procedures to maximize the likelihood of successful reproduction, especially with regard to nutrition, behavior, and social factors, (3) assessing and diagnosing departures from normal reproductive function, and (4) potentially developing strategies in assisted reproductive techniques (ARTs), should ARTs be required to achieve conservation, genetic, and/or husbandry endpoints for lion tamarins. In this chapter we review the contributions made to the science and conservation of lion tamarins from research addressing the endocrinology of reproduction. Most of the advances in this area have come through the use of noninvasive sample collection techniques that are clearly compatible with captive management strategies designed to maximize breeding and/or animal welfare and that do not interfere with the behavior of wild populations.

The chapter is organized as follows. First we briefly discuss methodological issues involved in the selection and development of these noninvasive protocols. Second, we provide information on the fundamental characteristics of female reproductive endocrinology, including the first comparative analysis of ovarian function in the genus *Leontopithecus,* with data from three of the four recognized species (no information is currently available on *L. caissara*). We also discuss the endocrine and (possibly) behavioral mechanisms that underlie seasonal breeding in both wild and captive populations of lion tamarins. Third, we present data on reproductive development in lion tamarins, with a focus primarily on females, since limited data are available for male lion tamarins. Fourth, we describe the apparent absence of social suppression of ovulatory function in subordinate females and reproductive function in males. Fifth, we review strategies developed for controlling reproduction in captive groups of lion tamarins, strategies that, for a variety of reasons, have focused to this point exclusively on regulating female reproductive function. Finally, we conclude with a description of new and exciting applications of noninvasive endocrine-monitoring techniques with wild populations of lion tamarins, efforts that are beginning to provide insights about the proximate mechanisms underlying the complex dynamics of lion tamarin social and behavioral ecology.

METHODOLOGICAL ISSUES IN THE NONINVASIVE STUDY OF FEMALE REPRODUCTIVE ENDOCRINOLOGY

Progress in understanding the reproductive biology of callitrichid primates, including lion tamarins, has lagged behind knowledge of the endocrinology of reproduction in other primate species for several reasons. First, the small body size of callitrichids (150–750 g) makes them difficult to handle for repeated blood sampling in order to measure plasma concentrations of pituitary gonadotropins and gonadal steroids. Second, repeated blood sampling, even of small volumes of blood (0.5 ml), can lead to significantly lower hematocrit values, even after as little as six blood draws in a 4-day period (French et al. 1992). Further, the disruption of capture and sampling can produce high levels of stress hormones (e.g., Smith and French 1997a, 1997b) that may, at a minimum, affect subsequent behavioral profiles or, at worst, deleteriously affect reproductive function through activation of the hypothalamic-pituitary-adrenal (HPA) axis. Finally, for many species the training and adaptation procedures required to produce animals that are relatively insensitive to handling stress are prohibitive and, in fact, incompatible with the husbandry protocols necessary to produce a successful captive breeding program. For all of these reasons, then, there is a premium on using techniques for assessing ovarian, testicular, and adrenal function that do not involve extensive blood sampling or induce stress.

Traditional endocrinology has relied on the measurement of hormonal levels in blood, for the primary reason that the concentration of hormones in the bloodstream reflects the level of activity at the target tissue (whether that tissue is brain, ovary, testes, or uterine epithelium). In contrast, the practice of noninvasive hormone-monitoring requires the measurement of steroid (or peptide) hormones in samples (such as urine, feces, or saliva) that can be collected without the need of a blood sample. Fortunately, the mechanisms of peptide and especially steroid metabolism render these forms of biological samples useful for estimating circulating levels of hormones (see Martin 1985; Norman and Litwack 1987; Schuster et al. 1976). The most common method to assess hormone concentration quantitatively is the use of competitive binding techniques, also referred to as *immunoassay.*

ENDOCRINE CHARACTERISTICS OF THE OVARIAN CYCLE AND PREGNANCY

Based on noninvasive sampling techniques, our knowledge of ovarian function in callitrichid primates now includes data on cotton-top tamarins (*Saguinus oedipus:* Brand 1981; French et al. 1983; Ziegler et al. 1987b), saddleback tamarins (*S. fuscicollis:* Heistermann and Hodges 1995), red-bellied tamarins (*S. labiatus:* Küderling et al. 1995), pied tamarins (*S. bicolor:* Heistermann et al. 1987), pygmy marmosets (*Cebuella pygmaea:* Ziegler et al. 1990), common marmosets (*Callithrix jacchus:* Hodges and Eastman 1984), and Wied's black tufted-ear marmosets (*Callithrix kuhlii:* French et al. 1996a). Information has been available on both steroid and gonadotropin excretion in golden lion tamarins for a number of years (French and Stribley 1985, 1987; French et al. 1992; Kleiman et al. 1978; Monfort et al. 1996), and these reports provide the bulk of the information presented in this section. In addition, we also present the first information on ovarian cycle dynamics in black lion tamarins (*L. chrysopygus*). A systematic comparison of the details of ovarian cycles in golden lion tamarins with those of other callitrichids reveals similarities and important differences among the species in steroid hormone metabolism and excretion, in the patterning of steroid excretion across the ovarian cycle and pregnancy, and in the resumption of fertility postpartum. A summary of the published data for four genera of callitrichids is contained in Table 6.1.

Fundamentals of Ovarian Cycles in *Leontopithecus*

The basic profile of ovarian cycles in lion tamarins, as determined by excretion patterns of urinary steroid metabolites, especially pregnanediol glucuronide (PdG) and estrone conjugates (E_1C), resembles those of other callitrichids. Urinary steroid cycles in golden lion tamarins (Figure 6.1) are characterized by low excretion rates of both steroid metabolites for approximately one-third of the cycle and elevated concentrations of steroids for two-thirds of the cycle. Similar patterns are observed when measuring changes in estrogen excretion in golden-headed lion tamarins and black lion tamarins (Figure 6.2; De Vleeschouwer et al. 2000a; J. A. French, unpublished data). Estimates of the duration of nonconceptive ovarian cycles come from several independent studies and yield similar results: 19.6 ± 1.9 days (n = 14 cycles, French and Stribley 1985) and 18.5 ± 0.3 days (n = 136 cycles,

Table 6.1

Reproductive Parameters in Callitrichid Primates

Species	Gestation (Days)	Interbirth Interval (Days)	Cycle Length (Days)	Litters per Year[a]	Postpartum Ovulation?	References
Leontopithecus chrysomelas	125.3	—	21.5 ± 2.5 21.1 ± 1.0 17.5[b]	1	Yes Nonconcept 17.3 ± 3.5 days	Chaoui and Hasler-Gallusser 1999; De Vleeschouwer et al. 2000a; Chapter 6 this volume
Leontopithecus chrysopygus	—	—	23.0 ± 2.0	1	Yes Nonconcept	Chapter 6 this volume
Leontopithecus rosalia	125.0	194.0	19.6 ± 1.9 18.5 ± 0.3	1	Yes Nonconcept	Baker and Woods 1992; French and Stribley 1985; Monfort et al. 1996; Chapter 6 this volume
Callithrix jacchus	144.0	154.0–158.0	30.0	2	Yes 10.5 days	Harding et al. 1982; Hearn 1983; Koenig et al. 1990; Tardif et al. 1984
Callithrix kuhlii	143.1	156.3 ± 2.9	24.9 ± 0.6	2	Yes 13.6 days	French et al. 1996a
Cebuella pygmaea	141.9	212.7	27.0[c]	2	Yes 15.6 days	Ziegler et al. 1990
Saguinus fuscicollis	149.7	167.0, 185.0	25.7 ± 1.0	1–2	Yes 17.4 days	Heistermann and Hodges 1995; Tardif et al. 1984
Saguinus imperator	—	262.0	—	1	—	Baker and Woods 1992
Saguinus oedipus	183.7	208.0, 235.0	23.6	1–2	Yes 19 days	Brand 1981; French 1984; French et al. 1983; Tardif et al. 1984; Ziegler et al. 1987b
Saguinus bicolor	160.0[c]	—	21.0[c]	—	—	Heistermann et al. 1987

Note: Values indicate data from single cases or mean ±s.e.m.

[a]Modal number of litters each year.

[b]Data from two cycles from a single hysterectomized female (ovaries intact).

[c]Data from a single female.

Monfort et al. 1996) in *L. rosalia* and 21.5 ± 2.5 days (n = 11 cycles, De Vleeschouwer et al. 2000a) in *L. chrysomelas*. Among callitrichids, then, the genus *Leontopithecus* has the shortest ovarian cycle, with most other species expressing nonconceptive cycle lengths greater than 23 days (Table 6.1).

Excretions of E_1C and PdG parallel each other throughout the ovarian cycle and reach peak and nadir concentrations at approximately the same time, suggesting that both steroids are derived from luteal synthesis and secretion. Golden lion tamarins excrete estrogens in two primary forms, estrone and estradiol. Like the metabolites excreted by tamarins of the genus *Saguinus* (e.g., *S. oedipus:* French et al. 1983; *S. fuscicollis:* Heistermann and Hodges 1995), the predominant metabolite excreted by female golden lion tamarins is estrone. Urine collected from females in varying stages of the reproductive cycle contains 15- to 25-fold higher concentrations of estrone than estradiol (French and Stribley 1985). Marmosets of both genera (*Cebuella* and *Callithrix*) excrete predominantly estradiol (French et al. 1996a; Hodges and Eastman 1984; Ziegler et al. 1990), which suggests a potentially important divergence in this biochemical process that may have bearing on issues of taxonomic status within this group (Pryce et al. 1995).

The biological significance of the sinusoidal steroid profiles is clarified when contrasted to the timing of the periovulatory gonadotropin surge. French et al. (unpublished data) used a gonadotropin assay that relies on a luteinizing hormone (LH) antibody with high cross-species generality (Matteri et al. 1987; validated for use in the genus *Leontopithecus* in French et al. 1992). Samples were collected for a period of 80 days from four female golden lion tamarins housed with a vasectomized male partner in facilities at the University of Nebraska in Omaha. Samples were analyzed for steroid conjugate and gonadotropin concentrations using enzyme immunoassay (EIA) and radioimmunoassay (RIA), respectively, and two examples of the resulting profiles are shown in Figure 6.1. Periovulatory spikes in LH excretion occur immediately prior to the rise in urinary steroids. As in other callitrichids, then, high levels of both estrogen and progesterone metabolites in lion tamarins are reflective of luteal, or postovulatory, endocrine events (see French et al. 1996a and Ziegler et al. 1987b, 1990, for examples from other species). Thus, high steroid excretion in females undergoing ovulatory cycles is useful for confirming that ovulation has occurred.

Little information is available regarding the endocrinology of pregnancy in lion tamarins. Kleiman et al. (1978) detected elevated concentrations of urinary chorionic gonadotropin beginning approximately 2 weeks after conception until 9 weeks prior to parturition. French and Stribley (1985) presented data on urinary estrogen excretion in a pregnant female in the immediate postpartum period through conception and early pregnancy. Following a short period of apparent anovula-

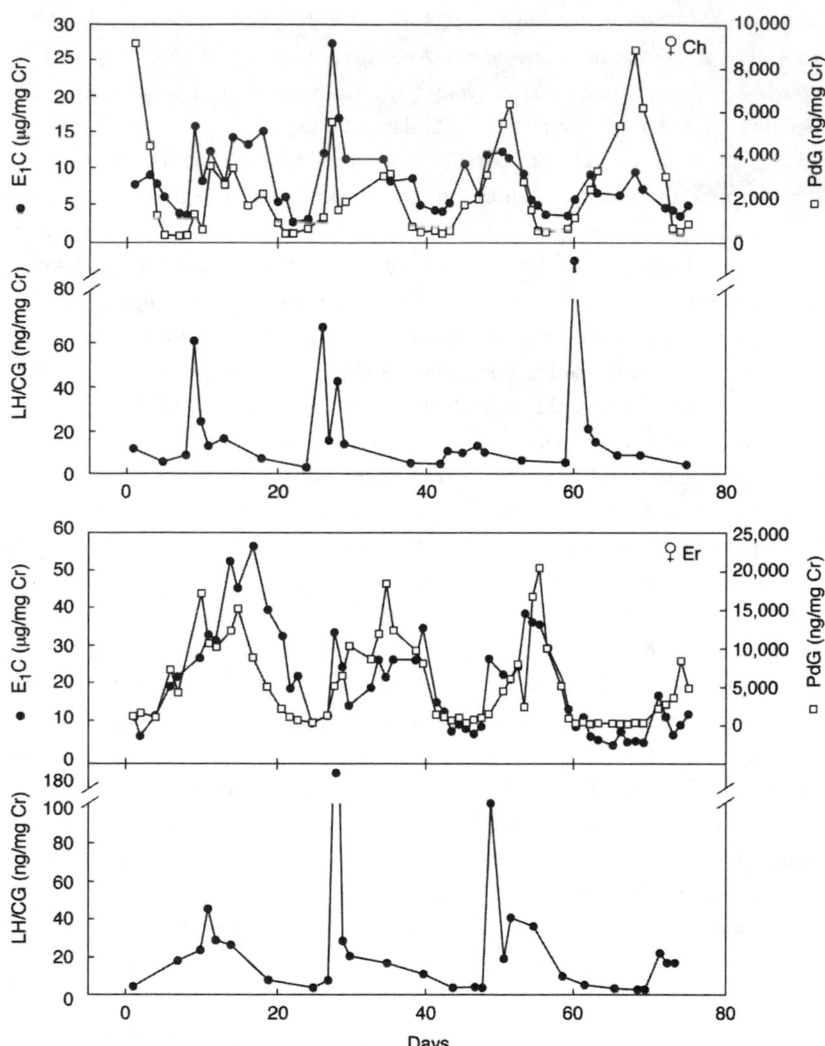

Figure 6.1. Successive ovarian cycles indexed by excreted ovarian steroids and gonadotropin in two female golden lion tamarins (Ch and Er) housed with a vasectomized male. Note the approximately 20-day periodicity of the ovarian cycle and the occurrence of the gonadotropin peak while steroid hormone concentrations are at nadir values. LH/CG = gonadotropins, E_1C = estrone conjugates, LH = luteinizing hormone, PdG = pregnanediol glucuronide, and Cr = creatinine.

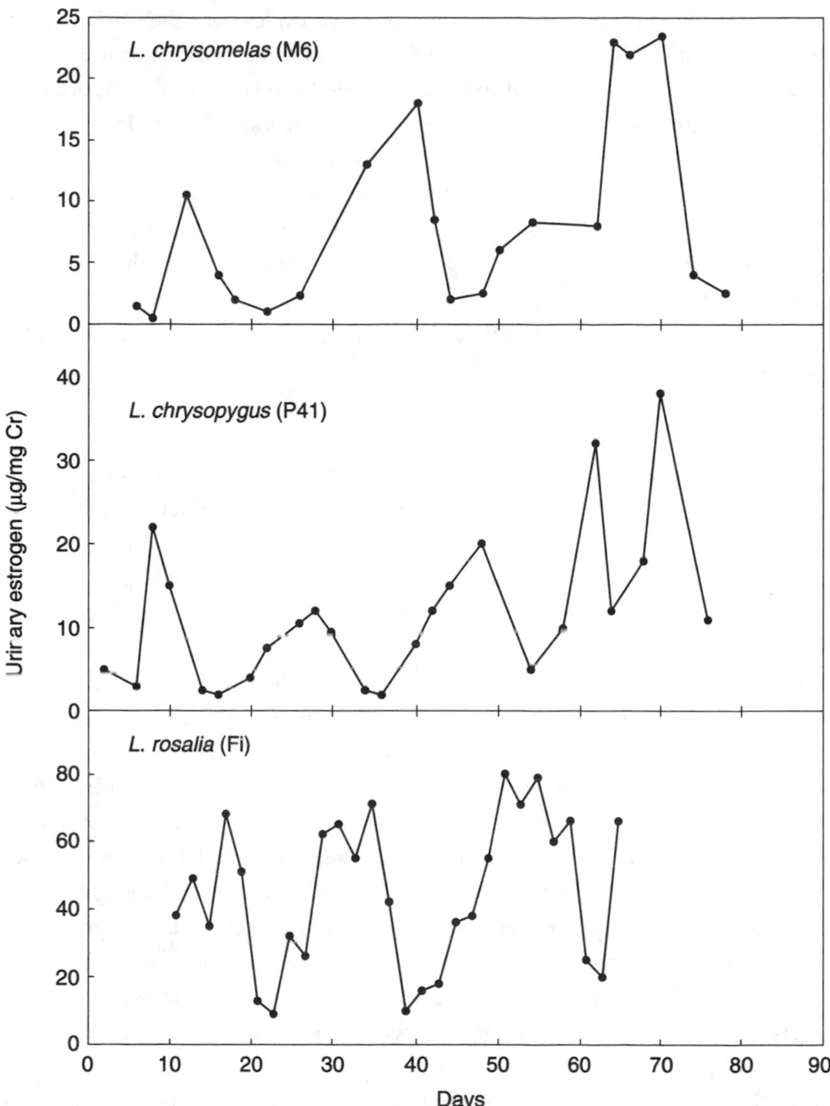

Figure 6.2. Comparative ovarian cycles in three representative females from three of the four species of lion tamarins. Samples for *Leontopithecus chrysomelas* and *L. chrysopygus* were collected from captive females housed at the Centro de Primatologia do Rio de Janeiro (CPRJ—Rio de Janeiro Primate Center), and the samples for *L. rosalia* were collected from a female housed with a vasectomized male at the Callitrichid Research Facility at the University of Nebraska in Omaha. Cr = creatinine.

tion, the female experienced three nonconceptive cycles over the 70 days postpartum. Ovulation and conception, based on the periovulatory nadir in estrogen excretion accompanied by a brief spike in estrogen excretion (see above), occurred on day 81 postpartum, and the female delivered triplets after a 118-day gestation. This estimate of gestation length is in agreement with estimates based on management and behavioral criteria, which place gestation in *L. rosalia* at 125 ± 5 days (Kleiman 1978c). De Vleeschouwer et al. (2000a) reported that the gestation of *L. chrysomelas* is very similar to that of *L. rosalia,* with a mean length of 125.5 ± 3 days (six pregnancies from a total of four females), and indicated that conception can be inferred from the fact that high levels of estrogen measured during the luteal phase of the ovarian cycle are being sustained (see Figure 1 in De Vleeschouwer et al. 2000a).

Comparative Aspects of Ovarian Cyclicity in Lion Tamarins

This section presents the first detailed information on reproductive endocrinology in black lion tamarins and compares these parameters with similar measures in golden and golden-headed lion tamarins. Ovarian cycles from the three species, measured with the same RIA procedures (as described in French and Stribley 1985), appear to be similar in both quantitative and qualitative aspects. Ovarian cycles in each species contain a phase of low steroid excretion, followed by a longer phase characterized by elevated steroid excretion (Figure 6.2). Figure 6.3 presents (1) a quantitative comparison of ovarian cycle parameters (i.e., ovarian cycle length) measured from successive ovulations and (2) mean concentrations of excreted estrogens (after hydrolysis and extraction) measured from multiple females and across multiple cycles per female. Neither cycle length (golden lion tamarins, 19.6 ± 1.9 days; golden-headed lion tamarins, 21.0 ± 1 days; black lion tamarins, 23.0 ± 2 days) nor estrogen excretion parameters differ significantly among species, indicating strong species similarities in this aspect of ovarian physiology. In addition, available data for lion tamarins suggest that females frequently undergo multiple nonconceptive ovarian cycles prior to conceiving at the onset of the breeding season. *Leontopithecus chrysomelas* and *L. chrysopygus* undergo an average of 1.2 and 2.3 nonconceptive ovulations, respectively, prior to conception (J. A. French and A. Pissinatti, unpublished data). This trait sets the genus *Leontopithecus* apart from all other callitrichid primates because postpartum ovulation and conception as early as 10 days and no later than 20 days after parturition is the rule in species in the other three genera in which return to fertility has been systematically addressed (French et al. 1996a; Hearn 1983; Heistermann and Hodges 1995; Ziegler et al. 1987b, 1990). There may be differences in the occurrence of nonconceptive ovulatory cycles between the northern and southern hemispheres because the incidence of longer interbirth intervals (suggesting nonconceptive

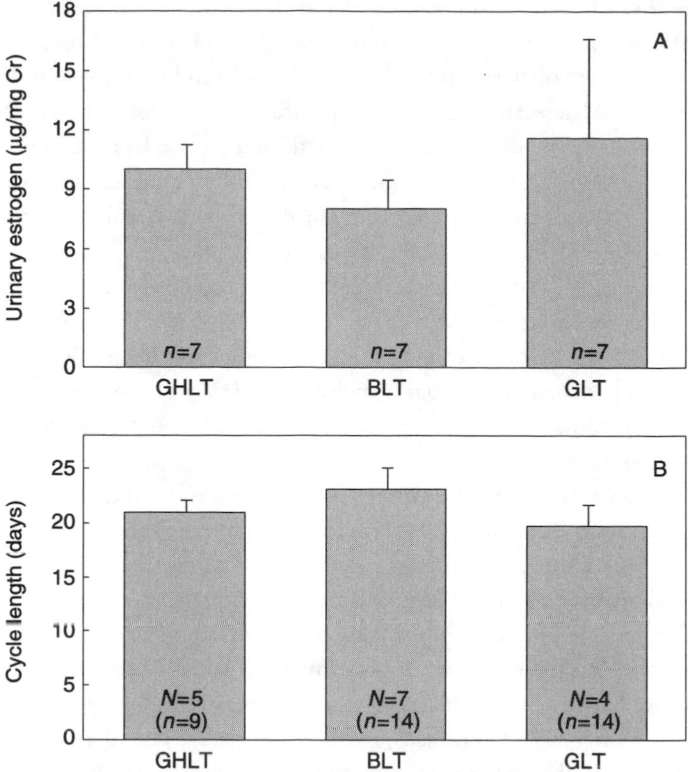

Figure 6.3. Cross-species comparison of two quantitative parameters of the ovarian cycle. Overall concentrations of urinary estrogen excretion (**A**) were assessed in seven nonpregnant females from each species, and mean cycle length (**B**) was determined for *N* females for each species, who contributed a total of *n* datapoints on cycle length. Parametric ANOVA revealed no significant differences among species in either parameter. GHLT = golden-headed lion tamarin, BLT = black lion tamarin, and GLT = golden lion tamarin. Cr = creatinine.

cycles) is more common in captive *L. chrysomelas* in South America than in North America or Europe (De Vleeschouwer 2000).

Seasonal Influences on Reproductive Function

Among callitrichids, lion tamarins display the most conspicuous seasonal pattern of reproduction in a wide range of environmental conditions, including in the wild (Dietz et al. 1994a) and in captivity in Brazil (Coimbra-Filho and Maia 1979a; De Vleeschouwer 2000; French et al. 1996b) and North America and Europe (De Vleeschouwer 2000; Kleiman et al. 1982). In North America and Europe, births are not evenly distributed over the year. There is a drop in the number of births in the northern winter, both for *L. rosalia* and *L. chrysomelas* (De Vleeschouwer 2000; Kleiman et al. 1982), and these populations do not show the conspicuous birth peak that is apparent for wild and captive Brazilian populations. Rather, populations in the north have an extended breeding season, lasting from about March to September, during which births are fairly evenly distributed over the months (De Vleeschouwer 2000; Kleiman et al. 1982).

In all three species housed at the Centro de Primatologia do Rio de Janeiro (CPRJ—Rio de Janeiro Primate Center), offspring production is limited to the months of August through March, and most offspring are produced between August and October (conceptions therefore occurring in the months of April through June). Almost no litters have been produced during the Southern Hemisphere winter months of April through July (French et al. 1996b). De Vleeschouwer (2000) confirmed this pattern for *L. chrysomelas* using data from CPRJ, the Rio de Janeiro Zoo, and the São Paulo Zoo and found photoperiod to be the single most important proximate factor explaining differences in the distribution of conceptions over the year. Temperature and rainfall did not explain the observed differences. In order to evaluate the potential contribution of seasonal anovulation to this phenomenon in lion tamarins, we evaluated ovarian function in female golden-headed and black lion tamarins housed at CPRJ during two periods (J. A. French and A. Pissinatti, unpublished data). The first was during the months of January through mid-March, when, according to the long-term colony records, females were reproductively inactive (the nonbreeding season). The second phase encompassed the period defined as the onset and peak period of reproductive activity in *Leontopithecus* species, namely, late March through mid-June, when the majority of conceptions occur (the breeding season). If seasonal breeding was produced by seasonally associated ovarian quiescence in females (produced by climactic or lactational factors), then we expected to see low and/or acyclic profiles in females during the first phase of the study. We collected urine samples from four female golden-headed lion tamarins and six female black lion tamarins and analyzed these samples for urinary estrone concentrations. All females were mature, multiparous, and living with a long-term pairmate.

Two of four golden-headed lion tamarins showed evidence of normal cycles in both the nonbreeding and breeding seasons, while the remaining two displayed low and noncyclic patterns of estrogen excretion in the nonbreeding season. In black lion tamarins, two of six females were ovulatory in the nonbreeding season, and the remaining four were acyclic. For both species, females in the breeding season displayed multiple nonconceptive cycles prior to conception. Representative

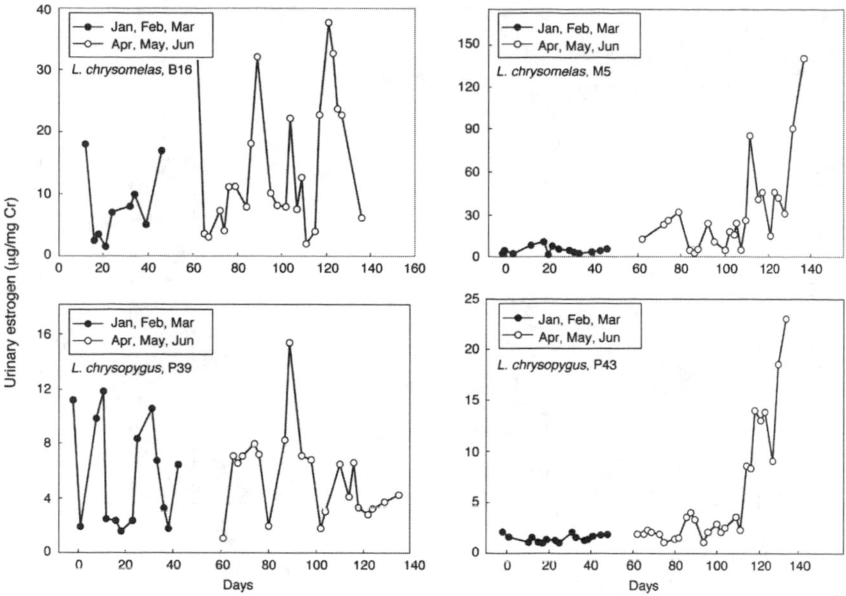

Figure 6.4. Representative cycles from two subsets of females of both *Leontopithecus chrysomelas* and *L. chrysopygus* during the period in which no conceptions have historically occurred (January, February, March = nonbreeding season) and during the period in which most conceptions occur (April, May, June = breeding season). The left column of the figure portrays endocrine profiles from females that suggest functional ovarian cyclicity during both nonconceptive and conceptive phases of the year, and the right column portrays females that show no evidence of ovulatory function during the non-conceptive phase but display the onset of ovulatory function as the breeding season progresses. Cr = creatinine.

cycles are presented in Figure 6.4. Unfortunately, these data do not clarify the mechanism(s) underlying the strikingly seasonal patterns of reproduction in this genus. Ovulatory function during the nonbreeding season is in fact quiescent in about half of the females studied, and it is only upon initiation of ovulatory cycling later in the year that these females become receptive to males and conceive litters. However, in some females of both species, ovulatory function is clearly *not* quiescent, yet these females fail to conceive as well. Behavioral data (Figure 6.5) reveal that while female solicitation behavior is slightly higher in the breeding than the nonbreeding season, the difference is not significant. In contrast, male mounting behavior does vary across seasons, with extremely low rates of male mounting in the nonbreeding season and significant elevations during the breed-

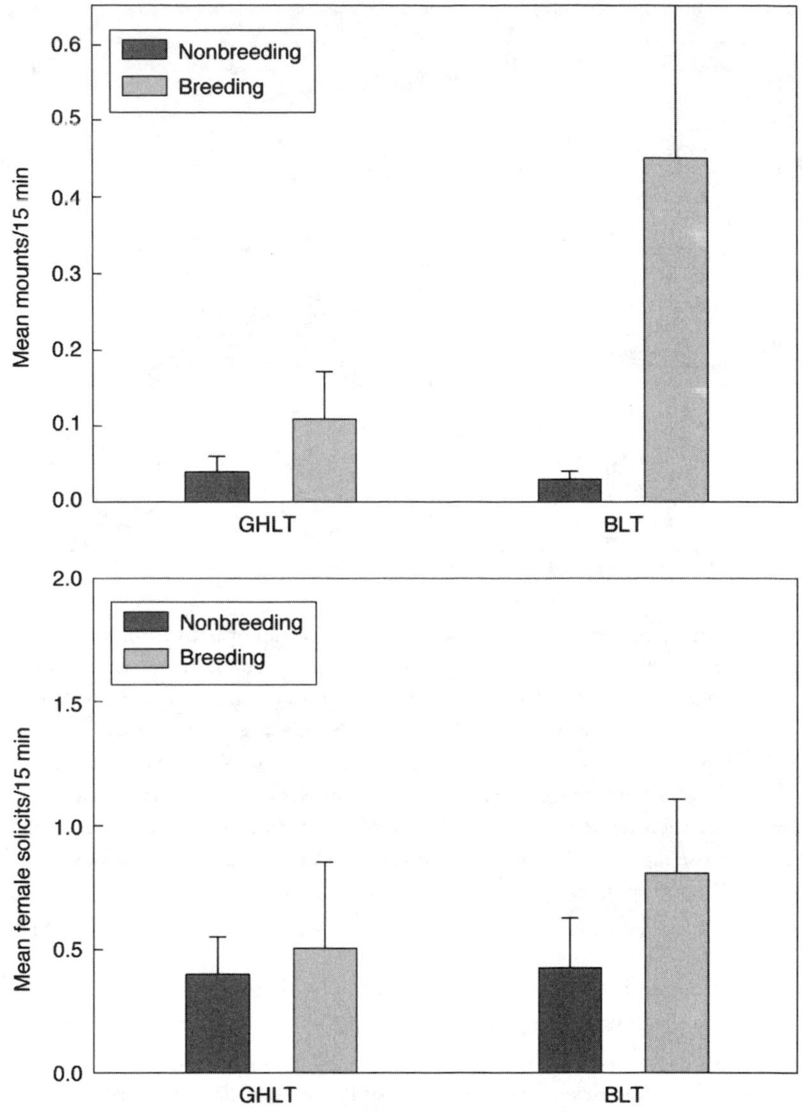

Figure 6.5. Behavioral indexes of male sexual activity (mounting behavior, top panel) and female sexual activity (sexual solicits, bottom panel) during the nonbreeding (January, February, March) and breeding (April, May, June) phases of the annual reproductive cycle. GHLT = golden-headed lion tamarin, and BLT = black lion tamarin.

ing season. Thus, the underlying mechanisms that produce the distinctly seasonal pattern of reproduction in lion tamarins may be independent of female gonadal status and may be regulated by seasonal variation in male reproductive function. Further, our data suggest a significant behavioral component underlying seasonal breeding in the genus *Leontopithecus*.

REPRODUCTIVE DEVELOPMENT AND SOCIAL EFFECTS ON REPRODUCTION

Reproductive Ontogeny and Reproductive Function in Nonbreeding Group Members

Among most callitrichid primates, social factors prominently influence the dynamics of female reproductive physiology and, hence, also the breeding structure and demography of both captive and wild populations. With a few notable exceptions (see review in French 1997), in the genera *Callithrix, Saguinus,* and *Cebuella* only a single female produces offspring in a given social group. Endocrine analyses reveal that daughters in natal groups and subordinates in groups of unrelated individuals exhibit reduced ovarian production of estrogen and progesterone, lower pituitary production of LH, and reduced hypothalamic sensitivity to positive feedback normally produced by rising estrogen concentrations (Abbott 1993).

For the golden lion tamarin, data from more than 25 years of captive management suggest that breeding activity, defined by the production of offspring, is also limited to a single adult female (e.g., Kleiman et al. 1982). The situation in the wild is less clear-cut because it has been found that some groups (approximately 10 percent) contained two adult females that showed morphological evidence of being parous (whether or not infants had been produced by both females in the season when the group was captured—Dietz and Baker 1993). Simultaneous production and rearing of offspring by multiple females in a single group has also been noted, albeit rarely, in wild golden lion tamarins (see Chapter 8 this volume). De Vleeschouwer et al. (2001) also documented a low degree of polygyny in captive groups of *L. chrysomelas*.

Despite similarities in breeding demography between the genus *Leontopithecus* and the other callitrichid primates, the mechanism in *Leontopithecus* that limits breeding to a single female is markedly different. In captive lion tamarins, daughters and subordinate females (at least those older than 16 months of age) display ovarian cycles that are indistinguishable from those of breeding adult females. French and Stribley (1987) and French et al. (1989) reported data on six female golden lion tamarins that were housed as daughters in natal family groups or as a subordinate in a female-female dyad. Daughters less than 12 months of age had low and acyclic profiles of estrogen excretion, but daughters over 17 months of age did not differ in levels of urinary estrogen excretion from those levels associated with cyclic dominant females, and their profiles suggested normal ovarian function. A systematic experimental analysis of social effects on ovarian function also revealed an absence of social modulation of ovarian function (Inglett 1993). Female golden lion tamarins were monitored in three social contexts: (1) as eldest daughters in natal family groups, (2) while housed alone in the absence of either family members or potential mates, and (3) after pairing with a gonadally intact but vasectomized unrelated adult male. Levels of urinary estrogen did not differ among the three social contexts, nor did temporal details of the ovarian cycle, such as overall cycle length (see French and Schaffner 2000). Similar findings have been reported for golden-headed lion tamarins. Three of four eldest daughters in large intact family groups exhibited ovarian cycles that were similar both quantitatively and qualitatively to those of adult females (De Vleeschouwer et al. 2000c; Van Elsacker et al. 1994; see Figure 6.6, later in this chapter). Two adult younger daughters housed in one of the groups did not show normal ovarian cycles (M. Heistermann and K. De Vleeschouwer, unpublished data). In a single semifree-ranging family of golden-headed lion tamarins, an older daughter (44 months of age) displayed normal ovarian cycles, whereas a younger daughter (18 months of age) appeared to be anovulatory, as indexed by fecal progesterone excretion (Chaoui and Hasler-Gallusser 1999). Finally, in four female-female golden lion tamarin dyads housed in the absence of adult males, both females displayed normal ovarian cycles (Monfort et al. 1996). The widespread absence of social influences on ovarian function in lion tamarins is in contrast to the relationship between social status and ovulatory status in other tamarins and marmosets, in which most daughters and subordinate females either are anovulatory (Abbott 1984, 1993; French et al. 1984) or experience ovarian insufficiency, such as low luteal progesterone and short luteal phases (Saltzman et al. 1994; Smith et al. 1997). In the absence of socially induced ovulatory dysfunction, singular breeding by females in most lion tamarin social groups thus appears to be maintained via behavioral mechanisms, including incest avoidance and high levels of female-female aggression, both within a group (De Vleeschouwer et al. 2001; Inglett et al. 1989; Kleiman 1979) and between groups (French and Inglett 1989). Little or nothing is known about male reproductive function and its potential regulation by social factors, although French et al. (1992) reported no significant differences in plasma testosterone concentrations in adult-aged sons housed in natal family groups, relative to males housed in same-sex dyads or with an adult female.

Synchronization of the Ovarian Cycle

In many group-living mammals, reproductively active females tend to synchronize ovarian cycles (e.g., R. E. Brown 1985). Because multiple females in groups of lion tamarins are capable of expressing normal ovarian function, we examined the temporal association in ovarian cycles among golden lion tamarin mothers and daughters and among breeding females living in cages in close spatial and olfactory (but not visual) proximity (French and Stribley 1985). We sampled urine from these females daily for a period of at least 100 days. Ovarian synchrony was determined by contrasting the discrepancy (in days) in the peak levels of estrogen excretion. While this marker is probably less reliable than using the estrogen nadir for alignment, in most cycles peak steroid excretion reliably occurs 4–5 days following the periovulatory period (see above). Statistical evaluation of synchrony involved contrasting observed peak discrepancy with the expected value (4.5 days), based on the assumption that there is a random distribution of cycles among females residing in a social group. For both mother-daughter pairs, and for nearby unrelated females, mean peak discrepancy was significantly less than expected by chance, which suggests a high degree of ovarian synchrony in females housed under these conditions.

Monfort et al. (1996) also tested ovarian synchrony by monitoring ovarian cycles for 16 months in four golden lion tamarin female pairs housed in the absence of males. They collected first-void urine from females temporarily separated from their female cage mate and analyzed these samples via RIA for PdG and E_1C. Statistical analyses were similar to those in French and Stribley (1985). Unlike the previous report, Monfort et al. (1996) found no evidence for ovarian synchrony beyond what would be expected by chance, either in cage mates or in females living in cages near other females. There were, however, several methodological differences in the studies. First, Monfort et al. (1996) used the nadir of PdG and E_1C excretion as the time point for indexing cycles, while French and Stribley (1985) used peak estrogen excretion. The former method appears to be better for aligning cycles because low steroid excretion is associated with the LH surge and, hence, the ovulatory phase of the cycle (French et al. 1992). Second, French and Stribley housed females in physical and olfactory contact with adult males, while the females studied by Monfort et al. had no access to males. Male-associated stimuli are critical for other socially mediated reproductive events in callitrichid primates (e.g., Widowski et al. 1990, 1992), and ovarian synchrony may be dependent upon appropriate levels of male stimulation. Third, the highest levels of ovarian synchrony in the French and Stribley data set were associated with mothers and daughters, and the Monfort et al. experiment studied only unrelated adult females. In any event, the potential for ovarian synchrony is a phenomenon worthy of further study.

CONTRACEPTION

It may be important to limit reproductive output in a captive breeding program for even the most critically endangered species. Limits on carrying capacity, overrepresentation of certain population "founders," and skewed population demographics are all sufficient reasons to justify manipulating and controlling reproduction in animals maintained as a managed population in captivity (Ballou 1996). The importance of regulating captive populations through contraceptive methodologies has been identified as crucial for at least two of the four species of lion tamarins (Chapter 4 this volume). For both *L. rosalia* and *L. chrysomelas,* captive population sizes are sufficient for maintaining genetic diversity for the near-term (Kleiman et al. 1990b). Long-term captive breeding plans for these species involve carefully controlled breeding of individuals in light of the representation of founder alleles in the captive population (see Chapter 4 this volume). For the remaining captive population, it is desirable to maintain normal female-male social groupings for both management and display purposes and to limit or eliminate the production of offspring. Leus (1999) summarized the goals and overall philosophies of the contraceptive program for *L. chrysomelas.*

Synthetic progestagen-like compounds are widely used to control ovarian function in a variety of mammals, including callitrichid primates (Ballou 1996). Implants of these compounds (most prominently melengestrol acetate—MGA) in a silicon-based substrate are done routinely in zoos and breeding facilities throughout the world (Asa et al. 1996). MGA implants have been used effectively in regulating breeding activity in golden lion and golden-headed lion tamarins. In the context of this chapter, we review the efficacy of these contraceptives and discuss the endocrinological, social, and behavioral consequences of female contraception via synthetic steroid implants.

Ballou (1996) reported the results of a fairly large study of the efficacy of MGA implants in the control of reproduction in golden lion tamarins. The preliminary data suggest a contraceptive efficacy of 97 percent (or, if one selects production of viable offspring as the measure, 100 percent efficacy) for MGA implants. Although no systematic behavioral observations were conducted as part of Ballou's study, the incidence of intrafamily conflict requiring management intervention in groups receiving implants was no greater than that in four control groups in which breeding females did not receive implants. Thus, the regulation of reproduction in *L. rosalia* through MGA implants is highly efficacious with minimal biomedical or biosocial concerns.

De Vleeschouwer and colleagues (De Vleeschouwer 2000; De Vleeschouwer et al. 2000b, 2000c; Van Elsacker et al. 1994) recently published a detailed analysis of contraception in golden-headed lion tamarins, primarily via MGA implants. Their analysis includes not only contraceptive efficacy assessments but also survey data from zoological parks on reproduction and group dynamics, detailed data on female reproductive endocrinology, and systematic observations on intragroup social interactions in four family groups at the Antwerp Zoo. As with golden lion tamarins, in golden-headed lion tamarins MGA implants were highly efficacious in preventing pregnancy. Of 48 cases in which the presence of the implant was confirmed throughout the study period, only 1 "unplanned" pregnancy was documented. Although the contraceptive efficacy was high, the potential for females to resume reproduction after the implant had expired or was removed appeared to be problematic. Of 26 cases followed for 2 or more years following implantation, only 5 females resumed reproduction. Reproductive failure in some cases may have been attributable to factors other than the implant (e.g., age or incest avoidance). Nonetheless the data suggest that return to fertility may be problematic for *L. chrysomelas* after MGA contraception. There was also the suggestion that the incidence of stillbirth was higher in females after contraception reversal than prior to contraception (however, the data set was not large).

The impact of the implants on reproduction in breeding females was of primary importance, but the impact of interfering with the dominant female's reproductive function on the ovarian function in the eldest daughters was equally important, so urine samples were collected for both mothers and eldest daughters. MGA implants had a dramatic effect on ovarian physiology. The top graphs of Figure 6.6 show profiles for the two mothers immediately prior to and following parturition and MGA implantation. Levels of E_1C were high immediately prepartum and exhibited the normal drop at the time of parturition. Implantation of the females with MGA eliminated the normal return to fertility and elevation of ovarian steroids that normally accompanies the weeks following parturition (De Vleeschouwer et al. 2000c; French and Stribley 1985). Urinary estrone conjugates remained at low and/or undetectable levels for the duration of the sampling period, providing endocrinological verification that MGA implants eliminate normal ovarian cyclicity. Neither of the females became pregnant during this time, in spite of the fact that sexual interactions were noted between the male and breeding female after implantation. Figure 6.6 also reveals, as described earlier, that adult *L. chrysomelas* daughters show normal ovarian cycles while housed in the presence of the natal family group and that ovarian dynamics in the daughters were not altered during the time that their mothers were implanted with MGA.

The survey that De Vleeschouwer and colleagues distributed among zoos indicated that daughters became pregnant and delivered offspring in 2 of 10 groups in which an adult daughter was present during MGA contraception of the mother (2 out of 25 daughters were potentially able to conceive). On a behavioral and social level, MGA implants had little disruptive effect on ongoing social behavior within family groups, at least in the short term. Sexual activity continued between the breeding adult male and female, although the temporal distribution of mounting by males and presenting by females was not limited to a short period of time, as occurred in pairs containing a normally cycling female. Social interactions between the breeding pair were unaltered after implantation. Neither fathers nor brothers directed significantly more social or sexual behavior toward the eldest daughters during the implantation of the breeding female, and interestingly,

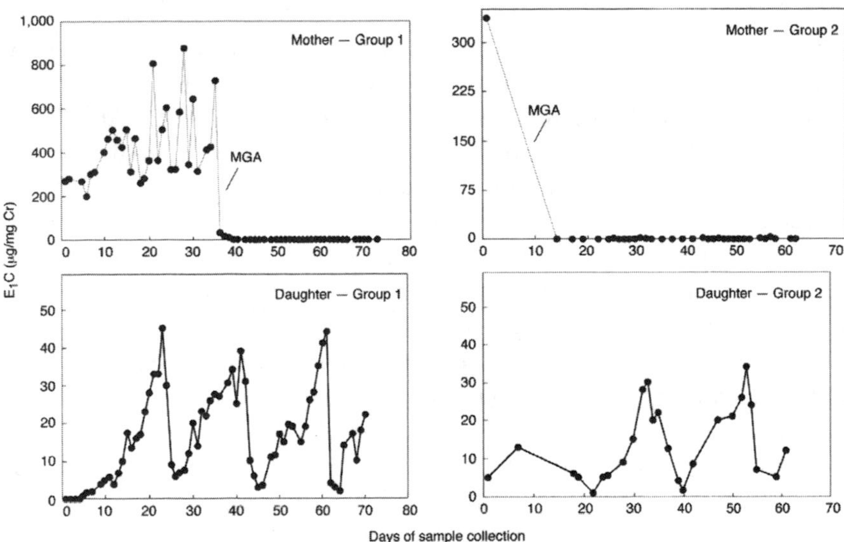

Figure 6.6. Patterns of urinary estrogen metabolite (E_1C) excretion in two family groups containing a breeding female and her adult-aged daughter. Breeding females received a contraceptive implant of melengestrol acetate (MGA) at the point indicated by the arrow, which effectively terminated ovarian activity. Daughters displayed ovarian cyclicity before mothers received their implants, and cycle dynamics did not appear to be affected when the mother received the implant (modified from Van Elsacker et al. 1994). Cr = creatinine.

mothers and daughters tended to exhibit more affiliative behavior with each other after implantation than before. Thus, there was a low incidence of incestuous mating when a large number of groups in various locations were surveyed, and MGA implantation does not appear to alter family social dynamics in a dramatic way in the short term; however, preliminary results indicate that long-term stability may be more problematic (De Vleeschouwer 2000). Further, the implantation of a breeding female does not alter within-group social dynamics, at least in groups with the demography studied here (large groups that can, under other circumstances, be highly susceptible to intragroup instability and fighting; Inglett et al. 1989). Decisions regarding the use of MGA as a contraceptive method should take into account the potential increased incidence of steroid-related endometrial decidualization (Murnane et al. 1996) and, because of the potential irreversibility of reproductive function, the importance of the individuals for later breeding purposes (De Vleeschouwer et al. 2000b). Alternative hormonal methods are Depo-Provera, a progestagen that needs to be injected approximately every 3 months. Sterilization and (ovario-) hysterectomy are also possible, although permanent, solutions. Little is know about the effects of any of these alternative methods.

There is one additional contraceptive regimen that can be useful for regulating female reproduction, but its use is much more labor-intensive than a single MGA implant and involves the repeated capture and restraint of females. Regular injections of analogues of prostaglandin 2α (e.g., cloprostenol, available in commercial form as Estrumate) is effective in lysing tissue of the corpus luteum, thus eliminating the primary source of progesterone that maintains early pregnancy (Summers et al. 1985). Monfort et al. (1996) showed that injections of 1.6 μg of cloprostenol are effective in eliminating luteal function within 2 days of the injection. Ovulation, as detected by increased PdG excretion, occurs 10 days following cloprostenol administration, and under normal housing conditions (i.e., an adult female paired with an unrelated male) females would become sexually receptive to their mate. Continued contraception with luteolytic agents like cloprostenol would thus require active monitoring of reproductive status of the female and intensive management to meet the endpoints of contraception.

It is clear that management of reproduction in captive animals is possible through a variety of routes. All of the methods described above, however, have drawbacks and potential problems that argue against their use for long-term reproductive management. Future research in the area of contraception should evaluate the use of compounds that do not involve steroid administration for their mode of action (e.g., gonadotropin-releasing hormone antagonists; Hodgen 1996). Further, methods that target male reproductive function, through physical, hormonal, or immunological blocks to male fertility, should receive a high priority. Male vasectomy and gonadectomy are effective (although permanent) contraceptive methods (De Vleeschouwer et al. 2000b; Inglett 1993).

MONITORING OF REPRODUCTIVE ENDOCRINOLOGY IN FREE-RANGING POPULATIONS

As this chapter demonstrates, our knowledge of the reproductive physiology of female captive lion tamarins is fairly well established. However, little is known regarding the endocrinological status of animals in the wild. This information is required for tests of critical hypotheses regarding the influence of habitat, ecology, and group demography on reproductive function and hence reproductive success. Further, the ability to predict the number and ages of breeding females in current and future generations, so critical for estimates of population growth or decline, requires detailed information on ovarian status.

Several investigators have had success in monitoring reproductive function in a wide variety of both captive and free-ranging primates by measuring levels of steroid metabolites excreted in the feces (Jurke et al. 1997; Pryce et al. 1994; Wasser et al. 1988; Ziegler et al. 1996). With regard to lion tamarins, two pilot studies have addressed the potential utility of these methods for assessing reproductive states in females. Ribeiro (1994) demonstrated that fecal progesterone can be detected in fecal samples collected from reintroduced but free-ranging golden lion tamarins and that females have concentrations of progesterone that are 10-fold higher than those in males. Samples collected from wild-born female lion tamarins in the Poço das Antas Biological Reserve also contained detectable progesterone, and Ribeiro identified profiles that may have been associated with cyclic ovarian function. These data clearly indicate that monitoring fecal progesterone excretion has the potential for aiding in the assessment and interpretation of reproductive patterns in wild lion tamarins.

More recently, Bales, Dietz and French initiated a program designed to assess reproductive function in wild lion tamarins using simplified extraction procedures and enzyme immunoassay of steroid metabolites as described by French et al. (1996a, submitted). Fecal steroid excretion is in fact closely related to the excretion of both PdG and E_1C in the urine, as shown in Figure 6.7. In the case of both steroids, there is especially good concurrence of the nadir in ovarian steroid excretion, which, as demonstrated above, is the critical marker for ovulation. Concentrations of steroid excretion in the feces during the luteal phase of the ovarian cycle tend to be more variable than steroid concentrations in the urine, but luteal elevations lasting approximately 10 days are still identifiable in the fecal

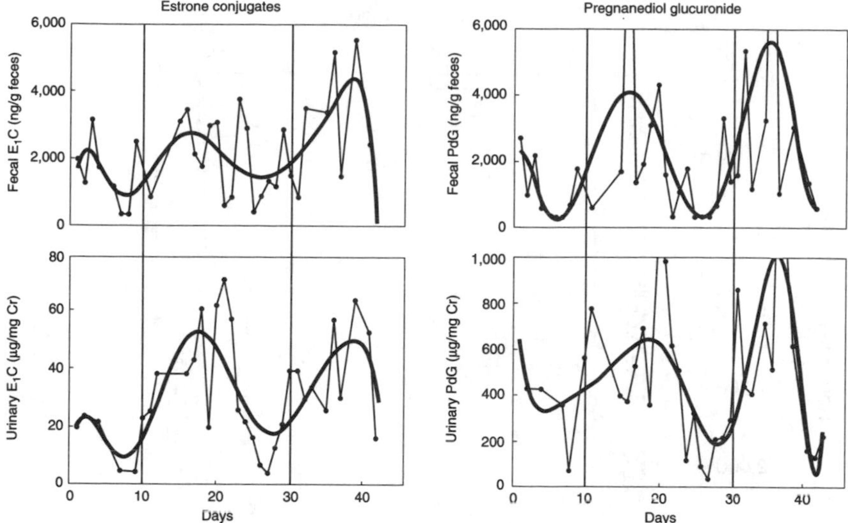

Figure 6.7. Steroid excretion profiles in paired fecal and urine samples from a singly housed golden lion tamarin. Dark lines represent best-fitting curves based on higher-order curve-fitting function. Peak and nadir time points are simultaneous in both estrogen and progesterone metabolite excretion profiles (modified from French et al., submitted). Cr = creatinine.

profiles. Although this work on a single female does not allow us to assess quantitatively the lag time between hormone excretion in the urine and its appearance in the feces, the similarity in the qualitative aspects of the profiles suggests that fecal steroid production is useful for marking significant reproductive events in female lion tamarins, such as pinpointing the time of ovulation.

The development and validation of the fecal hormone assays has led to significant advances in three areas, which are reviewed here (original data in Bales 2000; Bales et al. 2002; French et al., submitted). First, we are able to document significant reproductive events in wild golden lion tamarins. As an example, Figure 6.8 shows an ovarian hormone excretion profile for a single female, tracked over a 16-month period. Concentrations of E_1C and PdG are clearly elevated during each annual pregnancy and drop precipitously at the time of parturition. Although sampling frequency was lower during the "nonbreeding" season, the absolute concentrations and lack of conspicuous and long-duration elevations suggest ovarian quiescence during this time. Second, we have documented the reproductive status of daughters and subordinate females living in social groups (French et al., submitted). Not surprisingly, the social regulation of reproduction is considerably

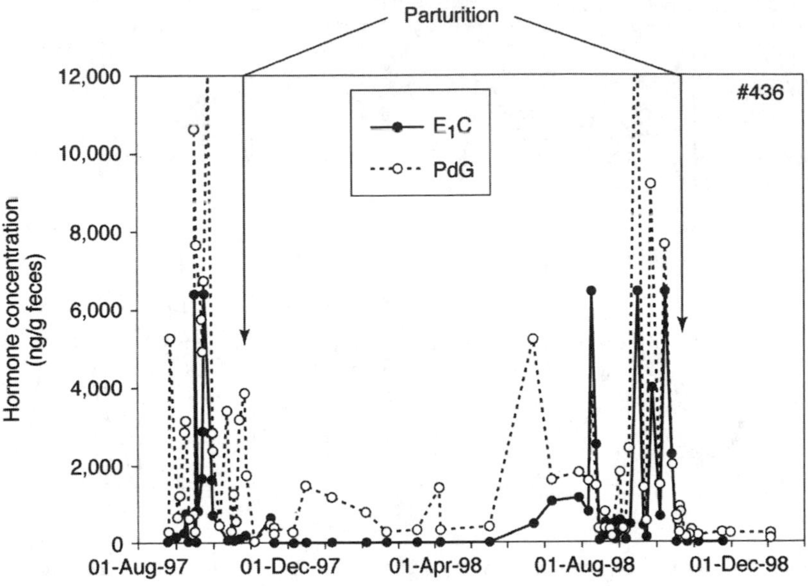

Figure 6.8. Fecal estrone conjugates (E_1C) and pregnanediol glucuronide (PdG) excretion over 16 months in wild-born breeding female (number 436) in the Poço das Antas Biological Reserve. Two litters were delivered at the times indicated on the figure, and clear endocrine manifestations of pregnancy are readily apparent in the profiles.

more complex in free-ranging groups than in captive groups. Age is a significant predictor of ovarian status, with younger females more likely to show low and acyclic patterns of steroid hormone excretion. Social status also is associated with ovarian status, independent of age, with the subordinate female in a dyad more likely to exhibit ovarian insufficiency. Finally, group demography accounts for some variation in ovarian status in subordinates/daughters, in that daughters living in intact natal groups are more likely to show ovarian insufficiency than females living with a stepfather or in a new social group. Bales et al. (2002) assessed the relative role of steroid hormones in regulating variation in maternal investment in offspring, both prenatally (assessed by infant weight) and postnatally (indexed by infant-carrying effort by the female). Endocrine variation accounted for maternal variation in prepartum investment (lower prepartum estrogen and higher prepartum cortisol predicted greater neonatal weight), whereas better maternal condition, larger litter size, and fewer alloparents were associated with greater postpartum maternal investment.

From these three examples, then, it is clear that fecal steroid assessment can play an important role in assessing fertility in both captive and wild lion tamarins and also in testing hypotheses in the wild regarding the ways in which social context and status translate into differential reproductive success in female lion tamarins. As of this moment there are no data available on fecal steroid excretion in any of the other three species of lion tamarins in the wild, but there is every reason to suspect that the methods established for golden lion tamarins will serve as useful tools for the assessment of reproductive status in the other species.

CONCLUSIONS

The development of noninvasive endocrine-monitoring methodologies has been critical for exploring reproductive function in lion tamarins, both in captivity and in the wild. The data reviewed in this chapter yield several important conclusions.

Leontopithecus displays considerable differences from the other callitrichid genera in details of female biology: it exhibits a shorter ovarian cycle, considerably shorter gestation, a higher rate of infertile postpartum ovulation, and a markedly seasonal pattern of reproduction.

There are only minor differences in the dynamics of ovarian cycles and pregnancy among the three species.

Singular breeding by females appears to be maintained predominantly via behavioral mechanisms, although recent field data hint at significant effects of social status and group demography on ovarian function in subordinate females.

Contraception based on manipulation of the ovarian cycle is well established, although the potential for negative consequences associated with MGA contraception remains a concern.

Little is known about male reproductive function in any species of lion tamarins and should be a priority for future research, especially in light of the potential that variation in male sexual interest and/or capacity may be important for shaping seasonal patterns of reproduction.

ACKNOWLEDGMENTS

We would like to first of all acknowledge the opportunity and access that Instituto Brasileiro do Meio Ambiente e dos Recursos Naturais Renováveis (IBAMA—Brazilian Institute of the Environment and Renewable Natural Resources) has provided for our work with *Leontopithecus*. The Conselho Nacional de Desenvolvimento Científico e Tecnológico (CNPq—Brazilian National Science Council) and the Associação Mico-

Leão-Dourado (AMLD—Golden Lion Tamarin Association) have also facilitated various portions of the work described in this chapter. We thank Tessa Smith, Linda van Elsacker, Steve Monfort, Elaine Ribeiro, and Alcides Pissinatti for access to data and comments on previous versions of the manuscript. Insightful comments from the volume editors, Devra Kleiman and Anthony Rylands, improved the chapter considerably. Jeffrey French thanks the following for financial support: the James M. Cattell Foundation, National Institutes of Health, and the National Science Foundation. Kristel De Vleeschouwer thanks the Royal Zoological Society of Antwerp, the University of Antwerp, and the Flemish Minister of Science for financial support. Karen Bales thanks the following for financial support: the National Science Foundation (NSF), the Biology of Small Populations, NSF Research Training Grant (BIR 9602266), Sigma Xi, TransBrasil, the Friends of the National Zoo (FONZ), and the University of Maryland.

MARIA CECÍLIA M. KIERULFF, BECKY E. RABOY,
PAULA PROCÓPIO DE OLIVEIRA, KIMRAN MILLER,
FERNANDO C. PASSOS, AND FABIANA PRADO

7
BEHAVIORAL ECOLOGY OF LION TAMARINS

The way in which animals distribute time among various activities has important consequences for survival and reproduction. For primates, many ecological and social factors can constrain time budgets (Bronikowski and Altmann 1996; Chapman 1990; Clutton-Brock and Harvey 1977; Defler 1995; Dunbar 1992; Isbell and Young 1993; Strier 1987; Watts 1988). Limited food resources (both spatially and temporally), the location of sleeping sites, predation pressure, and resource defense are all examples of ecological constraints. Social constraints on primates relate to the maintenance of group cohesiveness and the formation of social bonds. Intrinsic constraints include body size and metabolic demands. Those having small group sizes may be particularly influenced by predation pressure because there are fewer individuals to scan for predators or mob them. This may affect feeding efficiency and other essential activities.

In tropical environments, fluctuations in rainfall and temperature result in seasonal variation in food availability, forcing many species of vertebrates to shift to alternative diets, migrate into more favorable habitats, or otherwise experience periods of hunger (Peres 1994). The spatial and seasonal distribution of food influence time spent foraging, daily distance traveled, and patterns of home range use (Nunes 1995; O'Brien and Kinnaird 1997). Many studies of home range use by nonhuman primates have examined time budgets and habitat selection in relation to spatial and temporal variability in food distribution (Bronikowski and Altmann 1996; Clutton-Brock 1975; Dunbar 1992; Nunes 1995; O'Brien and Kinnaird 1997; Peres 1986a, 1986b; Stevenson et al. 1994; Stoner 1996; Watts 1991; Zhang 1995). Home range use may also be strongly influenced by social constraints, including the spatial arrangement of groups and population densities (Butynski 1990; Crockett and Eisenberg 1987; Ostro 1998; Peres 1986a, 1986b).

Lion tamarins (genus *Leontopithecus*), the largest members of the Callitrichidae, are represented by four distinct forms: *L. chrysomelas* (golden-headed lion tamarin), *L. rosalia* (golden lion tamarin), *L. caissara* (black-faced lion tamarin), and *L. chrysopygus* (black lion tamarin). They are endemic to the remaining fragments of the Atlantic coastal and semideciduous forest of southeastern Brazil. Ecological constraints on callitrichids, such as high predation pressure, may derive from their small size (Figure 7.1). To meet high metabolic demands, they must spend a large proportion of their time foraging and eating, which poses a conflict between resource acquisition and predator avoidance. Lion tamarins occur in four distinct regions in southeastern and south central Brazil, each of which differs in seasonality and the availability and distribution of food resources. These sometimes subtle differences result in varying population densities and behavioral parameters, such as activity budgets, home range size, day range length, and habitat use by the four species.

In this chapter, we review published and unpublished ecological and behavioral data collected during eight long-term studies of wild lion tamarins—as well as

Figure 7.1. Lion tamarins are preyed on by raptors, snakes, and small arboreal carnivores. Andrew Baker and Carlos Ruiz-Miranda hold a boa constrictor that was observed catching and consuming a wild female golden lion tamarin (Ruiz-Miranda points to swelling, indicating the site of the tamarin in the snake's gut). The boa was located the following day by radiotelemetry since the female tamarin was wearing a radiotransmitter. (Photo by Devra Kleiman)

additional data on population density and group composition collected during several short-term population surveys—with three objectives: (1) to provide an overview of time allocation patterns, ranging behavior, ecological strategies, and habitat preferences of lion tamarins, (2) to describe how these features of lion tamarin biology vary between populations and species, and (3) to consider the demographic and ecological factors that might underlie this variation.

FIELD SITES AND METHODOLOGY IN STUDIES OF LION TAMARIN BEHAVIORAL ECOLOGY

The data discussed here stem principally from seven distinct locations where lion tamarins have been studied in the wild (Table 7.1). Most of these sites consist of lowland evergreen tropical rainforest, although the two areas where *L. chrysopygus* was studied are semideciduous mesophytic forests. Despite the proposed dependence of lion tamarins on mature forest (Rylands 1996), most studies have been conducted in or near degraded areas, indeed most often in only partially forested mosaic habitats that include patches of mature and successional forest, agricultural land, and pastures. Many of these sites are also completely isolated from neighboring forests.

Most localities where lion tamarins have been studied experience a pronounced wet and dry season. However, the eastern portion of the range of *L. chrysomelas* is completely aseasonal (Rylands 1982, 1989a) and that of *L. caissara* has no conspicuous dry season, although there are seasonal fluctuations in rainfall (Prado 1999). All but one of the studies reviewed here were conducted on native populations in their original areas of occurrence. The exception is the *L. rosalia* population at União Biological Reserve, which was translocated from small forest fragments; the data included in the study at the União Biological Reserve were collected only after groups had established their home ranges (Kierulff 2000).

The studies of lion tamarins reviewed in this chapter lasted between 3 months and 3 years. All groups underwent a period of habituation prior to the collection of behavioral data. The number of groups investigated varied from one to eight (Table 7.2), and in all but one (Rylands 1989b), individuals were captured and fitted with radio collars to assist with locating the groups and to facilitate the habituation process.

Two different methods were used to record behavior systematically: focal animal sampling and instantaneous scan sampling (Table 7.2). In focal animal sampling, behaviors of individuals were either recorded continuously (Miller et al., in preparation) or instantaneously point sampled at minute intervals (Prado 1999; Valladares-Padua 1993). Behaviors selected for calculating time budgets were generally consistent across studies (Table 7.3). Discrepancies exist, however, in the definitions of foraging and feeding behaviors. Depending on the study, "foraging"

Table 7.1
Study Sites for *Leontopithecus* Species

Species and Study Site	Location	Size of Protected Area	Forest Structure	Forest Type
L. rosalia				
Poço das Antas Biological Reserve	22°31′ S 42°16′ W	63 km^2 (35 km^2 of forest)	Isolated forest remnant of mostly secondary forest fragmented by grasslands	Lowland moist forest (coastal)
União Biological Reserve	22°27′ S 42°02′ W	32 km^2 (24 km^2 of forest)	Isolated, well-preserved remnant of secondary and mature forest	Lowland moist forest (coastal)
L. chrysopygus				
Caetetus State Ecological Station	22°18′ S 49°35′ W	22 km^2	Isolated, well-preserved remnant of mostly mature forest, surrounded by agricultural land	Semideciduous mesophytic forest (inland)
Morro do Diabo State Park	22°30′ S 52°20′ W	341.56 km^2	Isolated forest remnant, mostly secondary, surrounded by agricultural land	Upland semi-deciduous mesophytic forest (inland)
L. chrysomelas				
Una Biological Reserve	15°09′ S 39°08′ W	70 km^2 (45 km^2 of mature forest)	Western portion of reserve highly fragmented; secondary and mature forest, contiguous along some borders	Lowland moist forest (coastal)
Lemos Maia Experimental Station	15°20′ S 39°5′ W	4.9 km^2 (3.9 km^2 forested)	Mostly isolated secondary and mature forest surrounded by agricultural land	Lowland moist forest (coastal)
L. caissara				
Superagüi National Park	25°15′ S 48°19′ W	34 km^2	Park made up of two islands. On one island, forest bordered by *restinga* (a coastal sandy habitat) and mangroves; on the other, secondary and mature forest	Lowland moist forest (coastal)

may or may not include consumption of food, and "feeding" may or may not include both plant and/or animal prey. Similarly, "stationary" may or may not include resting. Where possible, we indicate the precise definitions used in Table 7.3.

During data collection, information on vertical use of space, food types consumed, and foraging substrates was also recorded in most studies. Plant species used by lion tamarins were usually identified to genus or species level. The identification of animal prey to this level, however, was not possible.

The most common method for determining home range size in all of these studies was to estimate the location of groups at regular intervals by using marked trails as references, although in the earliest studies of *L. rosalia,* before groups were habituated, group positions were also estimated through triangulation of radio collar signals (Dietz et al. 1997; Peres 1986a, 1986b). To estimate home range size, researchers assigned group locations to particular quadrats that were summed to determine the shape and extent of a group's range. Differential use of certain areas within the home range was determined by calculating the percentage of time each quadrat was entered. More recent studies have used minimum convex polygon techniques to calculate home range (Dietz et al. 1997; Kierulff 2000; Prado 1999).

Day ranges were estimated by a variety of methods. One study tracked group movement by tracing the path of the group with a fishing line tied to trees; the line was later measured (Passos 1997a). In two other studies, group movement was traced by mapping the location of trees flagged at regular intervals (Keuroghlian 1990; Prado 1999). In all studies except Passos 1997a, day range length was taken as the cumulative straight-line distance between position estimates for full-day follows.

In some studies, the vegetation structure was classified and mapped (Kierulff 2000; Peres 1986a, 1986b; Rylands 1982, 1989b). Researchers were then able to identify preferences by comparing the different habitat types available with their use by lion tamarins. In a different approach, Valladares-Padua (1993) created an experimental design so that he could compare four groups of *L. chrysopygus,* each of which used a different type of habitat.

RESULTS FROM STUDIES OF BEHAVIORAL ECOLOGY OF LION TAMARINS

Time Budgets

Table 7.3 presents a summary of time allocation patterns found during eight long-term studies of lion tamarins in the wild. In several of the studies, lion tamarins were found to spend the majority of their time moving and traveling. In others,

Table 7.2

Summary of Methods for Studies of *Leontopithecus* Species

Species and Study Site	Study Period	Number of Groups	Total Observation Hours	Complete Days Observed	Behavioral Methods	Home Range Methods	Movement Methods	Source
L. rosalia								
Poço das Antas Biological Reserve	October 1983–May 1985	4–7	2,164	15–50 per group	Scan samples at 10-minute intervals.	Triangulation, estimate of position in marked 100 m^2 quadrats every 10 minutes Analysis: use of 0.25 ha quadrats	Cumulative step length between 10-minute position estimates	Dietz et al. 1997
Poço das Antas Biological Reserve	April–July 1984 May–August 1985	1	1,056	55	Scan samples at 15-minute intervals.	Triangulation, estimate of position in marked 100 m^2 quadrats every 15 minutes Analysis: use of 0.25 ha quadrats	Cumulative step length between 15-minute position estimates	Peres 1986b
União Biological Reserve	March 1997–February 1998	4	1,396	21–60 per group	Scan samples at 30-minute intervals and 10-minute focal with instantaneous point sampling every minute.	Estimate of position using detailed topographic maps with mapped trails every 15 minutes Analysis: minimum convex polygon model	Cumulative step length between 15-minute position estimates	Kierulff 2000
Poço das Antas Biological Reserve	March 1998–March 1999	8	374	—	15-minute continuous focal sampling	—	—	K. E. Miller and J. M. Dietz, unpublished data

L. chrysopygus								
Caetetus State Ecological Station	November–December 1988	1	164	—	Scan samples at 5-minute intervals (later clumped into 15-minute categories).	Flagging tape placed every 15 minutes. Path drawn on map by measuring angles and distances Analysis: use of 0.063 ha quadrats	Cumulative step length between markers placed every 15 minutes	Keuroghlian 1990
Morro do Diabo State Park	June 1988–August 1990	4	815 total	~24 days/group	10-minute focal with instantaneous point sampling every minute	Estimate of position in marked 50 m^2 quadrats every 20 minutes Analysis: use of 50 m^2 quadrats (including all quadrats within perimeter)	Cumulative step length between 20-minute position estimates	Valladares-Padua 1993
Caetetus State Ecological Station	August 1993–November 1995	1	550	—	Scan samples at 10-minute intervals.	Nylon cord tied to trees tracing group movement. Path drawn on map by measuring angles and distances Analysis: Use of 50 m^2 quadrats	Marked nylon cord every hour, measured cumulative distance between marks	Passos 1997a
L. chrysomelas								
Lemos Maia Experimental Station	April–November 1980	1	203.5	12 days	Scan samples at 5-minute intervals.	Estimate of position in marked 50 m^2 quadrats every 5 minutes Analysis: Use of 50 m^2 quadrats	Cumulative step length between 5-minute position estimates	Rylands 1982, 1989b

Continued on next page

Table 7.2 continued

Species and Study Site	Study Period	Number of Groups	Total Observation Hours	Complete Days Observed	Behavioral Methods	Home Range Methods	Movement Methods	Source
Una Biological Reserve	August 1992–August 1995	7	Minimum 192 hours/group	—	—	Position estimate 30-minute intervals Analysis: minimum convex polygon model MCPAAL software	—	Dietz et al. 1996
Una Biological Reserve	July 1998–December 2000	3	1,048 hours to 1,874 hours per group	45 to 92 days per group	Scan samples at 20-minute intervals.	—	Cumulative step length between 20-minute position estimates	B. E. Raboy and J. M. Dietz, unpublished data
L. caissara								
Superagüi National Park	April–December 1996	1	—	45 days	10-minute focal with instantaneous point sampling every minute	Flagging tape placed every 20 minutes. Path drawn on map by measuring angles and distances Georeferenced starting points with UTM Analysis: (1) use of 50 m^2 quadrats and (2) minimum convex polygon model	Cumulative step length between 20-minute position estimates	Prado 1999

Table 7.3

Percentage of Time Spent in Each Activity for *Leontopithecus* Species

Species and Study Site	Travel Mean ± S.D. (Range)	Stationary Mean ± S.D. (Range)	Feed Mean ± S.D. (Range)	Forage Mean ± S.D. (Range)	Other Mean ± S.D. (Range)	Source
L. rosalia						
Poço das Antas Biological Reserve	33.5 ± 9 (21.3–42.9)	17.6 ± 7.3 (7.4–24.7) resting + stationary	(15.3–21, groups R1, R2) plants + prey (17.9–29.1, groups A1, B) plants	19.7 ± 10.1 (9.2–30.2) foraging + feeding on prey	8.5 ± 4.3	Dietz et al. 1997[a]
Poço das Antas Biological Reserve	8	66	13 plants	7 foraging 3 feeding on prey	2	Miller et al., in preparation
União Biological Reserve	31.9 ± 1.6	30.9 ± 3.6 stationary 8.8 ± 2.7 resting	8.8 ± 2.2 plants	10.9 ± 1.8 foraging 1.6 ± 0.5 feeding on prey	6.9 ± 2.1	Kierulff 2000
L. chrysopygus						
Morro do Diabo State Park	(10–18)	(53–70) resting + stationary	(6–10)	(2–5)	(0.2–0.5)	Valladares-Padua 1993
Caetetus State Ecological Station	33.9	17.4 resting +stationary	29.9	15.7	2.9	Passos 1997a
L. chrysomelas						
Lemos Maia Experimental Station	43	20.2 resting + stationary +social	24.1	12.8	—	Rylands 1982
Una Biological Reserve	32.9	21 stationary (–3 resting)	10.5 plants	20 (foraging and feeding on prey)	13	B. E. Raboy and J. M. Dietz, unpublished data
L. caissara						
Superagüi National Park	56	1 resting	29.4 feeding + foraging	—	13.6	Prado 1999

[a]Pooled data from several studies.

lion tamarins were found to be more stationary or spend somewhat equal proportions of their time stationary and traveling. Feeding and/or foraging was the second most common behavior in most of the populations studied. In the Poço das Antas Biological Reserve, *L. rosalia* spent more time foraging for prey (20 percent) and feeding on fruit (23 percent) than *L. rosalia* in the União Biological Reserve (11 percent foraging and feeding on prey; 9 percent feeding on fruit) (Dietz et al. 1997; Kierulff 2000). The translocated groups of *L. rosalia* rested more and spent more time stationary (31 percent) in comparison with the Poço das Antas groups (18 percent). *Leontopithecus caissara* spent 56 percent of its time traveling and 1 percent stationary (Prado 1999). The percentage of time spent feeding was similar among the four species (Table 7.3).

Across study sites, lion tamarins were active for about 9–12 hours each day. They started their activities earlier and stopped later during the warm, rainy season, when the amount of daylight was greater, and they left the sleeping sites (Figure 7.2) later and entered them earlier during the cold, dry season, with shorter daylight hours. This trend may also relate to photoperiod because there are more hours of sunlight during the warmer months. Lion tamarins tended to leave their sleeping sites between 0500 to 0900 hours in the morning and entered their nighttime den sites between 1400 to 1800 hours.

In general, lion tamarins spent more time stationary during the middle of the day (between 1100 and 1400 hours) and traveled more in the early morning and late afternoon (Kierulff 2000; Passos 1997a; Peres 1986b; Prado 1999). Feeding on plants was concentrated between 0700 and 0800 hours (Kierulff 2000; Passos 1997a), and for *L. chrysopygus* a second peak was observed at 1500 h (Passos 1997a). Time spent foraging for animal prey was greatest in the morning and early afternoon, between 0900 and 1300 h (Kierulff 2000; Passos 1997a).

As already mentioned, lion tamarins showed similar seasonal differences in the allocation of time to their various activities. In the União Biological Reserve *L. rosalia* traveled more, foraged more, and fed more on plants during the cool, dry season and rested more and spent more time in social activities during the warm, wet season (Kierulff 2000). In the Caetetus State Ecological Station, *L. chrysopygus* spent more time traveling and foraging during the dry season and more time resting, feeding on plant material, and in social activities during the wet season (Keuroghlian and Passos 2001; Passos 1997a). In Morro do Diabo State Park, *L. chrysopygus* fed more on plants during the dry season and increased consumption of animal prey and exudates during the rainy season (Valladares-Padua 1993). During colder and drier months, *L. caissara* spent more time foraging and feeding and less time traveling, whereas during the warmer, wetter months, it rested more and spent more time in social activities (Prado 1999).

Figure 7.2. Black-faced lion tamarin in a tree hole. Lion tamarins use tree holes for sleeping at night. (Photo by Luiz Claudio Marigo)

Foraging and Feeding

All species of lion tamarins were found to have similar diets, which included ripe fruits, nectar, exudates, and small animals (see Figure 7.3 and Table 7.4). Fruit was the principal plant item eaten, followed by either nectar or exudates, depending on the population. Plant-feeding time was concentrated around a small number of important species. For *L. rosalia,* seven plant species accounted for 56 percent of plant-feeding time. For *L. chrysopygus,* five species accounted for 69 percent of plant-feeding time, and for *L. caissara,* only eight species accounted for 94 percent of the feeding observations (Table 7.4).

Fruits eaten by lion tamarins are typically small, soft, sweet, and have a lot of pulp, but large fruits are also occasionally consumed (Dietz et al. 1997; Kleiman et al. 1988; Passos 1999; Rylands 1982, 1989b). Ripe fruits are preferred over unripe (Dietz et al. 1997; Kierulff 2000; Passos 1999) and obtained more often from trees than from vines or lianas (Table 7.4). Melastomataceae is one of the most important families providing fruits for *L. rosalia* (Dietz et al. 1997; Kierulff 2000; Peres 1986a, 1986b) and *L. chrysomelas* (Rylands 1982), whereas Myrtaceae is the key family for *L. caissara* (Lorini and Persson 1994b; Prado 1999) and *L. chrysopygus* (Valladares-Padua 1993). *Leontopithecus chrysopygus* in the Caetetus State Eco-

Figure 7.3. Golden lion tamarin eating snail. (Photo by James Dietz)

logical Station did not commonly exploit either of these families but instead concentrated on fruits from the Palmae (Passos 1999).

Nectar is eaten by all species and appears to be an important seasonal food resource. In the Poço das Antas Biological Reserve, particularly when fruit was scarce, and during the dry season, *L. rosalia* groups spent 44 percent of their feeding time eating nectar of *Symphonia globulifera* (see Table 7.4 for averages). Miller et al. (in preparation) found that nectar from *S. globulifera* accounted for 10.4 percent of all time spent feeding on plant matter that could be identified. In the União Biological Reserve, 5 percent of the feedings observed for golden lion tamarins were on nectar of *S. globulifera* and bromeliads. Although there are no records of black lion tamarins eating nectar in Morro do Diabo State Park, they have been seen eating nectar from flowers of *Mabea fistulifera* in the Caetetus State Ecological Station during the dry season when fruit was scarce (Passos and Kim 1999; Valladares-Padua 1993). *Leontopithecus chrysomelas* was also observed feeding on nectar from both *S. globulifera* and *Manilkara* species (Sapotaceae) (Rylands 1982, 1989b), and *L. caissara* was seen eating nectar of the inflorescences of *Norantea brasiliensis* (Lorini and Persson 1994b).

All lion tamarin species have also been observed eating plant exudates (Dietz et al. 1997; Passos 1999; Peres 1986a, 1986b; Prado 1999; Valladares-Padua 1993), al-

Table 7.4

Microhabitats and Food Items Used by *Leontopithecus* Species

Species and Study Site	Diet (Percentage of Total)	Animal Prey Consumed	Microhabitats Most Exploited	Number of Plant Species	Principal Plant Species and Families	Source
L. rosalia						
Poço das Antas Biological Reserve	Ripe fruit (78) Animal prey (13.6) Exudates (1.5)	Invertebrates: arthropods Vertebrates: snails, nesting birds	Bromeliads, palm trees, woody crevices, dead leaves, epiphytic growth, tree bark, vine tangles, bamboo hollows, debris accumulated on the ground	63 species 23 families	Melastomataceae —*Clidemia* species —*Miconia* species Mimosoideae Guttiferae Palmae	Dietz et al. 1997; Peres 1986a
Poço das Antas Biological Reserve	Fruits (88.3) Nectar (10.4) Exudates (1.3)	Invertebrates: Orthoptera, spiders, caterpillars, other larvae, cockroaches Vertebrates: frogs, lizards	—	54 species	Twenty-one species accounted for 89 percent of the total time spent feeding.	Miller et al., in preparation

Continued on next page

Table 7.4 continued

Species and Study Site	Diet (Percentage of Total)	Animal Prey Consumed	Microhabitats Most Exploited	Number of Plant Species	Principal Plant Species and Families	Source
União Biological Reserve	Ripe fruit (79.5) Animal prey (15.4) Unripe fruit (0.1) Flowers/nectar (5)	Invertebrates: Blattaria, Coleoptera, Lepidoptera, Orthoptera, snails, spiders Vertebrates: tree frogs, fledgling birds	Wood crevices, tree bark, dead leaves (base of palm leaves), vine tangles, bromeliads and other epiphytes, palm trees, accumulated leaves, bamboo hollows, flowers, decaying coconuts	>80 species >21 families	Seven species represented 56 percent of the total feeding observations. Melastomataceae (23 percent) —*Miconia* species Sapotaceae (19 percent) —*Sarcaulus brasiliensis* Cecropiaceae (12 percent) —*Cecropia pachystachya* —*Cecropia hololeuca* —*Coussapoa* species Myrtaceae (11 percent) —*Eugenia* species Clusiaceae (5 percent) —*Symphonia globulifera*	Kierulff 2000
L. chrysopygus						
Morro do Diabo State Park	Fruit (78.5) Animal prey (13.5) Exudates (7.8)	Nonmobile invertebrates: 72 percent (cryptic adult cockroaches and larvae of Lepidoptera and Coleoptera) Mobile invertebrates: 27 percent (Orthoptera)	Natural tree cavities, cracked tree bark, other, similar foraging habitats	53 species 24 families	Myrtaceae (65 percent) Palmae (13 percent) —*Syagrus romanzoffiana*	Valladares-Padua 1993

		Others: snails, caterpillars, spiders, tree frogs (*Hyla* sp.), small lizards, fledgling birds, snakes				
Caetetus State Ecological Station	Fruits (73.4) Exudates (15.2) Animal prey (10.1) Nectar (1.1) Seeds (0.2)	Invertebrates (87 percent): beetles, butterflies, crickets, cockroaches, mantids Vertebrates: eggs and fledgling birds (13 percent), amphibians (11 percent)	Tree cavities, palm tree crowns, bamboo, vines, tree bark, epiphytes, *jequitibá* fruits, other	47 species 27 families	Five species represented 69 percent of the total. Palmae (29 percent) —*Syagrus romanzoffiona* Rhamnaceae (12 percent) —*Rhamnidium elaeocarpum* Ulmaceae (11 percent) —*Celtis pubescens* Moraceae (17 percent) —*Ficus trigona* —*Ficus* species	Passos 1999
L. chrysomelas						
Lemos Maia Experimental Station	Fruit (73.9–89.1) Animal prey (13.5–14.8) Flowers (1.43–3.75) Exudates (3–11.3)	Invertebrates: Blattaria, Coleoptera, Lepidoptera, Orthoptera, stick insects, tree snails, spiders	Epiphytic Bromeliaceae and Araceae, tree cavities, palm tree crowns, bamboo, vines, tree	23 species 16 families	Five families represented 61 percent of the plants used. Melastomataceae —*Miconia dodecandra*	Rylands 1982

Continued on next page

Table 7.4 continued

Species and Study Site	Diet (Percentage of Total)	Animal Prey Consumed	Microhabitats Most Exploited	Number of Plant Species	Principal Plant Species and Families	Source
		Vertebrates: tree frogs, lizards	bark, rotten wood, leaf litter piles		Apocynaceae Moraceae Rubiaceae Sapotaceae	
L. caissara						
Superagüi National Park	Fruits (74. 5) Fungus (12.9) Animal prey (10.3) Exudates (1.3) Flowers/nectar (1)	Invertebrates: mollusks, spiders, insects, crickets, cockroaches, stick insects, mantids, beetles Vertebrates: tree frogs	Wood crevices, palm tree leaves, bromeliads, clumps of flowers and fruits	30 species 17 families	Myrtaceae: 8 species accounted for 94 percent of total. Myrtaceae (31 percent) —*Myrcia acuminatissima* —*Malierea tomentosa* —*Myrcia multiflora* —*Psidium cattleianum* Anacarciaceae (20 percent) —*Tapirira guianensis* Clusiaceae (16 percent) —*Calophyllum brasiliensis* Myrsinaceae (16 percent) Palmae (13 percent) —*Syagrus romanzoffiana*	Lorini and Persson 1994b; Prado 1999

though exudate feeding accounts for only a small proportion of the total feeding bouts (Table 7.4). For *L. chrysomelas,* exudate feeding was found to be restricted to readily available gum exuded from the hanging seed pods of *Parkia pendula* (Rylands 1982, 1989b). For *L. chrysopygus,* feeding on tree exudates represented from 0 to 55 percent of feeding observations but was restricted mainly to the dry season, presumably as a consequence of the shortage of fruits (Passos 1999). *Leontopithecus chrysomelas* and *L. chrysopygus* include a greater proportion of gum in their diet than do *L. rosalia* or *L. caissara.* Lion tamarins also eat some less traditional items. Fungus is the second most common food item for *L. caissara* (Prado 1999), which also eats the leaf bases of small bromeliads (Lorini and Persson 1994b).

Small vertebrates and arthropods constitute the majority of animal prey consumed by lion tamarins. *Leontopithecus rosalia* captures sedentary and cryptic prey, such as adult orthopterans and larvae of Coleoptera and Lepidoptera (Dietz et al. 1997). Capture of mobile prey or flying insects is relatively rare.

The microhabitats where lion tamarins forage for prey vary across populations (Figures 7.4–7.7). In the Poço das Antas Biological Reserve, bromeliads, tree bark, and foliage of palm trees were most frequently exploited by *L. rosalia* (Table 7.4). In the União Biological Reserve, the most frequent microhabitats used by *L.*

Figure 7.4. Black-faced lion tamarin in bromeliad. Lion tamarins forage extensively in bromeliads for vertebrate and invertebrate prey. (Photo by Luiz Claudio Marigo)

Figure 7.5. Golden lion tamarin foraging for insects under tree bark. The tamarin is micromanipulating, using its long digits. (Photo by James Dietz)

rosalia were wood crevices and tree bark, followed by dead leaves, the bases of palm leaves, vine tangles, bromeliads and other epiphytes, palm trees, and clusters of dead leaves (Kierulff 2000; Table 7.4). Foraging in bromeliads accounted for nearly half of the foraging records for *L. chrysomelas* (Rylands 1982, 1989b). There are no bromeliads at Caetetus State Ecological Station and Morro do Diabo State Park. In Morro do Diabo State Park, *L. chrysopygus* foraged in natural tree cavities, cracked tree bark, and other similar sites (Valladares-Padua 1993; Table 7.4). In the Caetetus State Ecological Station, *L. chrysopygus* foraged for prey in tree cavities (41 percent), palm tree crowns (22 percent), bamboo (12 percent), vines (11 percent), bark (5 percent), seed pods of the *jequitibá* tree (4 percent), and epi-

Figure 7.6. Lion tamarins forage in palms. This golden lion tamarin is in a spiny palm. (© Ian Yeomans)

phytes (2 percent) (Keuroghlian and Passos 2001; Passos 1997a; Passos and Alho 2001; Table 7.4). *Leontopithecus caissara* foraged in wood crevices, palm tree leaves, and bromeliads (Prado 1999).

Use of Space

Comparative data on home range size, range use, and population density for various lion tamarin populations are presented in Tables 7.5 and 7.6. Range size and population density appear to be inversely related. The smallest home ranges were recorded for *L. rosalia* in the Poço das Antas Biological Reserve and for *L. chrysomelas* in the Lemos Maia Experimental Station, which also had the highest population densities. Population densities and home range sizes were intermediate for golden lion tamarins in the União Biological Reserve, golden-headed lion tamarins in the Una Biological Reserve, and black lion tamarins in Morro do Diabo State Park, whereas *L. chrysopygus* in the Caetetus State Ecological Station and *L. caissara* had the largest home ranges and the smallest population densities recorded to date.

The low population density of *L. rosalia* in the União Biological Reserve was likely a consequence of translocation. The absence of native golden lion tamarins at the release site allowed the groups to develop home ranges without the pressure of other groups. In this case, the home range used was probably influenced more by the resource base available than by competition between groups. Sup-

Figure 7.7. Golden lion tamarin eating an insect. Lion tamarins prize invertebrates and depend upon them for protein. (Photo by Jessie Cohen, Smithsonian National Zoological Park)

porting this idea is the fact that range size decreased as newly established groups were settled in adjacent areas (Kierulff 2000).

More generally, *Leontopithecus* species' population densities appear to be related to differences in habitat quality. The low population density and the large home ranges observed for black lion tamarins in Morro do Diabo State Park correlate with a low density of bromeliads, high seasonality of rainfall, and lower temperatures (Rylands 1993, 1996; Valladares-Padua 1993). In the Caetetus State Ecological Station, the regular use of spatially dispersed foraging sites is likely associated with the black lion tamarins' large home range requirements there (Keuroghlian and Passos 2001). The low density of golden lion tamarins in the Serra do Mar

Table 7.5
Population Densities (Ranges in Parentheses) of *Leontopithecus* Species

Species (Population) and Study Site	Individuals/km^2	Groups/km^2	Source
L. rosalia			
Poço das Antas Biological Reserve	12.0	1.96	Dietz et al. 1994a
Adjacent to Poço das Antas Biological Reserve	5.1	1.17	Kierulff 1993a, 1993b
Serra do Mar	2.2	0.39	
Centro Hípico	6.8	1.14	
Campos Novos	8.5	2.35	Kierulff 1993a, 1993b
União Biological Reserve	3.5	0.46	Kierulff 2000
L. chrysopygus			
Morro do Diabo State Park	3.7	0.73	Valladares-Padua 1993
Caetetus State Ecological Station	1.0	0.20	Passos 1997a
L. chrysomelas			
Una Biological Reserve	5.0	—	Dietz et al. 1994c, 1996
Lemos Maia Experimental Station	(5.0–17.0)	(0.90–3.00)	Rylands 1982, 1989b
L. caissara			
Various localities	1.5	0.30	Lorini and Persson 1994a

likely derived from poor habitat quality (Kierulff 1993a, 1993b; Kierulff and Rylands, submitted). In the Una Biological Reserve, the eastern half of the area is partially covered by relatively tall, mature forest unlike the more disturbed forests in the Lemos Maia Experimental Station, where *L. chrysomelas* was estimated to have higher densities (Table 7.2). At least in the case of *L. chrysomelas,* the presence of mature forests in the Una Biological Reserve may contribute to a lower population density when compared with Lemos Maia.

Two contrasting patterns of home range use emerge from studies of the genus *Leontopithecus:* lion tamarins spend most of their time either patrolling the edges of their home ranges or preferentially occupying core areas. These patterns are probably related to ecological factors such as habitat type and resource distribution, as well as social factors, including population density. In the Poço das Antas Biological Reserve, Peres (1989b) recorded golden lion tamarin groups concentrating much of their time in certain areas of the range periphery and rarely using the center. Peres (1986a, 1986b, 1989b) suggested that this was due to variation in habitat structure across the range, the distribution of food and the location of sleeping sites, and the influence of neighboring groups challenging home range borders. In the União Biological Reserve, with a low population density, groups spent more time in the center of their home ranges, exhibited very little territorial defense, and did not spend as much time patrolling territorial boundaries as groups did in the Poço das Antas Biological Reserve.

Three of the four *L. chrysopygus* groups monitored by Valladares-Padua (1993) showed a tendency for a greater use of the edges of their ranges, while one group concentrated its activities in the center. The differences in home range use by these groups were believed to be related to habitat structure, food distribution, and the location of sleeping sites (Valladares-Padua 1993). In the Caetetus State Ecological Station, a group of *L. chrysopygus* used the central area of its home range more intensively than the periphery; variation in food abundance and the location of sleeping sites in the central and peripheral regions of the territory were presumed to be the cause (Passos 1997a). *Leontopithecus caissara* was also seen to concentrate its activity in the center of its home range (Prado 1999).

Tree holes are the predominant sleeping sites for all species of lion tamarins. In the Poço das Antas Biological Reserve, sleeping sites of *L. rosalia* consisted primarily of tree holes and occasionally vine tangles, palm crowns, and bamboo thickets (Dietz et al. 1997). In the União Biological Reserve, one group used 35 different tree holes, while another used 41. The average number of tree holes used as sleeping sites among the four groups of black lion tamarins at Morro do Diabo State Park was 17 (range 16 to 19), and they rarely returned to a tree hole used on the previous night. All observed dens were located in or close to areas of intensive use, and dens had a clumped distribution in the four study areas (Valladares-Padua 1993). A group of *L. chrysomelas* studied by Rylands (1982) most frequently used holes in tree trunks, and only on two occasions used epiphytes. It used the same sleeping site on consecutive nights for up to 6 nights. One *L. caissara* group was seen sleeping in 16 different tree holes (Prado 1999).

Height preferences varied among study populations, likely reflecting differences in the availability of substrates on which to forage for prey, travel, rest, and feed on fruits (Table 7.6). *Leontopithecus rosalia* (see Coimbra-Filho and Mittermeier 1973; Kierulff 2000), *L. chrysopygus* in Morro do Diabo State Park (Valladares-Padua 1993), and *L. caissara* (see Prado 1999) all used the middle strata most, while *L. chrysomelas* (see Rylands 1989b) preferred higher levels and *L. chrysopygus* at the Caetetus State Ecological Station (Passos 1997a) used the lower and middle levels more frequently. *Leontopithecus rosalia,* which used mostly the midlevel of the forest (Coimbra-Filho and Mittermeier 1973), was observed going to the ground, both in captivity and in the wild, in order to forage and cross open areas (Dietz et al. 1997; Kierulff 2000; Kleiman et al. 1988). Golden lion tamarins in the União Biological Reserve showed a marked preference for the middle layers of the forest in all behavioral categories (Kierulff 2000). Valladares-Padua (1993) also observed that his four black lion tamarin study groups spent a considerable amount of time at the middle level of the forest and rarely went to the ground (Table 7.6). In the Caetetus State Ecological Station, *L. chrysopygus* spent the majority of time at 0 to 10 m and 10 to 20 m (Passos 1997a). Feeding on fruit and resting were the principal reasons for the use of the canopy, whereas moving and foraging occurred more often in the understory (Passos 1994; 1997a; Table 7.6). *Leontopithecus chrysomelas* spent more time higher in the forest, above 12 m in the majority of the records (80 percent), with foraging occurring generally between 13 and 19 m. The foraging level corresponded to that containing the highest abundance of large bromeliads, one of the preferred types of foraging sites, but nonbromeliad foraging was also largely restricted to this level (Rylands 1989b). *Leontopithecus caissara* showed a preference for the middle layers in all activities (Prado 1999).

Habitat Preferences

In the Poço das Antas Biological Reserve, *L. rosalia* preferred swamp and hilltop forest, avoided "gingers" (2 m–high vegetation laden with epiphytes), hillside forest, and pasture, and used corridors in accordance with their availability in the area. Among the described structural parameters of the forest, density and size of trees, canopy cover, and abundance of bromeliads and lianas were most relevant to habitat preference (Peres 1986b). In the União Biological Reserve, *L. rosalia* preferred swamp and lowland forest, where the abundance of lianas and bromeliads was high. As in the Poço das Antas Biological Reserve, it also avoided hillside forest, where the densities of microhabitats, such as palm crowns, bromeliads, and lianas, were lower.

The habitats that lion tamarins choose to occupy depend on the activities in which they are engaged. Peres (1986b) found different behavior patterns in each habitat used by *L. rosalia*. In general, the study groups used swamp forest when foraging for animal prey and used hilltops while sleeping and feeding on fruit, flowers, and exudates. Some fruit trees available in corridors of transitional forest were also visited during regular movements from hilltops to the lowlands or vice versa. Quadrats dominated by gingers were almost exclusively entered for bromeliad foraging, whereas those of hillside forests were used for occasional sleeping sites or while in transition to other vegetation types (Peres 1986b).

Conversely, the availability of widely dispersed and/or ephemeral resources may influence where lion tamarins choose to spend their time and the amount of time spent in an area. The spatial and temporal location of gum influenced the use

Table 7.6
Home Range Size, Average Day Range Length, and Heights Used in the Forest for *Leontopithecus* Species

Species and Study Site	Home Range Size (ha) Mean ± S.D. (Range)	Day Range (m) (Minimum–Maximum)	Preferred Height (m)	Source
L. rosalia				
Poço das Antas Biological Reserve	45 ± 16 (21–73)	1,339 ± 256 and 1,553 ± 406	Midlevel	Dietz et al. 1997; Miller et al., in preparation
União Biological Reserve	150 ± 72 (65–229)	1,873 ± 302 and 1,745 ± 484 (401–3,916)	7.2 ± 3.3 m	Kierulff 2000
L. chrysopygus				
Morro do Diabo State Park	138 (113–199)	1,725 (1,362–2,088)	8.5 ± 2.7 to 7.1 ± 2 m	Valladares-Padua 1993; Valladares-Padua and Cullen 1994
Caetetus State Ecological Station	277	1,904 ± 465 (1,164–3,103)	0–10 to 10–20 m	Passos 1997a
L. chrysomelas				
Lemos Maia Experimental Station	40	(1,410 to 2,175)	80 percent of observation records 12–30 m	Rylands 1982, 1989b
Una Biological Reserve	93 ± 17 118 ± 62	1,684 ± 582 to 2,044 ± 868	—	Dietz et al. 1997; B. E. Raboy and J. M. Dietz, unpublished data
L. caissara				
Superagüi - National Park	321	2,235 (1,082 to 3,398)	6–10 m	Prado 1999

of habitat by *L. chrysopygus*. In Morro do Diabo State Park, fruit feeding was associated with dry land forest, whereas gum feeding was associated with swamp forest (Albernaz 1997; Valladares-Padua 1993). The study group of Albernaz (1997) used a transition zone between dry land and swamp forest most frequently, and all of their sleeping sites were located there. Based on comparisons of plant species density and the composition of the different habitats within the home range of each group of *L. rosalia* in the União Biological Reserve, it was found that the density of trees species most commonly exploited by each group was the most important factor affecting habitat selection (Kierulff 2000). Time spent in each habitat throughout the year was related to the location of the principal tree species used each month. Seasonal differences in home range use by *L. chrysopygus* in the Caetetus State Ecological Station were also related to seasonal differences in the use of food resources (Passos 1997a).

COMPARISON OF THE BEHAVIORAL ECOLOGY OF LION TAMARINS

Time Budgets

Activity budgets vary according to a number of environmental factors, including diet and the distribution and abundance of food resources (Clutton-Brock 1977; Milton 1980; Oates 1987; Peres 1994; Rylands 1982, 1996). Intraspecific variation in time budgets has been observed in many other primate species and is often related to overall habitat difference, e.g., in species such as the red colobus (*Colobus badius*) (Marsh 1981), woolly monkey (*Lagothrix lagothricha*) (Defler 1995, Di Fiore and Rodman 2001; Stevenson et al. 1994), and muriqui (*Brachyteles arachnoides*) (Milton 1984; Strier 1987). Differences seen in the time budgets of the various lion tamarin species likewise appear related to differences in their habitats, especially given that the age and structure of the forest fragments in which different populations are found can result in significant differences between sites in food quality and quantity throughout the year.

In the Poço das Antas Biological Reserve, habitats used by *L. rosalia* consist mainly of degraded forest with little mature forest (Dietz et al. 1997), contrasting with forest in the União Biological Reserve where the majority of the forest is mature (Kierulff, personal observations). In the Poço das Antas Biological Reserve, *L. rosalia* groups spent more time feeding and foraging than in União. In the União Biological Reserve, the home range of the translocated groups was larger than that observed in the Poço das Antas Biological Reserve, but in both populations the groups spent similar percentages of time traveling, probably due to differences in food quality and abundance between the two forests. A high level of inactivity was observed in black lion tamarins in Morro do Diabo State Park relative to the other three species (Valladares-Padua 1993). This inactivity could derive from the fact that it was the only species studied in semideciduous forest, with a low frequency of bromeliads, high seasonality of rainfall, and low temperatures (Rylands 1996; Valladares-Padua 1993). This difference could also be due to variability in data collection techniques since black lion tamarins studied by Keuroghlian and Passos (2001) in Caetetus were much more active than those described by Valladares-Padua (1993).

Variation in time budgets is also influenced by population density, which affects home range size and the degree of territoriality exhibited among groups and, as a result, the amount of time spent moving. The time spent stationary and resting observed for the translocated lion tamarin groups in the União Biological Reserve was similar to that observed for *L. chrysopygus,* which has one of the lowest population densities recorded for the Callitrichidae (Valladares-Padua 1993; Table 7.5). Very little territorial defense was found in the *L. chrysopygus* groups in Morro do Diabo State Park (Valladares-Padua 1993) and the Caetetus State Ecological Station (Passos 1997a). In União, encounters between neighboring golden lion tamarin groups occurred on average once every 16.7 days, and less time was spent patroling the territorial boundaries (Kierulff 2000) than was spent by golden lion tamarin groups in the Poço das Antas Biological Reserve, where all the forested area was occupied by reproductive groups (Dietz et al. 1994a, 1997). Peres (1989b) observed that the ranges of two groups in the Poço das Antas Biological Reserve were contiguous with six and seven neighboring groups. Encounters between neighboring groups occurred once every 1.6 and 2.1 days. According to Peres (1989b), groups regularly move to, and remain at, encounter sites to reinforce boundaries in order to prevent neighbors from trespassing farther into the groups' territory.

Within each species of lion tamarin, intermonthly variation in time budgets is likely due to variation in food abundance, dispersion, and availability, as well as to seasonal differences in day length, temperature, and physiological constraints related to reproduction (Clutton-Brock 1977; Ménard and Vallet 1997; Morland 1993).

The type and availability of food are important factors determining differences in activity between seasons. Lion tamarins travel farther and forage more during the cool, dry season, which is associated with food shortage, a pattern also observed in other primates during periods of food scarcity (Ménard and Vallet 1997; Passamani 1998; and others). Insect abundance is higher during the rainy season, and insect abundance, richness, and diversity all increase during warmer months (Ferrari 1991; Janzen 1973). In the União Biological Reserve, *L. rosalia* maintained its prey consumption rate throughout the year by foraging more during the dry season, when insect abundance is lower. However, an increase in foraging related to a lower availability of fruits during the dry season was also observed for *L. chrysopygus* (Passos 1997a; Keuroghlian and Passos 2001).

Seasonal variations in temperature and day length are also important factors affecting, particularly, the two most southerly species, *L. chrysopygus* and *L. caissara.* When total time spent inactive, including both stationary and resting time, was considered for the golden lion tamarin groups in the União Biological Reserve, there was little variation over the year. From March to August (dry season), they spent 39 percent of their time stationary and resting combined and 43 percent from September to April (rainy season). When time spent resting (lying flat on a branch) was analyzed separately, however, it was clear that they adopted the prone position more often during the warm, wet season than in the cold, dry season, probably to regulate body temperature.

A similar pattern was found for *L. chrysopygus* in the Caetetus State Ecological Station. When time spent resting was analyzed separately, black lion tamarins were found to spend more time lying flat during the warm, rainy season (Passos 1997a). *Leontopithecus* species' fur is fine, long, and wispy and much less dense than might be expected, with nearly bare areas on the ventral surfaces. The long fur may represent a compromise between insulation against seasonally cool or cold temperatures and the need to lose heat when faced with the high ambient temperatures and high relative humidity that characterize the daylight hours during much of the year in these species' natural habitat (Thompson et al. 1994).

As mentioned, seasonal physiological factors may also influence time budgets. *Leontopithecus rosalia* mates between August and March, with most births occurring between September and November (Dietz et al. 1994a). Similar patterns of birth seasonality were also recorded for captive populations of three species of *Leontopithecus* in Brazil (French et al. 1996b; De Vleeschouwer 2000). In golden lion tamarins, lactation, the peak of which is the most energetically expensive period of reproduction (Thompson et al. 1990), occurs during the warm, wet season. At this time, females rest more often. As in other callitrichids, lion tamarins are cooperative breeders: all members of a group carry and provision the offspring of the group's reproductive female(s) (Baker 1991; Dietz and Baker 1993; Chapter 8 this volume). Golden lion tamarin infants are carried continuously for the first 3 weeks of life, but by 10 weeks of age infants average more than 90 percent of their time off carriers (Baker 1991; see also Chapter 9 this volume). During the rainy season (September to February), time spent traveling and feeding is reduced, possibly due to constraints linked to the birth of the offspring.

The proportion of time spent in each activity during the day is similar across *Leontopithecus* species. Feeding and moving are concentrated during the morning and afternoon hours, and resting peaks at midday—a pattern that is common in many species of primates and is determined by energetic requirements after the long period of nocturnal inactivity and the need to facilitate thermoregulation during the hot midday hours (Lindburg 1977; Oates 1987; Stevenson et al. 1994). As in other primates (Clutton-Brock 1977; Passamani 1998; and others), lion tamarins eat a lot of fruit in the first hours of the day to obtain a high energy supply, and they forage for insects, their principal source of protein, during the warmer hours in the middle of the day.

Lastly, it is important to note that differences in time budgets observed between the two golden lion tamarin populations studied, and between the four species, may also result from methodological differences such as the degree of habituation (for which there is no objective measure), in recording techniques, in the number of groups sampled, and in the duration of the studies and seasons included (see, for example, Ferrari and Rylands, 1994, Table 7.2). Long-term studies with consistent methodologies can more precisely distinguish between the subtle variables that influence the daily behavior and demography of the lion tamarins.

Foraging and Feeding

Lion tamarins are fauni-frugivores (Coimbra-Filho 1976a, 1976b, 1981; Coimbra-Filho and Mittermeier 1973; Dietz et al. 1997; Kleiman et al. 1988; Rylands 1986b, 1996; Valladares-Padua 1993) and have long, slender arms and elongated fingers enabling them to exploit microhabitats when searching for animal prey (Coimbra-Filho 1981; Hershkovitz 1977). Because lion tamarins forage in specific microhabitats as opposed to gleaning prey from foliage (Rylands 1986b, 1996), the distribution of animal prey available to lion tamarins is largely dependent upon the distribution of foraging microhabitats, which in turn is a function of microclimatic conditions and floristic composition (Dietz et al. 1997).

The dietary differences observed between the four *Leontopithecus* species result from differences in food availability at the study sites. During the dry season in the União Biological Reserve, when insect abundance is low, golden lion tamarin groups increase time spent foraging. During the wet season, when more trees produce fruit, the golden lion tamarins spend more time feeding on plant material. In the Poço das Antas Biological Reserve, the availability of ripe fruit is lowest during the dry season, and the golden lion tamarins rely heavily on nectar (Dietz et al. 1997). Nectar was also found to be important in the diet of *L. chrysomelas,* despite the lack of seasonality (Rylands 1986b). A number of forest-dwelling primates are known to feed on nectar, a rich source of simple sugars (Ferrari and Strier 1992; Peres 2000a). It has been described as an essential keystone resource for *Saguinus fuscicollis* and *S. imperator* in southeast Peru (Terborgh 1983; Terborgh and Stern 1987). In the União Biological Reserve, *L. rosalia* feed on *Symphonia globulifera* nectar between March and June, with a peak in April and May, when it is the principal food consumed.

Leontopithecus chrysopygus in Morro do Diabo State Park compensates for low fruit availability in the dry season by increasing the consumption of exudates and animal prey (Valladares-Padua 1993). In contrast, in the Caetetus State Ecological Station, black lion tamarins spend more time eating exudates and nectar during the dry season, whereas time spent feeding on animal prey remains constant throughout the year (Passos 1997a). Prado (1999) found that *L. caissara* complemented its diet with fungus during the dry season. Intense exploitation of a small number of fruit species is common in the genus *Leontopithecus* (Table 7.4) and has also been observed in other callitrichids (Garber 1993; Peres 2000a). The amount of feeding time devoted to the top five to eight plant species each season remains relatively constant, despite the differences in the food resources exploited by the species at different times of the year.

Use of Space

Patterns of home range use are influenced strongly by the temporal and spatial distribution and abundance of food resources as well as population density and the interactions between neighboring groups (Butynski 1990; Crockett and Eisenberg 1987; Ostro 1998; Peres 1986a). Home range sizes observed for different golden lion tamarin populations, and for each of the *Leontopithecus* species, were probably influenced by the different population densities at each site, in part a consequence of habitat quality. In more productive habitats, species tend to require smaller home ranges than in less productive habitats (Eisenberg 1980; Foster 1980).

Leontopithecus species and other large callitrichids, such as Amazonian tamarins (genus *Saguinus*), use the largest home ranges per unit of group biomass of all New World primates (Dietz et al. 1997; Peres 2000b). The similarity in spatial requirements between Amazonian tamarins and lion tamarins may be attributed not only to their relatively large size and group biomass but also to their convergent feeding ecology (Peres 2000b).

Two explanations have been proposed for the maintenance of large home range size in fauni-frugivorous primate species, in which ranging patterns and habitat use are food-dependent. Dawson (1979), Kinzey (1981), and O'Brien and Kinnaird (1997) suggested that home ranges should be large enough to allow for invertebrate foraging because this resource is evenly distributed and difficult to exploit. Ferrari (1988, 1991) advocated that this is a key determinant of range size in callitrichids in general. Terborgh (1983) and Rylands (1986b), on the other hand, argued that home range sizes must be large enough to encompass sufficient fruit resources during the periods of fruit scarcity. Whether fruit or animal prey is the limiting factor depends on the nature and seasonality of the habitats occupied.

Dietz et al. (1997) and Peres (2000b) concluded that the long daily path lengths and large ranges of *L. rosalia* result from foraging on animal prey that are locally abundant but patchily distributed. In the Poço das Antas Biological Reserve, the size of a given home range was negatively correlated with the proportion of swamp forest occurring within it. Swamp forest was the habitat that supported the highest densities of foraging microhabitats and animal prey (Dietz and Baker 1993; Dietz et al. 1997).

The differences in the daily path lengths observed for the lion tamarins are also correlated with home range size (Table 7.6). An increase in home range size of a golden lion tamarin group in the Poço das Antas Biological Reserve was associated with an increase in day range (Dietz et al. 1997). At the same time, the high population density in the Poço das Antas Biological Reserve is also believed to result in longer day ranges as a consequence of groups patrolling and defending their range against neighboring groups (Peres 2000b). With lower population densities and less intraspecific competition, variation in food distribution probably had a stronger effect on the ranging patterns observed for the translocated golden lion tamarin groups in the União Biological Reserve, as well as for black lion tamarins and black-headed lion tamarins. *Leontopithecus rosalia, L. chrysopygus,* and *L. caissara* were recorded using different parts of their home ranges during different times of the year and following a foraging strategy that has also been described for gray-cheeked mangabeys (*Cercocebus albigena*). Plant species used by mangabeys bear fruit in an asynchronous or nonseasonal cycle, and although their location varied throughout the year, these primates were able to maintain a relatively consistent dietary pattern by exploiting large home ranges and shifting to different areas of their range each month (Waser 1975, 1984, in Garber 1993). The specific areas used within the home ranges of lion tamarin groups were defined by which fruits were available during the year. Thus, the annual home range was correspondingly large in order to accommodate an annual cycle of fruit resources and animal prey.

Habitat Preferences

Lion tamarins are thought to have evolved to exploit the tall, mature vegetation of lowland forests (Coimbra-Filho 1978; Peres 1986a, 1986b; Rylands 1993, 1996), but with the widespread destruction and degradation of those habitats, lion tamarins today largely occupy secondary and successional forests. The differences in habitat structure among the forests in which contemporary *Leontopithecus* species' populations have been studied likely influence foraging strategies and patterns of strata and microhabitat use. For example, in mature forests, there are more tree holes (Coimbra-Filho 1978; Dietz et al., in preparation; Peres 1986a, 1986b), and the nonmobile prey consumed by golden lion tamarins are more common (Dietz et al., in preparation) than in secondary successional forests. However, the types of fruits that lion tamarins evidently prefer are more common in secondary successional forests (Peres 1986a), and the characteristic greater productivity of secondary habitats generates a great abundance and diversity of insects (Foster 1980). Rylands (1996) argued that, unlike the marmosets (*Callithrix* spp.) and tamarins (*Saguinus* spp.), the genus *Leontopithecus* is primarily adapted to mature forest and not dependent on successional habitats, but analysis of the habitat of the subpopulations of golden lion tamarins shows they are also able to occupy older secondary growth. In late successional or mature forests, the relative densities of golden lion tamarins are lower (Kierulff and Rylands, submitted). Dietz et al. (1994c) also suggested that the low densities observed for *L. chrysomelas* in the Una Biological Reserve are a consequence of its relatively well-preserved forest.

Valladares-Padua (1993) observed that the habitats used by four groups of black lion tamarins at Morro do Diabo State Park had characteristics analogous to both mature and secondary growth forest. He concluded that the black lion tamarin is flexible in its forest use, and the ecological constraints of secondary forests, such as the relative absence of the black lion tamarin's preferred foraging and nest sites, do not seem to be limiting factors (Valladares-Padua 1993). Because of the absence of epiphytic bromeliads at Morro do Diabo and low density at the Caetetus State Ecological Station, *L. chrysopygus* uses alternative foraging sites (Passos and Keuroghlian 1999; Valladares-Padua 1993) and increases the consumption of exudates when animal prey are less abundant (Valladares-Padua 1993).

CONCLUSIONS

The widespread fragmentation and degradation of the forests formerly occupied by lion tamarins are the key reasons they are so threatened today. The lion tamarins' adaptability is expressed in their ranging behavior and their use of alternative food resources in forests that are radically altered and very different in their seasonality and floristic composition. Behavioral modifications allow lion tamarins to survive in habitats with varying degrees of food availability and population densities. When population densities are high, they show a more intensive range use with an increase in peripheral patrolling; when population densities are lower, lion tamarins are able to exploit larger home ranges. Whether found at high or low population densities, many of the surviving populations of lion tamarins are confined to areas so small that the key to their survival rests with their capacity to resist inbreeding and the loss of genetic heterogeneity. Metapopulation management strategies are now indispensable for the conservation of *L. rosalia* and *L. chrysopygus,* and a full understanding of the variation and adaptability of their feeding and ranging behavior in different population sizes and densities is a vital component for the success of these strategies. Understanding the spatial and temporal dynamics of lion tamarin resource use, and how lion tamarins distribute their time among different activities in different habitats and at varying population densities, is key to formulating conservation practices for these species—practices that are now constrained by the small populations and few localities where lion tamarins still survive (Chapter 2 this volume).

ANDREW J. BAKER, KAREN BALES,
AND JAMES M. DIETZ

8
MATING SYSTEM AND GROUP DYNAMICS IN LION TAMARINS

The study of callitrichid social structure and mating systems has been one of the most dynamic areas of primatology over the past decade. The unusual social organization and behavior of the callitrichids have long attracted research attention (e.g., Kleiman 1978a; Rothe et al. 1978), but until the 1980s, most work focused on captive groups. Evidence from captive studies led to wide acceptance of the hypothesis of obligate monogamy (*sensu* Kleiman 1977c) for this family (e.g., Epple 1972, 1975a, 1975b; Kleiman 1978b, 1978c). Evidence supporting this hypothesis included a lack of pronounced sexual dimorphism in physical (Hershkovitz 1977) and some behavioral characteristics (e.g., Box 1978; Kleiman and Mack 1980), high levels of male parental investment (e.g., Box 1977; Hoage 1978), and aggressive responses by pair members toward same-sex conspecific strangers (e.g., Epple 1978; Sutcliffe and Poole 1984). Captive multifemale groups were often unstable (Epple 1975b; Hampton et al. 1966; Rothe 1975). Usually only one female per peer group (Abbott 1984) or family group (e.g., Epple 1972; Epple 1975a; French et al. 1984; Chapter 6 this volume) was reproductively active, and other females typically were physiologically suppressed (e.g., Abbott 1984; Evans and Hodges 1984; Ziegler et al. 1987a).

The evidence for monogamy was historically evaluated with the typical mammalian pattern of polygyny as the (sometimes implicit) alternative hypothesis. However, by the mid-1980s, surprising field data led to significant support for the hypothesis that wild callitrichids are often cooperatively polyandrous (e.g., Goldizen 1987a, 1987b, 1989; Sussman and Garber 1987; Sussman and Kinzey 1984). In cooperative polyandry, two or more males, each with substantial probability of paternity, communally rear the offspring of a single female. Cooperative polyandry had previously been reported only in a few bird species (for reviews see Brown 1987; Faaborg and Patterson 1981; Oring 1986), and the rarity of this system among vertebrates doubtless contributed to the surprise and interest around the new field data. Evidence offered in support of the cooperative polyandry hypothesis included the large proportion of multimale groups documented in some callitrichid populations (for reviews see Heymann 2000; Sussman and Garber 1987), low levels of agonistic behavior among same-group males (Goldizen 1989), and participation by multiple males in sexual behavior (Rylands 1986a; Terborgh and Goldizen 1985) and infant care (Garber 1997; Goldizen 1987b; Rylands 1986a; Savage et al. 1996b). Other authors (Baker et al. 1993; Ferrari and Lopes Ferrari 1989) have argued that the available data are equally consistent with a system characterized by one reproductive male and additional nonreproductive male helpers.

Recent field data have also challenged the ubiquity of the "one breeding female per group" pattern. Polygynous groups (i.e., groups in which more than one female produces young) have been suspected or verified in a variety of wild callitrichid populations (e.g., *Callithrix jacchus:* Digby and Ferrari 1994; Roda 1989; *C. aurita:* Coutinho and Corrêa 1996; *Leontopithecus rosalia:* Dietz and Baker 1993; *Saguinus fuscicollis:* Goldizen et al. 1996; *S. oedipus:* Savage et al. 1996a, 1996b). This finding has provided an even stronger contrast to the results of captive studies than the finding that there are potentially polyandrous groups: whereas two-male, one-female captive callitrichid trios have proven reasonably stable (e.g., Epple 1972; Hampton et al. 1966; Kleiman 1978c; Malaga 1985), polygynous reproduction in captive groups appears to be extremely rare (De Vleeschouwer et al. 2001; Price and McGrew 1991; Rothe and Koenig 1991).

Another area of controversy concerns the membership dynamics of callitrichid groups. Behavioral studies of captive groups led to expectations that wild groups would be resistant to immigration due to resident aggression toward potential immigrants. Thus, births were anticipated to be the major mode of recruitment, immigration was expected to be rare, and "families" composed largely of closely related individuals were expected as the typical social unit (e.g., Epple 1975b; Kleiman 1977c). Field data have been somewhat ambiguous on this point. Some studies (e.g., Dawson 1978; Garber et al. 1984; Scanlon et al. 1988; Valladares-Padua 1993) have suggested a high rate of immigration and have led some reviewers (Sussman and Garber 1987; Sussman and Kinzey 1984) to conclude that wild callitrichid groups are likely to be composed mostly of unrelated individuals. Other studies (e.g., Baker and Dietz 1996; Digby and Barreto 1993; Ferrari and Diego 1992; Goldizen and Terborgh 1989) have documented immigration rates and patterns that are consistent with a kinship-based family structure; several authors have argued that almost all data from habituated groups of identifiable individuals support this "kin-based" view of callitrichid social structure (see Baker 1991; Digby and Barreto 1993; Ferrari and Lopes Ferrari 1989; Price 1991).

Data on wild golden lion tamarins (*L. rosalia*) have contributed to the current discussion in all of these areas of controversy (Baker and Dietz 1996; Baker et al. 1993; Dietz and Baker 1993). The population of this species on the Poço das Antas Biological Reserve is, with regard to the number of study groups and duration of continuous monitoring, the most thoroughly studied of any wild callitrichid population. In this chapter, we review what has been published with regard to the mating system and group dynamics in this population, and we provide new data with regard to natal dispersal patterns, polygyny, and potential polyandry. We also review available comparative data from the other lion tamarin taxa.

FIELD SITE AND METHODOLOGY IN STUDIES OF GOLDEN LION TAMARINS

Methods used in studying golden lion tamarins have been detailed in several publications (Baker and Dietz 1996; Baker et al. 1993; Dietz and Baker 1993; Dietz et al. 1994a). Here, we summarize the relevant methods.

Golden lion tamarins were studied on the 6,300 ha Poço das Antas Biological Reserve in the state of Rio de Janeiro in southeastern coastal Brazil. The reserve is isolated from other forested lands along most of its perimeter and is itself a patchwork of grasslands and forests in early to late secondary succession, a result of clear cutting and selective cutting prior to 1973 (Brazil, MA/IBDF/FBCN 1981). The total golden lion tamarin population on the reserve has been estimated at from 290 (Kleiman et al. 1990a) to 347 (Ballou et al. 1998) individuals. Most golden lion tamarins on the reserve live in year-round territorial groups; the remainder as solitary floaters or in floating associations (i.e., nonterritorial individuals or groups that do not occupy fixed ranges but rather travel through or reside temporarily within the territories of other groups: J. M. Dietz and A. J. Baker, unpublished data). During the period for which data are presented in this chapter, almost all suitable habitat within the study area was occupied by territorial groups (J. M. Dietz and A. J. Baker, unpublished data).

Reproduction in this population is seasonal, with almost all births occurring from September through February (Dietz et al. 1994a). Females produce one or occasionally two litters per year, with a modal litter size of two (Dietz et al. 1994a). All group members care for infants (Baker 1991). No successful reproduction has been documented outside of territorial groups.

Study groups were selected for ease of habituation and proximity to the road and major trails crossing the reserve. All study groups were native in origin (i.e., not reintroduced captive-born animals), and no study groups were territorially adjacent to reintroduced groups.

Trapping of golden lion tamarin groups began in 1983 and continues to the present, with most study groups being trapped twice a year (Figure 8.1). Captured golden lion tamarins are anesthetized with ketamine hydrochloride, weighed, measured, examined for reproductive status, tattooed, and dye-marked for individual field identification using a Nyanzol D and hydrogen peroxide solution. We affixed radiotransmitters (Telonics, Inc., and Wildlife Materials, Inc.) on a bead-chain collar to one or more individuals per group under behavioral observation (Figure 8.2). Trapped golden lion tamarins were held overnight before release at the capture site the next morning.

Group Compositions

For some analyses, we relied on cross-sectional group composition data, with compositions "sampled" twice a year. We included only those compositions that were confirmed through either repeated captures or (usually) visual verification in the field. For other analyses, we used essentially continuous data on group composition. These data were from groups that were sufficiently habituated to observers to allow reliable censusing. Such groups were typically visited and checked

Figure 8.1. James Dietz setting traps for the semiannual trapping of golden lion tamarins in the Poço das Antas Biological Reserve. (Photo by Devra Kleiman)

Figure 8.2. Golden lion tamarin in the field wearing a radio collar. (Photo by Devra Kleiman)

at least once per week for any changes in group composition, specifically births, immigrations, dispersals, deaths, and disappearances. Using this census technique, events that we classed as "simultaneous" (e.g., two immigrations into the same group) may have in fact been separated in time by several days.

Golden lion tamarins 18 months old or older were considered adults. For golden lion tamarins of unknown birth date, those weighing at least 550 g were considered adults (Dietz et al. 1994a). We counted animals as natal to the group in which they were trapped if they were known to have been born in the group or if they weighed less than 450 g when first trapped with the group (nearly all recruitment of young animals occurred through births within the group: Baker and Dietz 1996, unpublished data).

Parentage of individuals born during the study was assigned through the following methods:

- *Maternity.* Maternity could in most cases be definitively assigned through behavioral observation (carrying and nursing during the first few days postpartum) and visual and physical exams (visible distension and/or manual palpation prepartum; milk production and/or nipple elongation postpartum). Occasionally, if two females delivered litters in close succession, we could not definitively determine which infants to assign to which female.

- *Paternity.* If only one male thought to be unrelated to the breeding female(s) was present in the group at the time of conception, he was assigned paternity for the infant(s). If two males thought to be unrelated to the breeding female(s) were present, paternity was assigned to the male that was dominant in behavioral interaction (see Baker et al. 1993) or had been seen to exhibit sexual behavior at the estimated time of conception. If the dominance relationship could not be determined and no conception period sexual behavior had been seen, paternity was assigned to the oldest male in the group. If no males thought to be unrelated to the breeding female(s) were present at conception, paternity was assigned to the oldest male in the group, who was typically the presumptive father of the breeding female in question. In these cases, any other males in the group were typically the presumptive brothers of the breeding female.

Terminology

In an attempt to avoid confusion, the following terms are used as defined:

- *Monogyny.* A mating system characterized by a single reproductive female per group. There may be one or more reproductive males per group.
- *Polygyny.* A system characterized by more than one reproductive female per group. There may be one or more reproductive males per group.
- *Pregnancy polygyny.* Groups in which more than one female becomes pregnant within a single breeding season. This term does not indicate how many females per group actually produce live young or successfully rear young.
- *Rearing polygyny.* Groups in which two or more females rear offspring within a single breeding season.
- *Potentially reproductive male.* An adult male not closely related to the current breeding female(s) in the group.
- *Potential polyandry (PPA).* Groups containing two or more potentially reproductive males.
- *Polyandry.* A system in which more than one male per group is likely to father offspring—in other words, to copulate with the reproductive female(s) during fertile periods.

- *Intact group*. A group containing at least one reproductive or potentially reproductive individual of each sex.
- *Established group*. A territorial group in which successful reproduction has taken place. This term is used when discussing immigration patterns. Results presented here refer to established groups only. Prereproductive groups show different patterns of immigration and are not discussed here.

GROUP DYNAMICS

Immigration

Immigration into established breeding groups of golden lion tamarins in the Poço das Antas Biological Reserve was rare (Baker and Dietz 1996), with groups experiencing a mean of 0.34 immigration events per year, or 1 event per 2.9 years (based on 70.4 group-years of data from 17 habituated groups). In 33 percent of immigration events, multiple individuals entered a group simultaneously. All events involving multiple individuals were male events; all but one of the eight multi-individual immigrating units were composed of close relatives (brothers or father-son[s]).

Eighteen of 24 immigration events were "replacements" of breeding individuals, in that they followed or closely preceded the death or emigration of the same-sex breeding individual(s) in the recipient group (Baker and Dietz 1996). In most cases of replacement, it could not be determined whether the immigrant had entered the group before the previous breeder(s) had died/emigrated or after. Where order of events could be determined, both sequences were documented. In 3 cases, all involving males, there was evidence of aggressive takeover by the immigrating individual(s).

Nonreplacement immigration events (i.e., events in which the immigrant joined rather than replaced a resident same-sex breeder) were unusual (Baker and Dietz 1996). Only 4 of 17 study groups experienced such events, and overall, groups experienced a mean of 0.09 events per year (or 1 event per 11.1 years). Five of the 6 nonreplacement immigration events involved males, and most cases occurred under unusual circumstances. Two male events were coincident with the departure of one potential breeding male from two groups—each group had contained two potential male breeders. Thus, in these cases, the immigrant replaced one male and joined the remaining original male. Two other male events occurred in groups that had experienced replacement-type immigrations only days before: the "joiner" male joined a "resident" male that had himself immigrated (as a "replacer") only a few days previously. The single female event might justifiably be excluded as an immigration event involving an established breeding group. In this case, the group, through attrition, had been reduced to a single adult male. The male was joined by an adult female and then, less than 1 month later, by a second adult female. This nonreplacement immigration event did not result in a stable two-female situation: the first immigrating female left the group about 6 weeks after the second female entered.

The sex ratio of immigrants was significantly biased, with 85 percent of 34 immigrants being male (Baker and Dietz 1996). A number of factors possibly contributed to male bias in immigration into established groups. First, as described, males sometimes immigrated in duos or larger units. Thus, a single "vacancy" sometimes provided an immigration opportunity for multiple males. Second, the death of a breeding female in a group containing an adult daughter was usually followed by dispersal of the current breeding male (the "daughter's" father) and the entry of a new male into the group, with the daughter thus "inheriting" her mother's breeding position (A. J. Baker and J. M. Dietz, unpublished data). In contrast, sons did not inherit a dead sire's breeding position; when breeding males died, the current breeding female remained on the territory and was joined by an immigrant male. Thus, the pattern of territorial inheritance within families decreased female immigration opportunities and increased male immigration opportunities. Third, males but not females were sometimes able to join groups already containing a same-sex breeding individual. Finally, there was evidence that male immigrants could aggressively displace resident males, thus "creating" an immigration opportunity, whereas there was no evidence for "opportunity creation" by females. Females may be able to enter groups only if the previous resident female has already died. The fact that potential male immigrants were attacked only by male residents, but that potential female immigrants were sometimes attacked by residents of both sexes (see below), may explain the apparently greater vulnerability of males to aggressive replacement.

Behavioral evidence suggested that aggression from resident golden lion tamarins was the major proximate factor discouraging immigration into established groups. Floating individuals and single-sex groups appearing near territorial groups were the objects of aggressive behavior (chasing) from the territorial residents in at least 80 percent of such encounters. Most such aggression was intrasexual in nature: male floaters were chased only by male group members, and female floaters were chased mostly by female group members. However, female floaters were in some cases chased by resident males as well as resident females. The largely intrasexual nature of resident aggression toward floaters and the typically replacement-style mode of immigration suggest strong intrasexual competition for group membership (and thus reproductive opportunities) in this population (Baker and Dietz 1996). Baker and Dietz (1996) concluded that female-female aggression plays an important role in the maintenance of monogyny in this species, a characteristic also noted in captive golden lion tamarins (Kleiman 1979). Baker and Dietz (1996) also discussed the potential role of males in the maintenance of this system. They hypothesized that male aggression toward potential female immigrants might reflect a form of mate choice—specifically a preference for current over alternative mates, within the constraints of female-enforced monogyny—rather than reflecting a male preference for monogyny itself.

Emigration

No extensive data on emigration patterns in lion tamarins have been published previously. Here we present a preliminary summary of data from golden lion tamarins in the Poço das Antas Biological Reserve. Seven habituated groups were monitored closely from May 1986 to January 1991. During this period, extensive efforts were made to locate animals that had disappeared from study groups, and study groups that had experienced a loss were tracked more frequently and for longer periods than usual, in hopes that emigrations could be confirmed through reappearance of the emigrants near the groups. Data from this period suggested that dispersal accounted for most postinfancy losses of natal individuals (Baker 1991). Of 27 losses of animals older than 10 weeks, 21 (78 percent) were confirmed as dispersals (through sighting or trapping of the animal postdispersal), only 2 (7 percent) were confirmed as deaths, and the fates of 4 animals (15 percent) could not be determined (i.e., they disappeared from the natal group and were not seen thereafter).

Most surviving golden lion tamarins eventually dispersed from natal groups. By 3 years of age, approximately 60 percent of surviving natal individuals in these study groups had dispersed, and by 4 years of age, only approximately 10 percent were still present in natal groups (Baker 1991). Most golden lion tamarins still in natal groups at 4 years of age had become breeders themselves, following the death or emigration of one or both parents (Baker 1991; J. M. Dietz and A. J. Baker, unpublished data; see sections on mating system below).

Data from 18 study groups over a 12-year period (1986–1998) provide additional information about dispersal patterns in this population. Mean age of natal dispersal was not substantially different for males and females (males: $x = 27.5$ months, range = 12–53 months, $N = 65$ known-age confirmed dispersers; females: $x = 24.1$ months, range = 16–38 months, $N = 30$ known-age confirmed dispersers). There was a tendency for both sexes to disperse with same-sex relatives or join same-sex relatives that had previously left the group. Forty of 77 confirmed male dispersers (51.9 percent) either left the natal group in the company of a same-sex sibling (or occasionally with multiple siblings and/or the presumed father or uncle) or joined a sibling that was already floating or was resident in another group. Only three male dispersers (3.9 percent) left with or joined a female sibling. Eighteen of 38 confirmed female dispersers (47.4 percent) left with or joined female siblings. No female dispersers left with or joined male relatives.

Male and female dispersers differed in terms of immediate postdispersal fate. Of male dispersers sighted within 30 days after leaving the natal group, more than half (37 of 70 = 52.9 percent) were found in territorial groups adjacent to their natal groups. These males moved directly into potential breeding situations and avoided the potentially dangerous period (Baker et al. 1993) of floating in a nonterritorial, solitary, or small group situation. In comparison, only 12.1 percent (4 of 33) of dispersing females transferred directly into adjacent groups. Thus, a much higher proportion of female than male dispersers passed through at least some period in nonterritorial, floating status.

While dispersal events were rarely witnessed, circumstantial evidence suggested that a substantial proportion were forced evictions (Baker 1991; A. J. Baker and J. M. Dietz, in preparation, unpublished data). It was not uncommon for recent emigrants to reappear near the natal group, and these emigrants were typically chased off, usually by same-sex adults.

MATING SYSTEM

Males

Studying golden lion tamarins in the Poço das Antas Biological Reserve, Baker et al. (1993) sampled group compositions twice a year over 4.5 years ($N = 19$ groups over varying periods of time). Approximately 70 percent of group samples contained more than one adult male. However, many of the adult males were natal males that were either sons or brothers of the breeding female and thus not likely to be reproductively active within the group (Abbott 1984; Kleiman 1978c). With these natal males excluded, about 40 percent of group samples contained more than one potential breeding male. Baker et al. (1993) concluded that a one-male breeding system was modal for this population but that a large minority of groups did contain two potential breeding males and were thus potentially genetically polyandrous.

Additional data are now available from this population. From May 1986 through October 1996, compositions were documented for 22 habituated territorial groups tracked for 11 to 125 months each (a total of 1,749 group months). Groups contained more than one potential breeding male in 805 group months (46 percent). We refer to these groups as *potentially polyandrous* (PPA).

A total of 46 distinct male duos or sets were documented. Coresident males were close relatives (brother-brother, father-son, or uncle-nephew) in 28 of 37 sets (75.7 percent) for which the relationship between males was known. In the 9 remaining duos of known relationship, the two males were born in different groups and thus were probably not close relatives. We refer to such duos, in which the males are from different natal groups, as *nonkin duos*.

The PPA groups arose in a variety of ways (Table 8.1). Most commonly, two males entered an already established breeding group more or less simultaneously, replacing the resident male. It was not always clear whether the former resident male had died/emigrated before or after the entry of the new males, but as mentioned above, there was evidence that at least some of these cases involved aggressive takeover (Baker and Dietz 1996). Alternatively, already coresident males were involved in the formation of new territorial groups. Further variations on PPA group origin occurred when an immigrant male entered a group already containing a breeding or potentially breeding male but joined rather than replaced the resident (see section on immigration). In some cases the resident was a male that had already reproduced in the group, in others it was a recent immigrant that had not yet reproduced in that established breeding group, and in yet others it was the founding male of a new territorial group in which no reproduction had yet taken place. Finally, father-son or uncle-nephew PPA sets arose within groups following female replacement: when a breeding female was replaced by an immigrant female, an adult son of the original breeding female became a potential breeding male within his natal group, thus forming a PPA male duo with his father or uncle.

We documented only five cases (out of 46 PPA male sets) in which more than two potential breeding males were simultaneously coresident within a group. Two of these arose when a father and sons or uncle and nephews dispersed together from the "natal" group and entered an established group (one case) or founded a new group (one case). Another case occurred as a result of female replacement in a group containing a father and multiple sons (although only one of the sons was adult at the time of female replacement). A fourth case also followed female replacement but occurred in a group in which the current breeding male was a recent immigrant and thus not related to the natal males. Thus the group contained the new female, a recently immigrated adult male, and four natal males (two adult at the time of female replacement) unrelated to the immigrant male. The final

Table 8.1

Origin of Potentially Polyandrous (PPA) Groups in Wild Golden Lion Tamarins

	Duo Already Together			Immigrant Joins Resident Male			
Male relationship	Duo Enters Established Group (= Replacement)	Duo Founds New Group	Female Replacement Results in PPA Group	Resident Male = Established Breeder	Resident Male = Recent Immigrant in Established Group	Resident Male = Founder of New Group in Which No Reproduction Has Yet Occurred	Unknown
Brother-brother	9	0	0	2[a]	1	2	0
Father-son(s) or uncle-nephew(s)	4	2	7[b]	2[b]	0	0	0
Unrelated	2	0	1	3	1	1	1
Unknown	3	1	0	0	0	0	4

Note: See text for details of origin categories.

[a]In one case, the immigrating sibling was born after the older sibling had already left the natal group, and thus the duo partners had not overlapped in the natal group.

[b]A single male-male duo is counted in both these categories. The father-son duo was first potentially polyandrous following female replacement in the son's natal group. The father later emigrated to a neighboring group and then after 17 months in the neighboring group rejoined his son in the original group.

case occurred when two adult males of unknown relationship simultaneously joined a group already containing an adult male.

The duration of male duos ranged from less than 1 month to greater than 68 months. In the latter case, the duo was already together when it was first observed, and thus the documented duration was a minimum. The mean duration for all male duos was 16.6 months. There was not a large difference in mean duration for kin-based versus nonkin duos (kin-based: x = 15.1 months, range = 0–45 months; nonkin: x = 12 months, range = 5–26 months). However, a number of the kin-based duos were still intact at the end of the study or when monitoring of the group ceased, so the "mean" for kin-based duos is an underestimate of actual mean duration. One analysis of duo duration suggests a longer mean duration for kin-based duos than for nonkin duos, but it barely misses statistical significance (Cox's F-test [Lee 1992, pp. 113–16], $T = 1.981$, $p = 0.058$). The longest documented duration of a male trio during the study was 11 months.

Male duos dissolved through the death or emigration of one or both of the males (Table 8.2). In many cases, the fate of the individual(s) could not be determined (i.e., the animal[s] disappeared from the group and was [were] not seen again). Emigrating males transferred to adjacent groups, became solitary floaters, or joined other floating individuals. Baker et al. (1993) reported that aggression between males in two-male groups was rare, but that status-indicating interactions (fights, displacements, submissive vocalizations, arch-walks, and mounts) were unidirectional or strongly biased within most male duos, indicating the existence of a predictable dominance relationship between the males. While both dominant and subordinate males exhibited sexual behavior with the reproductive female in most two-male groups in that study, dominant males were responsible for 94 percent of mounts, copulations, and episodes of consorting behavior during periods in which the breeding female was expected to be fertile. The authors concluded that, despite the frequency of polyandrous group composition and copulations by more than one male, reproduction in this population was typically monopolized by a single dominant male per group.

Baker et al. (1993) also constructed a mathematical model predicting lifetime reproductive success for male golden lion tamarins following different strategies. The model incorporated survival and reproductive rates in one-male and two-male groups and likelihood of successful dispersal and settling by a male leaving a two-male group. The model assumed that a single, behaviorally dominant male sired all infants in a two-male group. The results of the model suggested that a subordinate male, even if achieving no current paternity, would typically experience higher lifetime reproductive success by staying in his current group and waiting for the dominant male to die or for a vacancy in a neighboring territory than

Table 8.2
Status and Causes of Dissolution of Golden Lion Tamarin Male-Male Duos

	Kin-Based Duos	Nonkin Duos	Duos of Unknown Relatedness	Total
Intact (as of September 1996 or last contact)	7[a]	—	3[b]	10
One male transferred to neighbor	3[a]	—	—	3
One male returned to natal group	3[c, d]	—	—	3
One male emigrated to (nonterritorial) floater status	3[a]	5	—	8
One male died	1	1	1	3
One male suspected to die[e]	3	—	2	5
One male disappeared (died or emigrated)	6	3	3	12
Both males emigrated to floater status at same time	1	—	—	1
Both males suspected to die at same time	2	—	—	2
Both males disappeared at same time	1	—	—	1

[a]Two duos were each counted twice. One of these duos dissolved after one male transferred to a neighboring group, but it was reconstituted when the transferring male returned to the original group. Thus, this duo was counted in both "intact" and "one male transferred to neighbor" categories. The second double-counted duo dissolved when one male joined a floating female, but it was reconstituted when the male rejoined his original group. Thus, this duo was counted in both "intact" and "one male emigrated to floater status" categories.

[b]This category included one group in which the disappearance (and suspected death) of one male followed trapping and handling, in case the trapping and handling were responsible for the dissolution of the duo. This categorization follows the premise that the group was "intact" at the last preinterference contact.

[c]This category contained one duo in which the males might have been genetically unrelated. The younger male of the duo was born into a group that contained the older male of the duo as one potential sire and another male, unrelated to the first, as the other potential sire.

[d]This category included one case in which both males transferred to a neighboring group and thus were still intact as a duo. However, the group into which they immigrated was the natal group and contained the mother of one of the males. Thus, the duo ceased meeting our definition of a "potentially polyandrous" duo at that point.

[e]Males were counted in the "suspected to die" category under the following circumstances: the male's radio collar was found shortly after his disappearance and was still closed (i.e., it had not broken), or the male or males disappeared simultaneously with at least one other group member other than or in addition to the other member of the duo.

he would by leaving the group and searching as a floater for breeding opportunities elsewhere. The authors concluded that the observed stability of two-male groups did not require that both males have nonzero certainty of paternity. The increase in reproductive success accrued by "staying" was greater for a subordinate who was coresident with a brother compared with a subordinate coresident with an unrelated male, due to indirect fitness benefits gained by the former.

Females

The large majority of golden lion tamarin groups in the Poço das Antas Biological Reserve contained a single reproductive female, but polygyny was a regular occurrence in this population. Based on data collected between March 1984 and September 1991, Dietz and Baker (1993) classified 10.6 percent of group samples as polygynous (groups sampled for composition twice per year over the study period). A similar pattern emerges after incorporation of more recent data (May 1986–December 1995), using a different "sampling" regime (A. J. Baker and J. M. Dietz, in preparation). Two females were confirmed pregnant within the same breeding season ("pregnancy polygyny") in 19.1 percent of 141 group breeding seasons of 20 habituated groups (one group breeding season = a single group monitored over a single breeding season). Two females reared offspring to weaning ("rearing polygyny") in 11.3 percent of the group breeding seasons. Groups were potentially polygynous (i.e., contained more than one adult female) in 61 of 141 group breeding seasons, and thus rates of pregnancy polygyny and rearing polygyny were much higher when only potentially polygynous group breeding seasons were considered (44.3 percent and 26.2 percent respectively; A. J. Baker and J. M. Dietz, in preparation).

Mother-daughter sets accounted for over 75 percent of multifemale group breeding seasons and accounted for all cases, among duos of known relationship, of successful rearing by two females within a season (A. J. Baker and J. M. Dietz, in preparation). Sister-sister situations were less common and never resulted in successful polygynous rearing. Duos of unrelated females were less common and also never resulted in successful rearing by two females.

Dietz and Baker (1993) found that only a small proportion of natal females still resident with their mothers became pregnant but that the presence of an unrelated adult male in the group appeared to be associated with a higher incidence of pregnancy and thus a higher incidence of mother-daughter polygyny. Among females reaching 2 years of age in the natal group, pregnancy was much more likely to occur (50 percent of individual natal females versus 21 percent of individuals) if the natal female's father had been replaced or joined by an immigrant male.

With additional data as described above (A. J. Baker and J. M. Dietz, in preparation), interesting age-related effects have emerged. Among 2-year-old females (the age of earliest female reproduction documented in this population), the same pattern seen in the earlier study was evident: only a small percentage of 2-year-old females coresident with their mothers were known to conceive, and all confirmed conceptions occurred in groups containing at least one immigrant male unrelated to the natal female. However, different patterns have emerged for 3- and

4-year-old females still coresident with their mothers. Among these females, incidence of pregnancy exceeded 50 percent and appeared unaffected by the genetic relationship between the natal female and the resident male: a natal female still coresident with her father was as likely to conceive as a female coresident with an unrelated male.

When all resident males were closely related to natal females, natal female pregnancies resulted from either extragroup copulations or within-group, incestuous matings. Evidence regarding this issue was conflicting: both incestuous and extragroup sexual behaviors have been documented among natal females in this long-term study (A. J. Baker and J. M. Dietz, in preparation; Dietz et al. 2000). Although pregnancies resulting from incestuous matings appear to be rare in captive golden lion tamarins (French et al. 1989), they have been documented in intact captive groups of golden-headed lion tamarins (*L. chrysomelas*): (Chaoui and Hasler-Gallusser 1999; De Vleeschouwer et al. 2001; Price 1997) and other callitrichid species (e.g., *Saguinus oedipus:* Price and McGrew 1991; *Callithrix jacchus:* Rothe and Koenig 1991; *Cebuella pygmaea:* Schröpel 1998).

Some polygynous female (mother-daughter) duos broke down after a single breeding season, but a number lasted through two seasons or more (A. J. Baker and J. M. Dietz, in preparation). Three female duos each successfully reproduced for three consecutive breeding seasons. In one duo, both females became pregnant over five consecutive breeding seasons, although only one female successfully reared young in some of those seasons.

Typically, though not always, daughters in polygynous mother-daughter duos were subordinate to their mothers (Dietz and Baker 1993). Annual reproductive success for subordinate polygynous females was less than half that for either monogynously breeding females or dominant females in polygynous groups (Dietz and Baker 1993). The difference between subordinate and monogynous females was statistically significant, and that between subordinate and dominant females was nearly so ($p < 0.1$). Dominant females in polygynous groups experienced about 80 percent of the reproductive success of monogynous females, but this difference was not significant (Dietz and Baker 1993).

A demographic model from Dietz and Baker (1993) assessed costs and benefits of polygynous reproduction for both mothers and daughters. The results of the model suggested that "daughters" should always attempt to breed polygynously in their natal groups rather than disperse and attempt to find monogynous breeding opportunities outside the natal group. For "mothers," the model indicated that costs and benefits of allowing a daughter to breed within the natal group were approximately balanced. Costs were a small reduction in current reproduction and the possibility of a role reversal (the daughter becoming dominant to the mother), with a resulting greater reduction in reproductive success. "Mothers" benefited through the gains in inclusive fitness stemming from increased reproduction by their daughters. The authors hypothesized that the saturated nature of the habitat occupied by this population increased the incidence of polygyny by reducing opportunities for successful reproduction by daughters outside their natal groups.

While this model has not been re-evaluated using the additional, more recent data, a close balance between costs and benefits of polygyny for "mothers" is consistent with what is observed in wild groups. Some females "allow" adult daughters to remain and reproduce, whereas others apparently evict their daughters. Individual decisions may be based on variations in group membership, ecological conditions, or individual nutritional or health status.

Dietz and Baker (1993) evaluated potential ecological correlates of polygyny in the Poço das Antas Biological Reserve's population, including territory size and the density of fruit trees, potential sleeping sites, and bromeliads (used extensively by the golden lion tamarins as foraging sites) and the absolute abundance of these three resources per territory. The number of reproductive females per group was significantly correlated with the amount of swamp forest in the sampled territories, suggesting that some unidentified environmental resource(s) may influence the occurrence of polygyny in this population.

Further analyses of factors affecting components of reproductive success in female golden lion tamarins are currently underway. Bales et al. (2001) examined variables potentially affecting the first observable measure of female reproductive success, the number of live births. This model included the number of infants born to a female the previous breeding season, the number of infants weaned the previous breeding season, the female's age and body mass, the number of adult males and helpers in the group, and the inbreeding coefficient of the offspring. Female body mass and number of infants born the previous season were significantly positively correlated with number of live births. For second births within the same season, number of infants born the previous season and number of helpers were significantly positively correlated with number of live births (body mass was not included in this analysis due to sample size constraints). All other variables were not predictive of number of live births.

Bales et al. (2002) examined various hormonal, social, and historical factors affecting two further measures of investment that have a potential impact on reproductive success: infant birth weight and maternal care of infants. Higher infant birth weight was predicted by higher maternal cortisol levels and lower maternal levels of estrogen conjugates. In addition, singletons had higher birth weights than twins. Maternal carrying during week 1 showed very little variation and was not explainable by any variable in the model. However, during weeks 2 and 3 maternal carrying was positively correlated with maternal condition and litter size and negatively correlated with number of helpers per infant. Male infants tended to spend more time on the nipple during week 1 ($p = 0.06$). During weeks 2 and 3, time spent on the nipple was positively correlated with provisioning of the mother and negatively correlated with group size.

CONCLUSIONS ON THE MATING SYSTEMS AND GROUP DYNAMICS OF GOLDEN LION TAMARINS AND COMPARISONS WITH OTHER CALLITRICHIDS

Summary of Golden Lion Tamarin Group Dynamics and Mating System

Most golden lion tamarin groups were found to contain a single reproductive female. However, polygyny was a regular occurrence in this population. In all cases in which two females successfully reared young in the same season (and female-female relatedness was known), the two females were mother and daughter. Although pregnancy among young (2-year-old) daughters occurred only after male immigration, older daughters (≥3 years old) were likely to become pregnant whether or not an unrelated male was present in the group. It was not clear whether daughters in groups containing only male relatives were conceiving through extragroup copulations or through incestuous matings. A demographic model suggested that benefits to a reproductive female of allowing her daughter to remain in the group and reproduce were approximately balanced by costs, perhaps accounting for the apparent case-by-case nature of polygyny in this population. The two factors that appear to be most important to female ability to invest in offspring (through higher litter size and higher maternal care) are better female condition and larger group size.

In nearly half the documented group months, golden lion tamarin groups in this population were potentially polyandrous (i.e., they contained two—or occasionally more—adult males unrelated to the current breeding female). Male duos were often stable over multiple breeding seasons. Most but not all male duos were composed of related males, with brother-brother being the most common duo composition. Duos arose (1) during group formation, (2) in established groups through male immigration, either by a duo replacing previous male breeders or by a single immigrant joining a resident male, and (3) in established groups through female replacement, with the resident male and his adult son(s) or nephew(s) being potential sexual partners for the new female (Table 8.2). To date, available behavioral and demographic data from this population support the conclusion that potentially polyandrous groups are typically genetically monoandrous, with a single behaviorally dominant male monopolizing paternity.

Established, intact golden lion tamarin groups were typically closed to immigration. Thus, most recruitment resulted from births within the group. Most immigration events were replacements of breeding females or breeding/potentially breeding males. Most immigrations into established groups were by males; it appeared that there were fewer opportunities for females to enter established groups.

Most golden lion tamarins dispersed from their natal groups in young adulthood. Males and females did not differ greatly in mean age of dispersal. However, males tended to disperse together with a male relative, usually a sibling, while females tended to disperse alone. Additionally, as might be expected from the documented immigration patterns, the sexes differed quantitatively in terms of status immediately postdispersal: many males replaced or joined males in territories adjacent to the natal territory, whereas most females became floaters.

Comparisons with Other Lion Tamarins

The relatively low number of group years of data available on lion tamarin populations other than the golden lion tamarin and the restricted knowledge regarding within-group genetic relationships in study groups of the other taxa hamper quantitative comparisons of mating systems and group dynamics. In general, however, there are no data suggesting major differences among taxa, other than those that might stem from habitat saturation and habitat quality differences among populations (see below).

Group sizes overlap among lion tamarin taxa (Table 8.3). Groups containing multiple adult females have been documented in all four taxa (golden-headed lion tamarins: Dietz et al. 1994c, 1997; black: Valladares-Padua and Cullen 1994; black-faced: Valladares-Padua and Prado 1996), and groups containing multiple adult males have been documented for all but black-faced lion tamarins (golden-headed: Dietz et al. 1994c, 1997; black: Passos 1994, 1997a; Valladares-Padua and Cullen 1994). It seems likely that multimale black-faced lion tamarin groups will be reported as more groups are censused. Group size and composition data alone thus do not suggest major variations from the mating system and group dynamics reported for the Poço das Antas Biological Reserve's golden lion tamarins. Although black lion tamarins appear to occur in typically smaller groups than the other three species, small group size does not necessarily indicate a strictly monogamous system. Three-member potentially polyandrous groups are not uncommon in the Poço das Antas Biological Reserve, and polygyny, with both females rearing young, has occurred in groups that contained as few as three members (J. M. Dietz and A. J. Baker, unpublished data).

Table 8.3
Group Sizes Reported from Field Studies of Lion Tamarins

Taxon	Study	Number of Groups	Mean Group Size	Range
Golden lion tamarin	Dietz and Baker 1993	~20	5.40	2–11
Golden-headed lion	Rylands 1982	4	6.70	5–8
tamarin	Dietz et al. 1994c, 1996	7	5.00[a]	3–9[a]
Black lion tamarin	Passos 1994, 1997a	1	—	4–6[b]
	Passos 1994	3	4.30	3–6
	Carvalho and Carvalho 1989	9	3.60	2–7
	Valladares-Padua and Cullen 1994	4	4.75	3–7
	Albernaz 1997	1	—	4–5[b]
Black-faced lion tamarin	Valladares-Padua and Prado 1996	1	7.00	—

[a]These numbers exclude any infants present.

[b]In both these single-group studies, group size varied during the course of the study.

Polygyny has been documented in the population of golden-headed lion tamarins in the Una Biological Reserve (J. M. Dietz and B. E. Raboy, unpublished data), but to date, there have been no published reports of polygyny in the other lion tamarins. However, given the relatively small number of group breeding seasons of observations for the other species, polygyny might not have been detected even if other populations exhibit levels similar to that seen in the Poço das Antas Biological Reserve. With regard to male composition, relationships between coresident males were unknown or at least unreported in most cases (with the exception of information on golden-headed lion tamarins discussed below), and thus no conclusions can be reached with regard to the proportion of multimale groups that might represent potentially polyandrous situations.

A set of observations on golden-headed lion tamarins in the Una Biological Reserve provides the only published information on group dynamics for lion tamarins outside the Poço das Antas Biological Reserve. Dietz et al. (1996) documented natal dispersal by three females and four males. The three females all dispersed alone, whereas the males dispersed in two-sibling duos. The male duos each founded new groups adjacent to the natal group. In contrast, none of the female dispersers were known to settle immediately, and two of the three were known to become floaters upon leaving the natal group. This set of observations documents that some multimale groups of golden-headed lion tamarins in the Una Biological Reserve are potentially polyandrous. The observations also conform to patterns of natal dispersal seen in the Poço das Antas Biological Reserve: individuals of both sexes dispersed from natal groups; males dispersed with partners, whereas females dispersed alone; males settled on territories adjacent to the natal territory, whereas females became floaters.

Poço das Antas Biological Reserve: Potential Impacts of Habitat Saturation and Habitat Quality

Habitat saturation has been a recurring theme in discussing the social organization and mating system of the golden lion tamarin population in the Poço das Antas Biological Reserve. Dietz and Baker (1993) suggested that habitat saturation encourages polygyny through limiting breeding opportunities outside the natal group and thus increasing the net benefit of polygyny to both mothers and daughters in potentially polygynous groups. Similarly, Baker et al. (1993) argued that habitat saturation reduced the availability of male breeding opportunities, thus increasing the benefits of the "stay and help" strategy for subordinate males. Other authors have proposed habitat saturation as a key factor favoring polygyny (e.g., Digby and Ferrari 1994) and polyandry (Goldizen and Terborgh 1989; Rylands 1989a, 1996) in callitrichids.

There are multiple lines of evidence supporting the conclusion that the available habitat in the Poço das Antas Biological Reserve is indeed saturated. First, as noted above, the reserve is essentially isolated from other forested areas, and thus opportunities for golden lion tamarins to disperse over long distances or to marginal habitat are extremely limited. Second, within the study area, all but small isolated pockets of habitat were occupied by territorial groups. Third, many more animals dispersed from natal groups than successfully joined established groups or formed new groups. Finally, direct observation of radio-collared emigrant individuals showed that many spent long periods in floating, nonterritorial situations before eventually settling successfully or disappearing.

Habitat quality is another ecological variable to be considered in interpreting the results from the Poço das Antas Biological Reserve. Several authors (e.g., Rylands 1986b, 1987; Rylands and Faria 1993; Terborgh 1983) have concluded that secondary and edge forests provide richer habitat for callitrichids than does mature forest, due to more favorable resource abundance and distribution patterns and perhaps greater concealment and locomotion opportunities. Rylands (1989a) argued that a monogynous mating system arose in an environment dominated by (relatively resource-poor) mature forests, with conditions that typically did not allow a male to defend resources sufficient for more than one female and her offspring. To explain the fact that documented callitrichid group sizes are often larger than would be predicted under this scenario, Rylands pointed out that human activity has greatly increased patches of secondary forest in recent centuries, both in quantity and in area. Secondary forest patches may encompass many contiguous callitrichid territories, resulting in entire callitrichid populations occupying rich habitats. A similar hypothesis could be advanced with regard to levels of polygyny documented in current callitrichid populations—that human activities have resulted in large patches of high-quality habitat, allowing an increase in the incidence of polygyny.

Could this hypothesis be relevant to golden lion tamarins in the Poço das Antas Biological Reserve? All golden lion tamarin habitats in the reserve are in a secondary state. Dietz and Baker (1993) noted that successful reproduction occurred on even the smallest, least resource-rich territories in their study area. Perhaps the high incidence of polygyny in this population is the result of a combination of habitat saturation and high habitat quality. A golden lion tamarin population living in undegraded primary forest might allow the testing of these ideas, but such a population does not exist. However, the captive population may provide an interesting contrast to the population in the Poço das Antas Biological Reserve. In comparison with the reserve, a zoo "habitat" is very rich in terms of food resources, resulting in higher body weights and higher reproductive rates among captive animals (wild animals: Dietz et al. 1994a; captive animals: Hoage 1982; Kleiman et al. 1982). However, this higher plane of nutrition has not resulted in an obviously higher incidence of polygyny in the zoo environment. Detectable pregnancies among daughters in intact golden lion tamarin family groups are very rare (e.g., there are no documented cases among 265 pregnancies in the U.S. captive population from 1980 to 1984) (Ballou 1985). For golden-headed lion tamarins, De Vleeschouwer et al. (2001) documented only five cases of (mother-daughter) polygyny in intact family groups in the captive population during the period of 1984 to 1998. Two additional cases of polygyny occurred in groups in which the resident males were unrelated to a mother-daughter or sister-sister duo (one case each). Different approaches to analysis make it difficult to compare the data for the Poço das Antas Biological Reserve fully with these data for captive golden-headed lion tamarins, but further analysis of data for both this population and the captive golden lion tamarin population may be fruitful.

Several captive/wild contrasts suggest that a lower level of polygyny should be expected in captive lion tamarin groups. Under the less physically confined conditions of the natural environment, physiological and/or behavioral suppression of daughters by their mothers may break down more often than in zoo settings. Additionally, adult daughters in wild groups have the opportunity to achieve outbred matings after male replacement or through extragroup copulations, an option not available to most natal females in a captive situation. Finally, captive lion tamarins may be responding to the perceived probability of reproductive success for natal females outside of the natal group (i.e., level of habitat saturation). A nonvolant species with large territory size may assess saturation through intrusion rate by floaters, which typically are unmated individuals "looking" for reproductive positions. Intrusion rate for a captive group is typically zero. Groups in an institution maintaining multiple families may have neighbors but not intruders. Thus, a mother in a zoo lion tamarin group may not "allow" a daughter to reproduce within the natal group or a daughter may "choose" to avoid incestuous pregnancies because both perceive the habitat as unsaturated and the chances of successful reproduction outside the group as high. More detailed analysis of captive records, including housing and social conditions, may allow testing of hypotheses about these issues.

Kinship in Golden Lion Tamarin Society

Golden lion tamarin groups in the Poço das Antas Biological Reserve were found to be largely kin based. A typical group was composed of a single breeding female, one or two potentially breeding males, and offspring of the breeding individuals (note that our information on paternity of offspring is based on behavioral data alone; if, through genetic work, extragroup copulations prove to result in a significant number of pregnancies, we will have to reconsider our conclusions). Nearly all natal individuals were closely related to at least one of the current breeders in the group and thus also to future offspring of the current breeders. It was extremely rare for two unrelated adult females to be coresident in a group, and "rearing" polygyny (with two females successfully rearing young) occurred, as far as we know, only in mother-daughter situations. In terms of coresident potential male breeders, three-quarters of such male duos were composed of close kin. Clearly, the picture of immigration-based groups composed largely of unrelated individuals, as proposed in some reviews on callitrichids (Sussman and Garber 1987; Sussman and Kinzey 1984), is not appropriate for this population.

The observed patterns suggest that kinship has played a major role in the evolution of golden lion tamarin society. Additionally, kinship ties are likely to affect individual behavior and may be of significance for understanding such areas as infant care behavior by nonreproductive individuals (see Baker 1991; Tardif 1996), behavioral tolerance toward group mates, and intergroup interactions. Certainly the typically high degree of within-group relatedness we documented provides an appropriate context for the evolution of behaviors based on kinship and inclusive fitness (see Hamilton 1964; Trivers 1971; West-Eberhard 1975).

Comparison of Male and Female Life Histories

Sex differences in patterns of intrasexual gregariousness and tolerance were found to result in different modal life histories for male versus female golden lion tamarins in the Poço das Antas Biological Reserve. Same-sex associations were markedly more pervasive in male life histories. While both sexes showed a tendency toward same-sex duo formation at natal dispersal, males were also sometimes able to join groups that already contained a breeding male, whereas intact territorial groups appeared to be completely closed to female immigrants. Additionally, male duos were often stable over multiple breeding seasons. Female duos were typically not stable, with the exception of mother-daughter duos that developed "in situ."

Male and female life histories also differed in terms of patterns of dispersal and territorial inheritance. A much higher proportion of dispersing males than females moved directly into potential breeding positions. Overall, far more males than females successfully joined nonnatal territorial groups. In contrast, females were more likely to inherit natal territories. In summary, most reproductively successful females first reproduced within their natal groups, whereas the vast majority of males first achieved reproductive positions outside their natal groups (J. M. Dietz and A. J. Baker, unpublished data). It is important to note that although inheritance was the most frequent route to breeding success for females, but was relatively unimportant for males, the sexes did not differ greatly in their level of philopatry. The majority of individuals of both sexes dispersed. Female inheritance was relatively important only because few dispersing females succeeded in reproducing outside their natal groups.

Implications for Population Management

Golden lion tamarin conservation efforts have included frequent transfers of individuals between forest patches and between captive and free-ranging populations (see Chapters 12 and 13 this volume). It seems likely that such management strategies will be used in the future for some or all of the lion tamarin species (see Chapter 14 this volume). Based on the information presented here on golden lion tamarin social dynamics, we offer several observations that may be useful in planning such transfers:

1. Given the high reproductive rate that characterizes the callitrichids (Rylands 1989a, 1996) on the one hand, and the increasing fragmentation of lion tamarin habitat on the other, it seems likely that most wild lion tamarin populations will remain at or near habitat saturation. We expect that an individual lion tamarin, especially a female, introduced into saturated habitat will have low probability of survival and reproduction. "Effective" migration and exchange of genes between free-living populations may be most easily accomplished through the simultaneous swapping of entire territorial groups. We recommend two-male groups as the best candidates for such swaps. Two-male groups may have a competitive advantage over challengers compared with one-male groups. Addition-

ally, a male duo will provide insurance against the loss of one of the males. Because of group permeability patterns, females are not recommended as candidates for release in a saturated habitat.

2. Short of moving intact breeding groups, male duos may be the best candidates for release in saturated (or probably any) habitat. In addition to the likely survivorship advantage described above, duos may benefit from increased vigilance compared with a lone male.
3. Our results suggest that insertion of individual adults into existing territorial groups may be possible, especially if the same-sex breeding individual of a group is removed prior to the attempted introduction. Males are the preferred candidates, since groups experience less destabilization after male replacement than after female replacement. Such insertion attempts are more likely to succeed if they involve male duos rather than single individuals.
4. Given the potential advantages of two-male groups, formation of such groups in zoos and breeding centers may be used as preparation for reintroduction programs.

ACKNOWLEDGMENTS

This research was supported by grants to Devra Kleiman, coordinator of the Smithsonian Institution/National Zoological Park Golden Lion Tamarin Conservation Program, from the Smithsonian Institution International Environmental Sciences Program, Friends of the National Zoo, World Wildlife Fund (U.S.), and the National Geographic Society; by National Science Foundation (NSF) grant BNS 8616480 to James Dietz; by NSF grants BNS 9008161 and 9318900 to James Dietz and Andrew Baker; and by a Sigma Xi grant, an NSF Graduate Fellowship, and a Smithsonian Institution Predoctoral Fellowship to Andrew Baker. We greatly appreciate the dedication of field assistants O. J. Narciso, E. Teixeira, S. de Mello, J. Ramos, and A. de Oliveira. We also appreciate the support of J. D. Ballou, B. Beck, M. I. Castro, M. C. M. Kierulff, D. G. Kleiman, A. Martins, P. Procópio de Oliveira, D. Pessamilio, D. M. Rambaldi, C. Ruiz Miranda, and other members of the golden lion tamarin team. Finally, we thank the state and federal authorities of the Instituto Brasileiro do Meio Ambiente e dos Recursos Naturais Renováveis (IBAMA—Brazilian Institute of the Environment and Renewable Natural Resources) and the Conselho Nacional de Desenvolvimento Científico e Tecnológico (CNPq—Brazilian National Science Council) for permission to work in the Poço das Antas Biological Reserve and with the golden lion tamarins.

SUZETTE D. TARDIF, CRISTINA V. SANTOS,
ANDREW J. BAKER, LINDA VAN ELSACKER,
ANNA T. C. FEISTNER, DEVRA G. KLEIMAN,
CARLOS R. RUIZ-MIRANDA,
ANTÔNIO CHRISTIAN DE A. MOURA,
FERNANDO C. PASSOS, ELUNED C. PRICE,
LISA G. RAPAPORT, AND KRISTEL DE VLEESCHOUWER

9
INFANT CARE IN LION TAMARINS

Cooperative infant care, in which numerous group members in addition to the mother transport infants and provision them with solid food, is a hallmark of the callitrichid primates (marmosets and tamarins). Much of callitrichid socioecology is hypothesized to be shaped by this unusual infant care pattern (Caine 1993; Goldizen 1987b; Goldizen et al. 1996; Kleiman 1985; Kleiman 1977c; Sussman and Garber 1987; Tardif and Bales 1997).

This chapter reviews what is known regarding the characteristics of cooperative infant care in three of the lion tamarins, *Leontopithecus rosalia* (the golden lion tamarin), *L. chrysomelas* (the golden-headed lion tamarin), and *L. chrysopygus* (the black lion tamarin). Studies of both captive and free-ranging animals are reviewed and compared. Where differences exist between the species or between captive and free-ranging populations, the potential relationship of these differences to environmental variables is examined. Finally, a general comparison of cooperative infant care in lion tamarins with that seen in other callitrichid primates is provided, with an emphasis on how such a comparison may provide insights into the role of cooperative infant care in the overall socioecology of each group.

INFANT CARE

Infant Transport

By far the best studied of the lion tamarin species is the golden lion tamarin. Detailed analyses of infant care by captive golden lion tamarins have been conducted by Hoage (1978, 1982), Kleiman (1984b, 1985, unpublished data), and Santos et al.

(1997, unpublished data). In captivity, infants are physically transported for the majority of the time up to week 8. Transport of infants is reported by Hoage (1978) and Kleiman (unpublished data) to cease around week 11 to 12 (Figure 9.1). However, Santos et al. (1997) reported infants still being transported approximately 53 percent of the time during weeks 10 to 12. Comparison of the populations studied reveals that infant transport by adult helpers may account for the discrepancy. The population studied by Santos et al. (1997, unpublished data) contained more groups with adult helpers than those populations studied by either Hoage (1978) or Kleiman (unpublished data). The presence of adult helpers that actively participate in infant transport appears to increase the likelihood that infants will be carried more frequently at later ages, given that the adult helpers are frequently infant carriers at later ages. Even in a population with infrequent infant transport at later ages (Kleiman, unpublished data), there was a relationship between the participation of helpers and more frequent transport of infants at later ages (Figure 9.2).

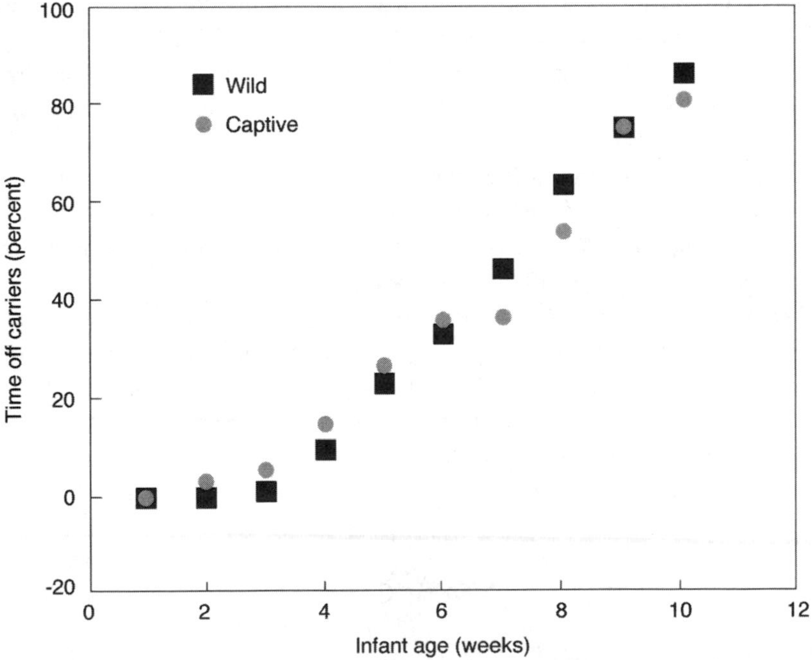

Figure 9.1. Comparison of percentage of time infants were off carriers (by week) for free-ranging wild and captive golden lion tamarins (captive data from D. G. Kleiman, unpublished data; wild data from Baker 1991).

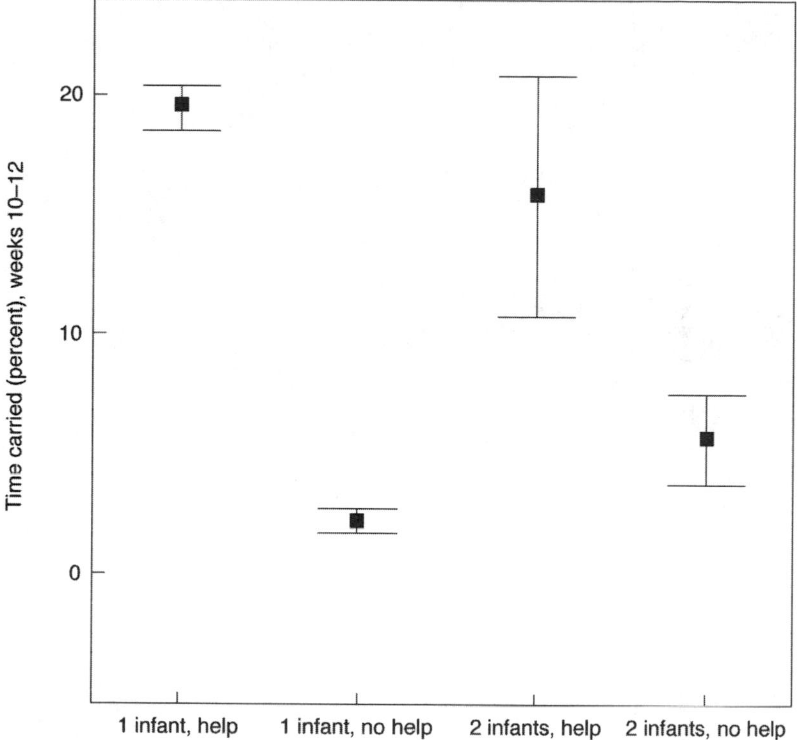

Figure 9.2. Comparison of mean (±S.E.) time singleton and twin infants were carried between 10 and 12 weeks of age for golden lion tamarin groups with and without helpers (from D. G. Kleiman, unpublished data).

The infant transport behavior of free-ranging golden lion tamarins has been examined by Baker (1991). The basic pattern of infant transport is quite similar to that observed in captivity. Figure 9.1 illustrates a similar pattern of infant transport during the first 10 weeks for the captive lion tamarins studied by Kleiman (unpublished data) and free-ranging tamarins (Baker 1991). Figure 9.3 illustrates the percentage of time that mothers and adult males carried infants during weeks 10 to 12. The frequency with which wild and captive mothers carried infants at later ages did not differ; however, transport by wild versus captive adult males did. The relaxed energetic demands of captivity compared with the free-ranging setting could well explain the longer periods of infant transport in some captive populations than in the free-ranging groups. However, it is interesting to note that this "captivity effect" was observed in adult males but not in mothers. These differences suggest that the infant-care behavior of nonmothers (i.e., fathers and helpers)

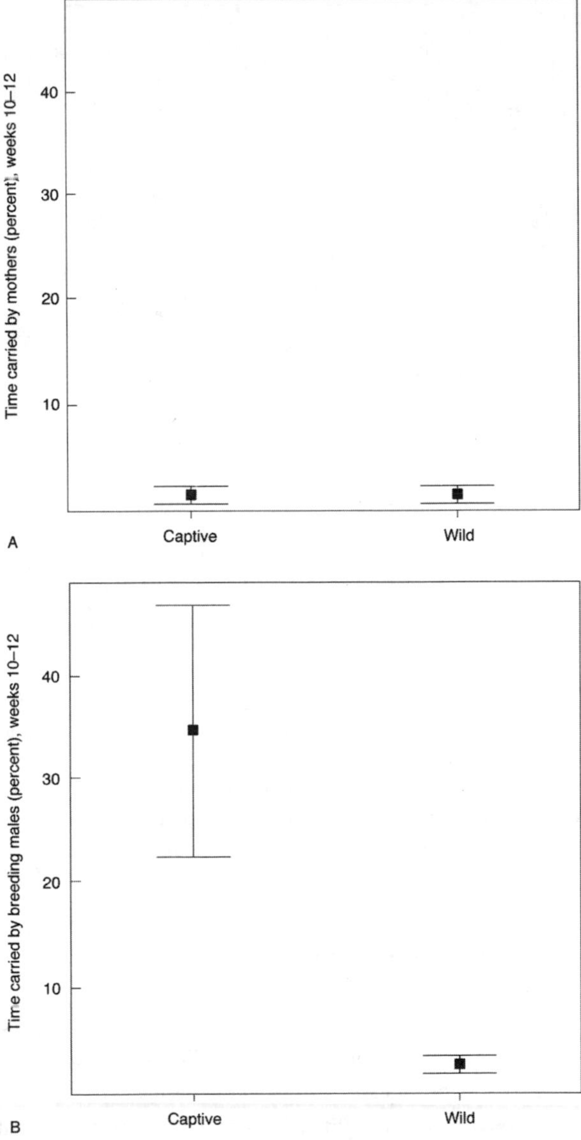

Figure 9.3. Comparison of mean time spent carrying infants 10 to 12 weeks old by mothers (**A**) and adult males (**B**) in one captive and one free-ranging population of golden lion tamarins. Adult males were defined as those over 18 months of age (captive data from C. V. Santos, unpublished data; wild data from Baker 1991).

may be more malleable or subject to environmental influences than is that of the mother, a result similar to that observed for other callitrichid species (McGrew 1988; Tardif 1996; Tardif et al. 1990; Wamboldt et al. 1988). The lack of a captivity effect on mothers might be related to the fact that the mother is the only group member who has the significant energetic cost of lactation, regardless of location.

Golden lion tamarin mothers are the primary infant carriers for the first 3 weeks; they carry significantly more than both fathers and other group members during this time. Mothers display a precipitous decline in transporting infants after week 3, while fathers and other group members continue to transport infants as, or more, often than before and show a more gradual decline (Hoage 1978, 1982; D. G. Kleiman, unpublished data; Santos et al. 1997). The temporal pattern of involvement of mothers, breeding males, and helpers in infant transport appears similar in free-ranging and captive golden lion tamarin groups, with mothers the primary caregivers for the first 3 weeks (Baker 1991). A similar pattern of maternal and paternal involvement in infant transport in captive versus free-ranging populations has also been found for *Callithrix jacchus* (Yamamoto et al. 1996).

Breeding males often play a prominent role in infant transport during late infancy by carrying infants more than either the mother or nonreproductive individuals. It has been proposed that such infant care may serve as a courtship strategy in callitrichid primates (Ferrari 1992; Price 1990; Rylands 1982, 1996). However, studies of free-ranging golden lion tamarins find no relation between a male's participation in infant transport and his access to the breeding female (Baker et al. 1993).

Recent analyses of data on captive populations also raise questions regarding the likelihood of infant transport as courtship in other callitrichid genera (Jurke et al. 1995—*Callimico* spp.; Tardif and Bales 1997—*Saguinus* and *Callithrix* spp.). Rather, breeding males might benefit from relinquishing infant care to other individuals. In free-ranging groups of golden lion tamarins, the number of helpers predicts reproductive tenure of males but not females: being relieved of infant carrying increased male reproductive tenure (Bales et al. 2000). An overview of carrying patterns in captive groups for three of the lion tamarins, in addition to other callitrichids, indicates that paternal effort is more likely to be reduced by helper presence than maternal effort (Bales et al. 2000).

Participation of nonreproductive group members in transport of infants is highly variable (Baker 1991; Hoage 1982; C. V. Santos, unpublished data), but overall, helpers appear to be less involved in infant transport than is either parent. In general, if helpers are the previous offspring of the breeding pair and are less than a year old, their participation in infant transport amounts to less than 10 percent of total transport time (Hoage 1982; D. G. Kleiman, unpublished data).

However, adult helpers (those over 12 months old) may carry substantially more. In the captive population studied by Santos (unpublished data), some adult helpers carried up to 50 percent of the time during weeks 10 and 12.

Some differences in the care-giving behavior of male and female helpers have also been noted. Hoage (1982) observed that juvenile females began carrying infants earlier and had a fairly consistent, low level of involvement in transporting infants in captive golden lion tamarins. Male juveniles began carrying later and displayed a concentrated period of carrying that overlapped with the highest levels of infant transport displayed by adult males. Baker (1991) found that female helpers of all ages carried more than male helpers in free-ranging golden lion tamarins. Baker (1991) also provided a review of the role that direct and indirect fitness benefits may have played in shaping the sex differences in helper involvement in infant transport in lion tamarins. Also, as mentioned previously, transport by helpers may delay the infant's onset of independence from transport (Figure 9.2).

The basic pattern of infant transport is similar in the three lion tamarin species that have been studied in captivity. Mothers are the predominant carriers of infants during the early postpartum period. Some studies report that mothers are the predominant carriers in weeks 1 through 3 in black lion tamarins and golden-headed lion tamarins, as they are in golden lion tamarins (Santos et al. 1997; Figure 9.4). However, Oliveira et al. (1999) reported that mothers were the predominant carriers in golden-headed lion tamarins for only the first 3 days and suggested that differences in this maternal behavior among the lion tamarin species may be related to ecological differences (e.g., larger home ranges for golden-headed lion tamarins than for golden lion tamarins). Van Elsacker et al. (1992) reported differences in the participation of mothers and fathers in early infant transport between two groups of golden-headed lion tamarins of similar composition, a finding substantiated by further research (De Vleeschouwer 2000; L. Van Elsacker and K. De Vleeschouwer, unpublished data). Maternal efforts either decreased sharply after the first week and remained low or decreased more gradually over the next weeks. Mothers seemed to control their energetic expenses by adjusting their investment in infant carrying. Fathers readily compensated for reduced maternal effort. The actions of helpers did not affect either maternal or paternal carrying effort. Santos (unpublished data) found that helpers participated more in golden-headed lion tamarin groups than in golden lion or black lion tamarin groups. This difference was not due to a difference in number of helpers.

There may be significant differences between the lion tamarin species in the process whereby infants become independent of transport. Figure 9.5 illustrates the percentage of time that infants were off carriers in three lion tamarin species housed under similar conditions (C. V. Santos, unpublished data). Golden lion

Figure 9.4. Golden-headed lion tamarin carrying two young of advanced age. (Photo by Vincent Sodaro, Jr., Chicago Zoological Society)

tamarins and golden-headed lion tamarins show a similar pattern, with infants being independent about 25 percent of the time during weeks 7 through 9 and 52 percent of the time during weeks 10 through 12. Black lion tamarin infants were independent significantly earlier, being off carriers an average of 51 percent of the time during weeks 7 through 9 and 87 percent of the time during weeks 10 through 12. Given the previous finding that the actions of adult helpers can lead to infants being transported at later ages, analysis of those groups with adult helpers was conducted. When only groups with adult helpers were compared, black lion tamarin infants were still independent earlier. It also appeared that male helpers participated in infant transport significantly more than did females in black lion tamarins, but this was not the case in golden lion tamarins or golden-headed lion tamarins (C. V. Santos, unpublished data). Observations of one free-ranging group of black lion tamarins suggest a rate of development of independence from transport that is similar to that seen in free-ranging golden lion tamarins (F. C. Passos, unpublished, versus Baker 1991). Van Elsacker and De Vleeschouwer (De Vleeschouwer 2000; L. Van Elsacker and K. De Vleeschouwer, unpublished data) reported infant golden-headed lion tamarins in captive groups as being independent about 75 percent of the time by week 7, which is substantially higher than Santos's findings and comparable to those for black lion tamarins. Therefore,

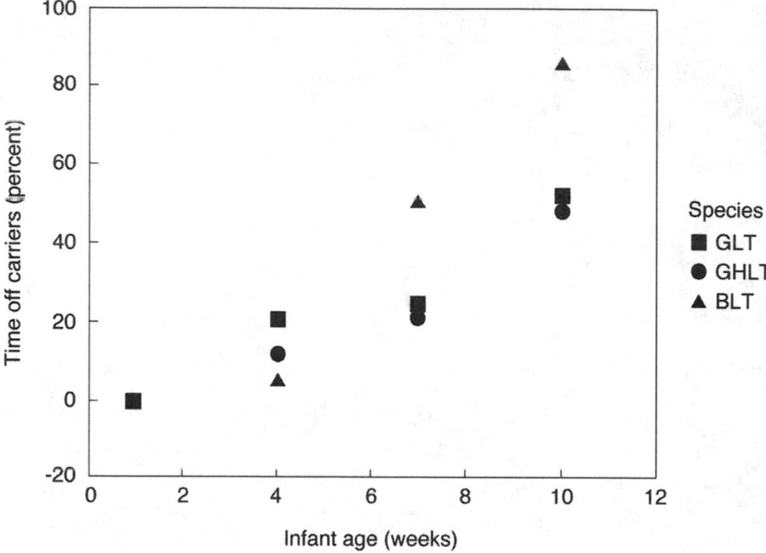

Figure 9.5. Comparison of percentage of time infants were off carriers (by week) for the golden lion tamarin (GLT), the golden-headed lion tamarin (GHLT), and the black lion tamarin (BLT) in a similar captive setting (data from C. V. Santos, unpublished).

although there is some suggestion of differing rates of independence from transport among the lion tamarin species, more data are necessary to demonstrate this effect conclusively.

Food Provisioning

In addition to physical transport, infant caregiving in callitrichid primates includes provisioning infants and juveniles with solid food. Interactions over food take several forms (Feistner 1985; Feistner and Chamove 1986; Ruiz-Miranda et al. 1999), and definitions of those interactions relative to the behavior of both the food possessor and the potential recipient may differ from one study to another. Therefore, we provide the following description of terminology to be used here. *Food transfer* refers to possession of food exchanging from one individual to another and may fall into one of three categories: begging, offering, or stealing.

Food begging occurs when one individual closely approaches an individual who has food and attempts to take it. Infants and juveniles often call loudly (mostly rasps and trills) in this context (Chapter 10 this volume). In contrast, *food offering,* which is much rarer, is initiated by the food possessor. The offering individual adopts a characteristic posture, emits a particular call, and often makes eye contact with the potential recipient who responds by approaching and taking food (Hoage 1982; Figure 9.6). This suite of traits tends to be associated with a lack of resistance to giving up food (Brown and Mack 1978; Hoage 1982; Ruiz-Miranda et al. 1999). Begging attempts may be refused by a possessor turning away, running away, or swatting at the begging individual. Alternatively, the possessor may simply drop the food. When an individual takes a food item without prior offering or begging, this is classed as *food stealing.*

Food sharing, which can be defined as the voluntary transfer of defensible food items by food-motivated individuals (Feistner and McGrew 1989), includes both successful begging and offering of food. Shared food is most likely to be live prey

Figure 9.6. Golden lion tamarin showing characteristic posture of food offering and food sharing. (From a drawing by C. Dorsey Rathbun from super 8 mm film sequences taken by R. J. Hoage in spring 1974)

or foods that require skill to process, though other items might also be shared (Feistner and Chamove 1986; Price and Feistner 1993; Ruiz-Miranda et al. 1999). Food items may be transferred between all group members, with the most frequent recipients generally being the youngest members, but in the wild, the beneficiaries of food transfers are mostly immatures and sometimes pregnant females (Ruiz-Miranda et al. 1999). Infants initially receive most of their solid food through such provisioning and only gradually become independent foragers and feeders (Hoage 1982). In captivity and in the wild, parents, subadults, and natal adults all provide food to infants and juveniles. Adult helpers and parents relinquish food to immatures at similar rates (Rapaport 1997; Ruiz-Miranda et al. 1999; C. V. Santos, unpublished data).

In lion tamarins, food transfers are generally first seen when the infants are 5 weeks old (Hoage 1982; Feistner and Price 2000; E. C. Price and A. T. C. Feistner, unpublished data) and continue for an extended period. Observations of captive golden-headed lion tamarins (E. C. Price and A. T. C. Feistner, unpublished data) and black lion tamarins (Feistner and Price 2000) indicate that food transfers peak at 12 and 9 weeks respectively. The frequency of transfers remains high in both species until about 5 months of age and then declines gradually. Even at 6 months of age, juveniles still receive items regularly from others, and black lion tamarins experience 65 percent begging success. Observations of wild golden lion tamarins suggest that the provisioning period may extend to 21 months of age (Ruiz-Miranda et al. 1999).

Other workers have found similar results, with transfers providing a high proportion of food items eaten by infants during weeks 1 through 12: 41 percent of food for golden lion tamarins (C. V. Santos, unpublished data), 52 to 56.4 percent of food for golden-headed lion tamarins (C. de A. Moura, unpublished data; E. C. Price and A. T. C. Feistner, unpublished data; C. V. Santos, unpublished data), and 55 percent of food for black lion tamarins (A. T. C. Feistner and E. C. Price, unpublished data). Hoage (1982) found that captive infant golden lion tamarins up to the age of 16 weeks received 90 percent of their food from others. This figure then declined to 15 percent for juveniles and subadults up to the age of a year.

The developmental pattern of food transfers appears similar in three of the four species (Figure 9.7), but offering may be less frequent in golden-headed lion tamarins than in the other species (Hoage 1982; E. C. Price and A. T. C. Feistner, unpublished data). During the peak sharing period (9–21 weeks), 0 to 18.2 percent (mean = 5.3 percent) of shared items received by captive infant black lion tamarins were from offering, whereas in golden-headed lion tamarins, on average only 0.8 percent of items shared were offered (range = 0–6.3 percent). In Hoage's (1982) study of captives, approximately 12 to 30 percent of shares appeared to be

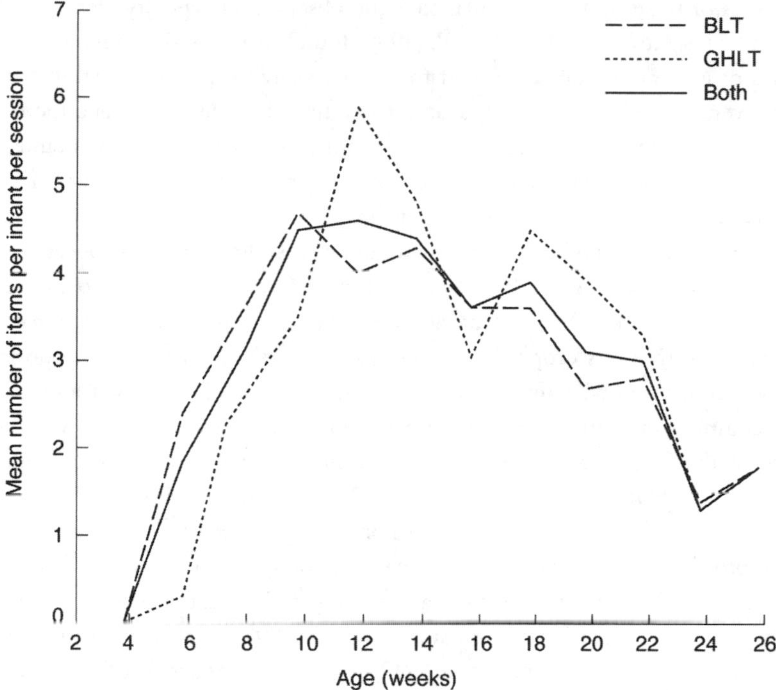

Figure 9.7. Development of food sharing in black lion tamarins (BLTs) and golden-headed lion tamarins (GHLTs), as shown by the mean number of items shared (successful begging plus offering) per observation session with infants aged 4 to 26 weeks (Feistner and Price 2000; A. T. C. Feistner and E. C. Price, unpublished data).

offers, and observations of wild and reintroduced captive-born golden lion tamarins indicate that offering may be present in 11 percent and 24 percent of the transfers, respectively (Ruiz-Miranda et al. 1999; but see Rapaport and Ruiz-Miranda, in press). Compared with food transfers in nonprovisioned groups, those reintroduced animals in provisioned groups showed less begging and also less frequent vocalizations by recipients. However, before it can be decided whether or not this apparent species difference is genuine, further work is needed to take into account possible effects of group size, housing conditions, methodological differences of studies, and other factors that may affect data collected on stealing, sharing, and offering.

In free-ranging groups, transfer of food items between adults and infants occurs much less frequently than in captive groups (Baker 1991); however, the characteristics of this transfer are similar to those observed in captivity. Perhaps relevant to this point, Ruiz-Miranda et al. (1999) found that provisioned reintroduced groups of golden lion tamarins most frequently shared the commercial diet they were given. The most commonly shared food items between adults and their dependent offspring in free-ranging groups of nonprovisioned golden lion tamarins are vertebrate and invertebrate prey (Ruiz-Miranda et al. 1999), with 51.1 percent of transferred food items being prey.

The ontogeny of food transfers in wild and captive lion tamarins appears to be similar. Ruiz-Miranda et al. (1999) observed that lion tamarins from 1 to 9 months of age received a relatively constant rate of food from transfers, after which receipt of food from transfers dropped dramatically. There seem to be some important differences between wild and captive animals in the dynamics of food transfers. In free-ranging groups the recipients were almost always infants, there was no evidence of the food circulation observed in captivity, and transfers among adults were exclusively to pregnant females (Ruiz-Miranda et al. 1999).

Experimental work on three captive lion tamarin species, as well as on wild and reintroduced golden lion tamarins, has provided data to test several hypotheses to explain the unusually frequent and widespread occurrence of food sharing in the genus *Leontopithecus* (Moura and Langguth 1997; Price and Feistner 1993; Rapaport 1999; Ruiz-Miranda et al. 1999). One hypothesis is that food transfers ensure that infants receive sufficient food even if it is difficult for them to acquire it themselves. Reducing the ease with which infants could reach a food source led to infants feeding themselves less but receiving more food from other group members, who were more likely to respond to infants' begging attempts by sharing (Moura and Langguth 1997; Price and Feistner 1993).

A second possible explanation is that sharing ensures that infants receive foods that are rare but important in the diet. When items were present singly rather than all at once, food sharing did indeed increase, while self-feeding by infants decreased.

A third hypothesis is that sharing helps teach infants an appropriate diet. If this was the case, foods to which infants have not previously been exposed would be transferred more frequently than familiar foods, and one would expect some degree of neophobia with regard to novel foods. Price and Feistner (1993) found that although infants did feed themselves less when the food item was novel, adults were less likely to share novel foods than familiar foods, and infants would take a novel food for themselves the first time they tried it rather than wait until an adult shared or offered it. Price and Feistner (1993) interpreted these as not supporting the hypothesis that food sharing teaches infants which foods they should eat. Moura and Langguth (1997), on the other hand, found decreases in both food sharing and self-feeding when novel foods were fed to golden-headed lion tamarins, and they offered the proposal that these decreases could be considered a strategy to avoid ingestion of toxins from unknown foods.

Rapaport (1999) found that although older infant and juvenile golden lion tamarins were less likely to reject novel foods acquired from other groups members, compared with those acquired through self-feeding, they, like the infants in Price and Feistner's (1993) study, occasionally picked up novel food for themselves and ate it. Rapaport (1999) offered her study subjects foods in three categories: familiar to all group members, novel to all group members, and novel to immatures but sampled several months previously by all adults in a given group. The adult golden lion tamarins transferred to immatures foods in the last category (novel to immatures but known to adults) and foods that were novel to all more frequently than foods that were familiar to all group members. Therefore, in contrast to Price and Feistner's (1993) results, Rapaport found that adults are more likely to give infant and juvenile golden lion tamarins foods that are novel, rather than familiar, to the young.

The results of these studies may not be as contradictory as they might first appear. First, Price and Feistner's (1993) novel food category consisted of both foods that were completely new to all group members and foods with which some adults were familiar. However, we know that novel foods are less likely to be picked up and eaten than known foods (Moura and Langguth 1997; Rapaport 1999). Moreover, Rapaport (1999) deleted from her analysis instances in which foods were only briefly picked up and then dropped; most of these rejections were of novel foods. Thus, only "accepted" novel foods were transferred more frequently to young. On the other hand, not only were foods that were known to adults, but novel to young, more frequently accepted (and thus available for food transfer) but also they were more frequently transferred to young group members. Thus, adult food transfer response to young group members may differ according to whether or not the adults have sampled the foods before, whether or not they find the foods acceptable, and the amount of time that has passed since a given food was last eaten.

These combined results also suggest a possible dual role for provisioning of young lion tamarins that is dependent upon age (Rapaport 1999). Young infants, who sustain a high rate of growth yet are inexperienced foragers, may primarily receive items that are high in lipids or protein and that they would be unlikely to acquire on their own. In contrast, older immatures may selectively receive from adults items that they have not previously sampled. Thus, provisioning of older juveniles may function to inform the recipients about diet. Whether age-related change in the function of provisioning is an artifact of captivity unique to golden lion tamarins or is a general feature of the lion tamarin or callitrichid caregiving strategy remains to be seen.

Other Infant Care Behaviors

Besides infant transport and food provisioning, lion tamarin parents and helpers provide other significant support to their young during development. Kleiman and Malcolm (1981) detailed the types of direct and indirect parental care in mammals that may be seen in fathers and other helpers. All of these are regular features of lion tamarin infant care behavior, and many are features also of adult social interactions. Parents and helpers huddle with young at night and during daily rest periods, thus providing additional warmth and reducing energetic costs to the young through heat loss. Parents and helpers specifically target the young for protection. If an infant or juvenile falls to the ground, an adult will retrieve it. In case of a threat to the group, parents and helpers will pick up and carry juveniles that are locomoting independently the majority of the time. Adults groom infants and juveniles during rest periods. They are also vigilant during play bouts of infants and juveniles, and at least in captivity, fathers and other helpers may participate in play bouts, thus providing some socialization experience for the young.

DIFFERENCES BETWEEN LION TAMARINS AND OTHER CALLITRICHID GENERA

While the general features of infant care in all four callitrichid genera (*Leontopithecus, Saguinus, Callithrix,* and *Cebuella*) are similar, there are some differences that might be informative as to how infant care demands shape—and are shaped by—the species' socioecology.

Relative Involvement of Mothers versus Nonmothers in Infant Transport

It appears that, in general, lion tamarin mothers play a more significant role in early infant care than do mothers in the other genera. Mothers are the primary infant carriers in most lion tamarin groups for the first 3 weeks, both in captive and free-ranging groups (however, see Oliveira et al. 1999). In *Saguinus, Callithrix,* and *Cebuella,* it appears that mothers play a predominant role in transporting infants for only the first few days. For all of these genera, fathers are reported to carry as frequently and sometimes more frequently than do mothers during the period of early infancy (Cleveland and Snowdon 1984; Digby 1995; Epple 1975a; Goldizen 1987b; Santos et al. 1997; Tardif et al. 1986, 1990; Yamamoto et al. 1996). It has been proposed that lion tamarin mothers may play a more prominent role in transporting young due to the relatively smaller burden that the infants represent relative to adult body weight (Hoage 1978; Kleiman 1977c). However, comparisons of birth weight and growth data across the genera suggest that there are not marked differences in the relative weight burden or energetic burden that the infants represent (Tardif et al. 1993). However, in the wild, marmosets (*Callithrix* and *Cebuella* spp.) frequently give birth twice a year, while tamarins (*Leontopithecus* and *Saguinus* spp.) usually have only one litter per year (Dietz et al. 1994a; French et al. 1996b; Rylands 1996), although multiple litters in captivity are common in the latter as well (De Vleeschouwer 2000; Kleiman et al. 1982). Thus, reproduction in marmosets is potentially more expensive, given that periods of high energy demand (lactation) are more closely spaced and gestation and lactation may overlap in time. Such factors might produce selection pressures that would explain the differences in maternal care between lion tamarins and marmosets. For captive golden-headed lion tamarins, Van Elsacker and De Vleeschouwer (De Vleeschouwer 2000; L. Van Elsacker and K. De Vleeschouwer, unpublished data) found that maternal carrying effort was related to the likelihood of conception at the first postpartum ovulation: mothers investing more in infant carrying delayed conception. For free-ranging golden lion tamarins, Bales (2000; Bales et al. 2002) found that maternal carrying effort was affected by the mother's condition, litter size, and group size: mothers seemed to invest more in infant carrying because they could (being in good condition) or because they had to (due to a low number of helpers). Infants of provisioned mothers (of reintroduced groups) also spent more time nursing. Both results suggest energetic constraints on the level of reproductive investment in lion tamarins as well. In addition, an argument based on differences in energetic demands alone cannot explain the difference between lion tamarins and the genus *Saguinus*, given that *Saguinus* species also usually reproduce only once a year.

On a proximate level, more extensive involvement of lion tamarin mothers in carrying infants might be due to the behavior of the mother, the infant, or other potential carriers. Hypotheses regarding the potential role of each of these players are as follows:

1. Lion tamarin mothers may carry more during early infancy because they are more tolerant of transporting infants than are mothers in other genera. Alternately, lion tamarin mothers might be less tolerant of helpers; data from other callitrichids (*Callithrix* spp.) suggest that mothers may play a selective, active role in determining who has access to the infants (Albuquerque 1999). Data on both free-ranging and captive *Saguinus* and *Callithrix* species suggest that transporting infants is not compatible with a variety of other behaviors, including travel, foraging, and sexual interactions (Digby and Barreto 1996; Goldizen 1987b; Price 1992a; Tardif and Bales 1997). This incompatibility may be particularly important for mothers during lactation, when the time required for foraging and feed-

ing is greatly increased (e.g., *Saguinus fuscicollis:* Goldizen 1987b). Perhaps lion tamarins do not display such marked incompatibility between infant transport and foraging activities. Such a difference between lion tamarins and the other callitrichid genera would raise interesting questions regarding the role of foraging and antipredator strategies in each genus and how those roles have shaped the evolution of its overall social structure.

If lion tamarin mothers are able to meet the energetic demands of early lactation more easily than are other callitrichid mothers, perhaps through larger energy reserves, then perhaps lion tamarin mothers face less of a tradeoff between infant transport and other activities, such as foraging. The observations of Van Elsacker et al. (De Vleeschouwer 2000; Van Elsacker et al. 1992, unpublished data) suggest participation of lion tamarin mothers in infant transport may be mediated by energetic concerns. For example, Kleiman (1980) found that captive female golden lion tamarins with larger litters (twins and triplets) first transferred young to other group members earlier than females with singletons. Females thus seem to control their energetic expenses by adjusting their investment in infant carrying. Males readily compensate for reduced levels of maternal carrying, and this has an impact on the mothers' condition. Equally, maternal condition influences level of maternal carrying in free-ranging groups of golden lion tamarins as well (Bales 2000; Bales et al. 2002).

2. The interest of nonmothers in infants may be more limited in lion tamarins than in other callitrichid genera. Observations by Kleiman (unpublished data) suggest that lack of interest is unlikely to be the reason for the relatively late involvement of breeding males and helpers. Potential helpers, including breeding males, have been observed attempting to retrieve young infants but being rebuffed by the mother. The observations of Van Elsacker et al. (De Vleeschouwer 2000; L. Van Elsacker and K. De Vleeschouwer, unpublished data; Van Elsacker et al. 1992) suggest that the behavior of the mother is critical in determining who has access to the infant during the earliest period. Characteristics of the maturing infants, including their behavior, may also be important factors in determining infant transport patterns. The decrease in maternal care of infants begins at a point close to when weaning begins, so increased activity of the infants may also play a part in reducing the mother's role and increasing the role of others in infant transport.
3. Compared with infants in the genera *Callithrix* and *Saguinus,* lion tamarin infants may be behaviorally more altricial during early infancy and

thus not transferred as often from mothers during the earliest weeks. However, comparison of average ages of reaching developmental milestones suggests that lion tamarins do not differ from other callitrichid genera (Yamamoto 1993).

Role of Food Transfers

Food transfer in response to infant begging has been observed in all marmosets and tamarins studied so far. However, lion tamarins differ from most other callitrichids in the degree to which the voluntary donation of food by possessors (the food offering) to infants occurs. Other species known to offer food include wild *Callithrix flaviceps* (Ferrari 1987), captive *C. jacchus* (Simek 1988; Tardif et al. 1993), captive *Cebuella pygmaea* (Feistner and Price 1991), and captive *Saguinus bicolor* (Price and Feistner 2001). Work on captive *S. oedipus* (Feistner and Price 1990) has shown that offering is particularly prominent in this species and accounts for an average of 28 percent of food transfers to infants; in *S. bicolor*, in contrast, offering is quite rare (Price and Feistner 2001). Offering was also observed much less frequently in *Callithrix* species than in *S. oedipus* (see Simek 1988). Also, as cited previously, the rate of food offering is higher and the rate of development of independent feeding is much slower in lion tamarins than in *Saguinus* or *Callithrix* species (Figure 9.8).

Currently only two *Saguinus* and two *Callithrix* species have been confirmed to offer food. It seems likely that in these genera interspecific differences may be important. For example, although intensively studied under a variety of conditions, *S. fuscicollis* has not been observed offering, although food transfers occur through begging and stealing. In contrast, three species of the genus *Leontopithecus* have been shown to offer food. Intergeneric morphological and ecological differences may affect the occurrence of food transfer behavior.

The lion tamarins have elongated fingers (Hershkovitz 1977) and a highly manipulative form of extractive foraging (Rylands 1989b; Thomas 1995). All lion tamarin diets in the wild include a high proportion of insects and small vertebrates; searching for these items takes 13 to 14 percent of the lion tamarins' activity budgets (golden lion tamarins: Kleiman et al. 1986; Peres 1989b; golden-headed lion tamarins: Rylands 1989b; black lion tamarins: Carvalho et al. 1989; Keuroghlian and Passos 2001; Passos 1997a; see Chapter 7 this volume for a summary). Black-faced lion tamarins (*L. caissara*) also feed on insects and small vertebrates (Prado 1999; Valladares-Padua and Prado 1996). These items require skill to locate, catch, subdue or kill, and manipulate during consumption, and infants have to learn these skills gradually. For some of the larger, more vigorous prey, it is likely that the infants' only access to such items is through sharing live prey

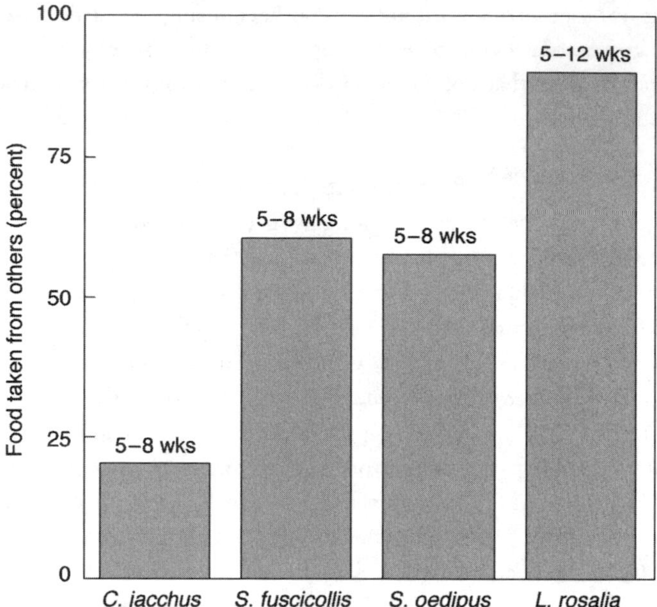

Figure 9.8. Comparison of proportion of food taken by infants from others in different callitrichid species: *Callithrix jacchus* (infants age 5–8 weeks), *Saguinus fuscicollis* (5–8 weeks), *S. oedipus* (5–8 weeks), and *Leontopithecus rosalia* (5–12 weeks) (data from Tardif et al. 1993; except for *L. rosalia,* which is from Hoage 1982).

caught by others; for example, "wild adult black lion tamarins were frequently observed sharing their prey with the infants, especially with large prey such as anurans, cerambycid coleopterans (long-horned beetles) or tettigoniid orthopterans (katydids)" (Passos 1992, 1997a; Passos and Keuroghlian 1999). Thus sharing, and in particular offering, may play an extremely important role where there is a high dependence on relatively large prey in the diet. Relevant to this point, recent evidence suggests that wild golden lion tamarin parents may demonstrate the location of large hidden prey to their inexperienced juveniles, in essence, tutoring their young (Rapaport and Ruiz-Miranda, in press).

Various studies have shown that food transfer in marmosets and tamarins is increased for highly preferred rare items, including live prey, relative to readily available fruit (Brown and Mack 1978; Feistner and Chamove 1986; Ferrari 1987; Izawa 1978; Neyman 1980). Experimental work with captive cotton-top tamarins (*S. oedipus*) has explored the effect of adult motivation on food-sharing behavior (Feistner and Chamove 1986). This research has shown that the more highly motivated adults and older siblings are toward food, the more transfers to infants occur. When there was preferred food, infants begged significantly more frequently, and their success rate increased. Thus motivation appears to be an important factor during food sharing. Adults could share everything, or they could share the most familiar or most common items. However, none of these would result in the infants preferentially getting the best food items. The strategy employed seems to be one of matching sharing with high motivational level. This ensures that infants receive ample food, the best items, and also rare items of high quality. Although not explicitly studied in lion tamarins, a similar strategy may be in operation. This hypothesis is supported by observations that preferred and rare food items are shared preferentially (Price and Feistner 1993).

In summary, unlike other primates for which food transfer is largely between mothers and infants, in callitrichids most group members transfer food to infants. In lion tamarins, offering also occurs and may account for a large proportion of transfers. This behavior thus plays a significant part in infant care. Further work on this important and characteristic behavior is clearly needed to explore its ecological, social, and taxonomic correlates.

While these comparisons suggest some interesting differences both within the genus *Leontopithecus* and among the four callitrichid genera, further research on a wider variety of populations, in both captive and free-ranging settings, is necessary to clarify the true nature of these differences. Though tentative, such comparisons are valuable in that they suggest fruitful areas for additional research. Two such areas that merit further research are (1) the role that energetic costs of, and tradeoffs between, foraging, vigilance, and infant-care activities might play in determining the role of mothers versus nonmothers in infant care and (2) the role of diet and foraging strategy as a determinant of rates of behavioral development, food provisioning, and independence.

CONCLUSIONS

This chapter reviews what is known regarding the characteristics of cooperative infant care in three of the lion tamarins, *L. rosalia* (golden lion tamarin), *L. chrysomelas* (golden-headed lion tamarin) and *L. chrysopygus* (black lion tamarin). Infants are physically transported for the majority of the time for the first 8 weeks and are largely independent by weeks 10 to 12 in all three species. There is some suggestion that infant independence might proceed at a faster pace in black lion tamarins, but further research is necessary to confirm this difference. The presence of adult helpers delays independence from transport. Mothers are the predominant infant carriers for the first 3 weeks, after which adult males are the primary carriers. Food transfers to infants from other group members generally begin at about 5 weeks of age and continue at a fairly constant level until infants are about 9 months old. During this period, captive infants obtain approximately 50 percent of their food from food transfers. Experimental examination of food sharing suggests that it ensures that infants receive adequate amounts of foods that are difficult to locate or acquire and that sharing may help to ensure the proper response to novel food, particularly for older infants and juveniles. A general comparison of cooperative infant care in lion tamarins with that seen in other callitrichid primates reveals two areas of difference. First, the more extensive involvement of mothers in early infant transport in lion tamarins suggests that those factors defining costs of infant transport to the mother might be different for lion tamarins than for other callitrichid genera. Second, the more extensive provisioning of lion tamarin infants might be related to the role of large prey in the diet. The overall effect of enhanced provisioning on growth and development merits further examination.

ACKNOWLEDGMENTS

Suzette Tardif acknowledges the support of National Institutes of Health (Grants R01-RR02022 and P51-RR013986). Anna Feistner and Eluned Price thank the callitrichid staff at the Jersey Zoo, Durrell Wildlife Conservation Trust, for its help with food-sharing studies. Devra Kleiman's captive golden lion tamarin studies were supported by National Institute of Mental Health Grant 27241 and the Smithsonian National Zoological Park (SNZP). Carlos Ruiz-Miranda, Devra Kleiman, and Andrew Baker gratefully acknowledge the support of the Associação Mico-Leão-Dourado (AMLD—Golden Lion Tamarin Association); Conselho Nacional de Desenvolvimento Científico e Tecnológico (CNPq—Brazilian National Science Council); Frankfurt Zoological Society Fund for Threatened Species; Scholarly Studies Program and International Environmental Sciences Program of the Smithsonian Institution; Friends of the National Zoo, Washington, D.C.; the National Science Foundation (Grants BNS 9008161 and 9318900 to James Dietz and Andrew Baker); the World Wildlife Fund (U.S.); the National Geographic Society; and the Centro de Primatologia do Rio de Janeiro/Fundação Estadual de Engenharia do Meio Ambiente (CPRJ/FEEMA—Rio de Janeiro Primate Center/ State Foundation for Environmental Engineering). The Instituto Brasileiro do Meio Ambiente e dos Recursos Naturais Renováveis (IBAMA—Brazilian Institute of the Environment and Renewable Natural Resources) provided financial and logistical support for the research at the Poço das Antas Biological Reserve. Lisa Rapaport would like to thank the SNZP and Riverbanks Zoo for their logistical support and cooperation. Lisa Rapaport was funded through a National Science Foundation Dissertation Improvement grant (BNS 9204342), Sigma Xi's Grants-in-Aid of Research, the Roger Williams Park Zoo's Sophie Danforth Conservation Biology Fund, and a Smithsonian Institution Graduate Student Fellowship. Funding for Linda Van Elsacker and Kristel De Vleeschouwer was made available by the Flemish Ministry of Science, through the Royal Zoological Society of Antwerp and the University of Antwerp.

10
CONSPICUOUSNESS AND COMPLEXITY: THEMES IN LION TAMARIN COMMUNICATION

Animal communication studies reveal that the communication patterns of a species (i.e., the preferred sensory modalities and the structure and function of signals and displays) are correlated with the types of intraspecific social interactions, the resulting social structure, and even the kinds of predators encountered. Communication is also constrained by phylogeny and environmental characteristics (Bateson 1994; Bradbury and Vehrencamp 1998; Hauser 1997).

Lion tamarins inhabit both primary and secondary tropical forests (Coimbra-Filho 1969; Dietz et al. 1997; Rylands 1993). The structure of the Brazilian Atlantic Forest makes visual communication at a distance relatively inefficient for such activities as maintaining group cohesion, warning about predators, and signaling about food sources. Lion tamarins rely on effective vocal communication for these functions. For New World primates, auditory communication is integral to species and individual identity, group cohesion, intragroup social dynamics, group spacing, parent-offspring communication, foraging, and antipredator behavior (Boinski 1991; Boinski et al. 1994; Halloy and Kleiman 1994; Kleiman et al. 1988; Ruiz-Miranda et al. 1999; Snowdon 1988, 1989b). Vocalizations are also important as taxonomic characters (Mendes 1997; Omedes 1979; Snowdon 1993; Snowdon et al. 1986). Olfactory communication, on the other hand, is believed to mediate territorial behavior, individual recognition, and reproduction (Epple et al. 1993). Visual communication is used during territorial encounters and social and reproductive interactions within the group (Kleiman et al. 1988).

Knowledge of a species' communication patterns can support conservation efforts because a basic understanding of the behavioral mechanisms related to socioecology, survival, and reproduction is necessary to manage a species in the wild and in captivity (Clemmons and Buchholz 1997; Dietz 1997; Kleiman 1980, 1994; Koontz and Roush 1996; Maple and Finlay 1989; Sutherland 1998).

The study of auditory communication and bioacoustics has direct practical applications too: it can assist in censusing, mapping, and monitoring changes in population structure, as well as sexing individuals through recordings and playbacks (Baptista and Gaunt 1997; Kierulff and Procópio de Oliveira 1996; Kierulff et al. 1997; Pinto 1994). The development of a successful breeding and captive management program also requires detailed knowledge of a species' social and reproductive behavior (Carlstead 1996; Hediger 1964; Kleiman 1980, 1994; Price 1984; Snowdon 1989a). Finally, programs directed at reintroduction should focus on behavior studies because there is compelling evidence that the captive environment has profound influences on the expression of behaviors directly related to survival and reproduction (Arnold 1995; Belyaev 1979; Castro et al. 1998; Kleiman 1989; Lickliter and Ness 1990; Miller et al. 1998; Price 1984; Renner and Rosenzweig 1987). The low survival rate of reintroduced captive-born animals (Beck and Castro 1994; Chapter 13 this volume) suggests that we must expand our knowledge of the species' behavior and ecology so as to modify captive conditions and improve reintroduction techniques.

There is surprisingly little information available describing the effects of captivity on communication or expressive behavior. The few studies that directly compare captive and wild primates suggest that the function (perception and usage), but not the form or rate of use, of social and expressive behaviors differs between captive-born and wild-living primates (Kummer and Kurt 1965; Mason 1985; Rowell 1967). Direct comparisons between captive and wild primates are obviously confounded by ecological differences between them, and differences between them can derive from ontogenetic processes or just reflect differences in context and opportunities. Reintroduction programs, such as that for the golden lion tamarin, offer a unique opportunity to explore the effects of captive rearing on behavioral development by allowing the researcher to observe captive-born and wild-born animals in the same environment.

This chapter reviews previous work on communication in lion tamarins, presents new information on vocal, as well as olfactory and visual, communication, and reviews the practical applications of research on communication for conservation. Information is presented and discussed in terms of important sources of variation such as context, sex, and age. We stress the differences between captive-born and wild-born animals by looking at the form, usage, and function of communication patterns.

METHODOLOGY FOR STUDYING LION TAMARIN COMMUNICATION

This chapter presents both published results and original data. Details on methods and techniques used can be obtained from publications by Boinski et al. (1994), Halloy and Kleiman (1994), and Ruiz-Miranda et al. (1999, in preparation). We obtained vocalization and behavioral data from recordings and observations of captive golden lion tamarins at the free-ranging exhibit of the Smithsonian National Zoological Park (SNZP) (Bronikowski et al. 1989), of captive-born golden lion tamarins that were reintroduced at Fazenda Rio Vermelho (Rio Vermelho Ranch) (Beck 1991; Beck et al. 1986b; Chapter 13 this volume), and of wild animals at the Poço das Antas Biological Reserve. We refer to animals from the Poço das Antas Biological Reserve as wild (19 adults and 33 immatures), from the Fazenda Rio Vermelho as reintroduced (14 adults and 8 immatures), and from SNZP as captive (8 adults). The animals were all individually marked.

Recordings of vocalizations were typically made from within 10 m of the target animal and during 20-minute-long focal observations. Vocalizations were recorded using a Sony Professional Walkman and a Marantz PMD 430 cassette recorder with metal tapes with a Sennheiser ME80 or ME88 microphone. They were categorized using a Kay Elemetrics Sonagram and analyzed with Canary 1.2.1 in a Macintosh Quadra at the Sound and Video Analysis Laboratory of the SNZP. No filters were used during recording or analyses. Details on bioacoustic measurements are presented elsewhere (Halloy and Kleiman 1994; Ruiz-Miranda et al. 1999, in preparation). The function of sounds was inferred by scoring the occurrence of a sound during different behaviors (Green 1979) or by calculating vocal rates under diverse circumstances (Boinski et al. 1994; Ruiz-Miranda et al. 1999). We calculated vocal rates by counting the total number of calls emitted and dividing it by the observation time (minus time out of sight) for each individual.

AUDITORY COMMUNICATION

The Acoustic Repertoire of *Leontopithecus*

Except for one short study of captive *Leontopithecus chrysomelas* by Haazen (1988), the only species for which acoustic communication has been studied in captivity and in the wild is *L. rosalia* (see Benz 1993; Benz et al. 1990, 1992; Boinski et al. 1994; Green 1979; Halloy and Kleiman 1994; Kleiman et al. 1988; McLanahan and Green 1978; Ruiz-Miranda et al., in preparation). We believe the vocal repertoires of *L. rosalia, L. chrysomelas,* and *L. chrysopygus* are similar, but there are no data on differences in usage, except for a study of the long call (Snowdon et al. 1986).

The vocalizations of golden lion tamarins have been categorized as belonging to six discrete categories (see Kleiman et al. 1988; Table 10.1). Within these there are recognizable variants and graded variation. Visual classification of sonograms reveals a repertoire of 15 to 21 vocalizations. Sonograms for these vocalizations can be found in Figure 10.1 and in various publications (Boinski et al. 1994; Green 1979; Halloy and Kleiman 1994; McLanahan and Green 1978; Ruiz-Miranda et al. 1999). One feature of the repertoire is that lion tamarins do not emit the J call used by other callitrichids for intragroup long-distance communication (Snowdon 1989b) and also used by *Callimico* and *Saimiri* species (Cleveland and Snowdon 1982; Elowson et al. 1992; Pola and Snowdon 1975; Schott 1975; Snowdon 1989b). This indicates that an ancestral character of the Callitrichidae has been lost in *Leontopithecus* species.

Intragroup Auditory Communication

A striking characteristic of golden lion tamarins is that they are highly vocal and mix call types (Table 10.2; see also Boinski et al. 1994; Green 1979). Golden lion tamarins, especially immature animals, vocalize in bouts (C. R. Ruiz-Miranda, unpublished data), which can be monotypic (one vocalization type) or multitypic (more than one vocalization type).

THE FUNCTIONS OF VOCALIZATIONS Field observations suggest that intragroup calls are used to coordinate group activities (Boinski et al. 1994). The different call types and their variants are used in specific contexts and correlate with individual behavior (Table 10.1); individual activity appears to be the referent of most vocalization types. However, the referent can also be an external object or event such as food or the appearance of a predator.

The vocalizations most frequently emitted by adults are clucks (Figure 10.1). Straight clucks (clucks with a sharp descending change in acoustic frequency) are emitted during encounters between groups as part of the three-phrase long call (Halloy and Kleiman 1994) but are disassociated from long calls when holding ground, when chasing intruders, and during predator mobbing (Castro 1990).

Captive studies strongly suggest that clucking rate is associated with the quality of a food item ("preference score") (Benz 1993). Experiments with captive golden lion tamarins have also demonstrated a weak association between acoustic measures (i.e., downslope) and food type (Benz 1993; Benz et al. 1992). It is likely that a species that eats over 80 plant species and many species of invertebrate and

Table 10.1

The Vocal Repertoire of Adult and Young Golden Lion Tamarins

Vocalization Type[a]	Vocalization Variants[b]	Frequency Range[c] (kHz)	Duration[d] (seconds)	Oscillations or Syllables[e]	Individual Behavior[f]	Function[g]
Tonal	Whine	3–6	0.4	0	Reaction to predator presence, periphery of the group startled	Alarm
	Peep	5–7	0.08	0	Behavior of solitary young after being retrieved, finding food	Affiliation
Clucks	Chevron	—	—	0–1	Reaction to presence of novelty, foraging	Ambivalence
	Straight	4–11	0.09	0	Encountering neighbors and intruders, chases, mobbing predators	Aggression
Trills	Tsick	4–8	0.04	1	Foraging for prey, food transfers	Recruitment, localization of caller
	Trill	4.2–9.8	0.18–0.44	3–14	Locomotion, begging, encounters, foraging	Multiple
	Trill B	4.3–10.6	0.24	1–4	Responding to long call	Identity, localization
Atonal	Rasp	4–10	0.37 (0.6)	0	Stationary, soliciting, playing, grooming	Immature begging call
	Screech	1–7	0.23	0	Fighting, agonistic behavior	Repellent
Multisyllable	Short call	5.4–7.4	1.1 (0.4)	3–5	Traveling, at group periphery	Group coordination
	Long calls					
	Two phrase	5.2–10.3	3.4 (0.5)	12–17	Separated from group, auditory encounters	Group cohesion, intergroup spacing
	Three phrase	4.8–10.1	5.2 (0.8)	16–22	Auditory and visual intergroup encounters	Intergroup spacing
Combination	Trill-rasp	4.6–9.7	0.56 (0.4)	?	Young begging	Seeking attention/protection/food
	Trill-whine	—	—	—	Foraging at group periphery	Startling predator, inducing withdrawal
	Cluck-whine	—	—	—	Mobbing predators, encounters	Aggression/defense

Sources: Measures were taken from Green 1979; Ruiz-Miranda et al. 1999; Ruiz-Miranda et al., in preparation. Functions and individual behavior were also obtained from Boinski et al. 1994.

[a] *Vocalization type* refers to major categories of vocalizations that are similar acoustically.

[b] *Vocalization variants* refers to acoustically different variations that have been consistently recognized by various authors.

[c] *Frequency range* refers to the average range of frequencies between the lowest and highest frequency of a call type.

[d] *Duration* refers to the average duration for calls within that call type. When duration varies considerably among the variants of a call type, a range is presented instead of a mean value.

[e] *Oscillations or syllables* refers to the number of sharp frequency modulations (forming a wavelike structure) as seen in a sonogram.

[f, g] *Individual behavior* and *function* note the situations and contexts for and presumed function of the vocalization type and its variants.

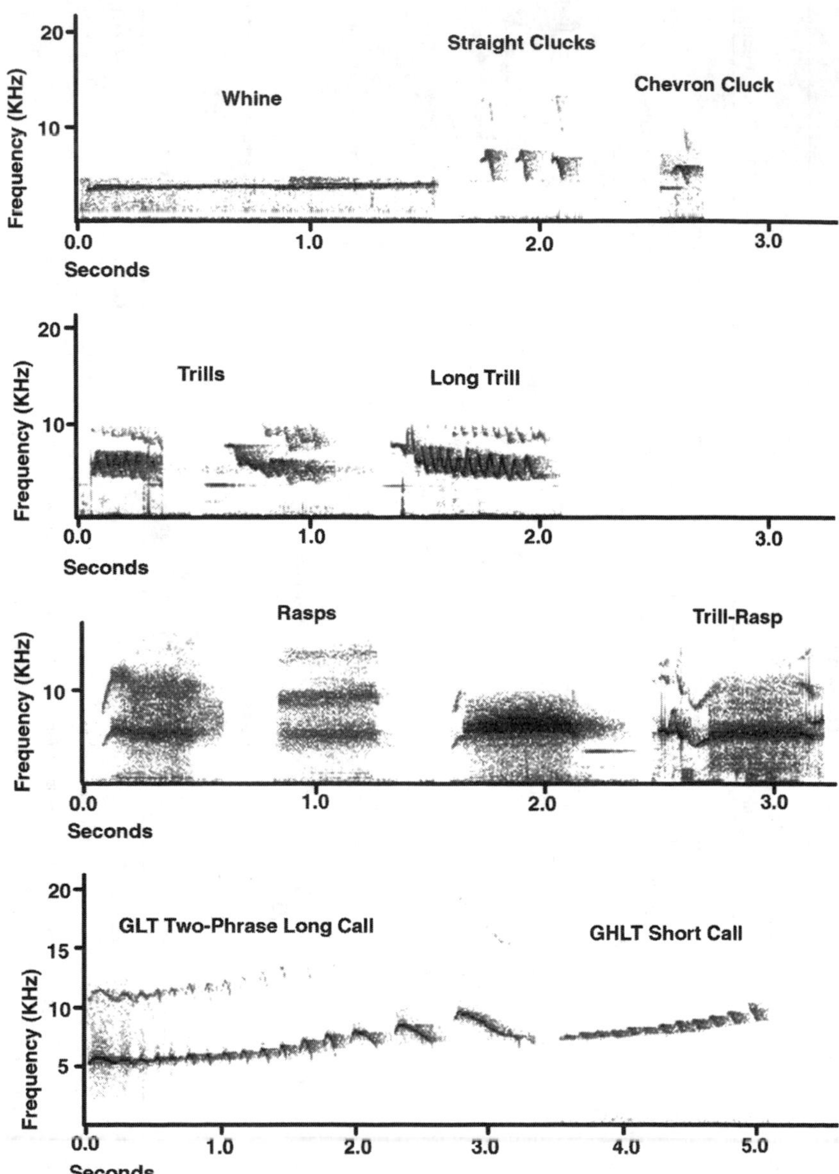

Figure 10.1. Sample sonograms of vocalizations from wild golden lion tamarins (GLTs) and of a short call from a golden-headed lion tamarin (GHLT). The calls were emitted by different individuals. The rasps were obtained from different juveniles. (Redrawn by Stephen D. Nash/Conservation International)

Table 10.2
Comparison of Rates of Emission of Vocalizations of Wild Male and Female Golden Lion Tamarins

Sex	Vocal Rate	Clucks	Trills	Whines	Short Calls
Males	3.6 (1.6)[a]	1.3 (0.91)	1.0 (0.5)[a]	0.20 (0.1)[a]	0.05 (0.04)
Females	1.8 (1.2)	0.9 (0.60)	0.4 (0.3)	0.05 (0.1)	0.06 (0.06)

Note: Rates of emission are mean rates for seven males and seven females. The numbers in parentheses are the standard deviations.

[a]Numbers indicate sex differences in vocal rates, trills, and whines that are statistically significant ($p < 0.05$). Obtained from a Student's t-test for unpaired samples.

vertebrate prey in the wild (Dietz et al. 1997; Lima et al. 1995) would evolve signals for a few major food categories such as large prey or plentiful fruit.

Lion tamarins share food, especially with weanlings and juveniles (Feistner and Price 2000; Chapter 9 this volume). Boinski et al. (1994) noted that clucking animals were not approached by listeners. However, adult lion tamarins also emit a "food-offering call" that consists of variable tonal sounds and clucks (Figure 10.2) and appears to signal willingness to share (Brown and Mack 1978; Hoage 1977; Ruiz-Miranda et al. 1999). Young lion tamarins hurriedly approach an individual emitting these calls. Further studies are necessary to reveal any possible relationship between food-offering calls and food quality.

Trills are the second most common category of vocalizations emitted by wild golden lion tamarins (Figure 10.1); variants of trills may have different functions (Green 1979; Kleiman et al. 1988; McLanahan and Green 1978; Table 10.1). Wild golden lion tamarins emit short-duration trills prior to leaping to another branch in which they will often forage (Boinski et al. 1994). Longer trills are used during locomotion and agonistic encounters. Medium and long trills are used by juveniles while soliciting food from adults (C. R. Ruiz-Miranda, unpublished data). A particular trill (Trill B) is often emitted in response to long calls by a group member. Trills emitted in bouts often vary in duration and frequency measures, and further studies are needed to understand the apparent complexity of their use.

Captive experiments and field observations have shown that lion tamarins use different vocalizations in response to terrestrial and aerial predators (Castro 1990; Castro et al. 1998; C. R. Ruiz-Miranda, personal observations). Castro (1990) presented predator models (snake and hawk) to captive animals at the Centro de Primatologia do Rio de Janeiro (CPRJ—Rio de Janeiro Primate Center) and found that whines, trills, clucks, and even short calls were used by lion tamarins in re-

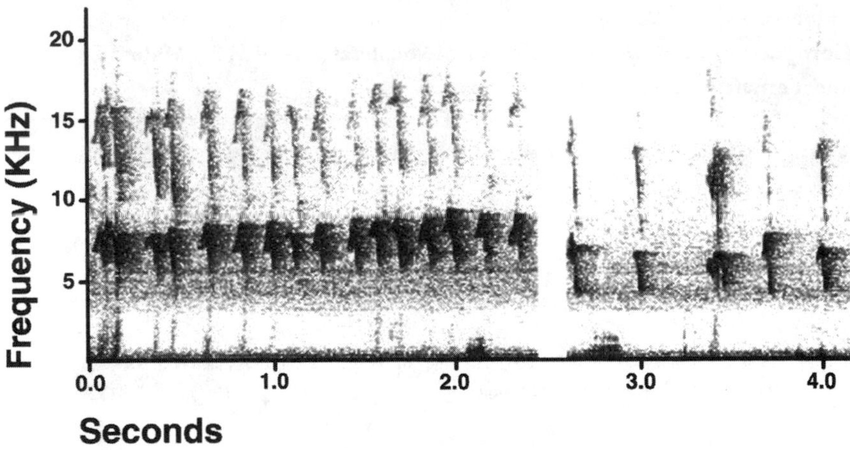

Figure 10.2. Sonogram of a food-offering call emitted by an adult golden lion tamarin prior to a food transfer. (Redrawn by Stephen D. Nash/Conservation International)

sponse to terrestrial predator models and that a short burst of clucks was used in response to an aerial predator model. Wild golden lion tamarins respond to terrestrial and arboreal predators (e.g., snakes and tayras [*Eira barbara*]) by mobbing them.

Not all predators are mobbed, however. Brown capuchin monkeys (*Cebus nigritus*) have been observed attacking young golden lion tamarins (C. R. Ruiz-Miranda, personal observations), but instead of mobbing, the golden lion tamarins usually just retreat quietly and quickly. The response to aerial predator warning calls is different from the response to warnings of terrestrial predators. Upon hearing a short burst of clucks, wild golden lion tamarins may drop or climb down to the ground, move to the opposite side of the tree trunk, or hide while looking up. If a bird of prey perches in a nearby branch, golden lion tamarins may mob it as they do a terrestrial or arboreal predator. These behaviors, observed in captive-born and wild-born lion tamarins (Castro 1990; C. R. Ruiz-Miranda, personal observations), are similar to those observed for other callitrichids (Ferrari and Lopes Ferrari 1990).

SEX DIFFERENCES Captive and field studies have demonstrated sex differences in overall vocal rate (Table 10.2) and the rate and acoustic structure of some specific calls. Sex differences have been observed in captive golden lion tamarins for the rate of emission and use of trills and nontonal vocalizations (McLanahan and Green 1978). Males emit trills more often than females, particularly when vigilant, when leaping and foraging, and during solo activities. When demonstrating social behaviors, females emit twice as many trills as males. Females also emit significantly more atonal sounds, such as snoughs (nonvocal atonal sounds) (Green 1979) and rasps (atonal vocalizations), (Figure 10.1) than males, and their snoughs and rasps occur during social behaviors involving contact, whereas males emit rasps mostly during solo activities. For captive animals, Benz et al. (1990) found sex differences in acoustic structure for chevron clucks (called "chirps" in their paper) and peeps ("chucks" in their paper) but not for trills.

For wild animals, the vocal behavior of the sexes may change seasonally. Recording sessions in the wet season (October–February) indicated that golden lion tamarin males had higher vocal rates than females overall, most especially for trills and whines (Table 10.2). While recording during the peak of the dry season (July), Boinski et al. (1994), however, found no sex difference in the rate of emission or usage for any type of vocalization.

AGE DIFFERENCES Immature reintroduced and wild golden lion tamarins vocalize at rates that are up to three times higher than those of adults (C. R. Ruiz-Miranda, unpublished data). We recorded bouts that ranged from 1 vocalization (<1 second long) to up to 400 consecutive vocalizations (16 minutes long). We also recorded large differences in the rate of vocalizing of twins and triplets (Table 10.3). Bout structure is also different between young and adults. Immature wild golden lion tamarins use a significantly higher proportion of multitypic bouts than adults. Animals as young as 1 month of age already can produce the adult vocal repertoire, but they mostly rasp and trill conspicuously for their first 9 months (C. R. Ruiz-Miranda, unpublished data; Figure 10.3). Rasps are loud (measured over 80 dB at 10 m), atonal, broad-banded vocalizations (Ruiz-Miranda et al. 1999; Figure 10.1). Young animals produce trills with an open mouth; the trills are loud, broad banded, and of varied duration (Ruiz-Miranda et al. 1999; Figure 10.1). The trills and rasps are often produced together. Age differences in vocal rate and repertoire disappear after golden lion tamarins reach 1 year of age (Boinski et al. 1994; C. R. Ruiz-Miranda, unpublished data).

Why should young signal so conspicuously and why does the acoustic communication pattern change with age so abruptly? Immature lion tamarins, like many mammals, emit noisy, conspicuous signals when interacting with their parents, even after weaning. Up to 2 months old, golden lion tamarins vocalize loudly when left alone on a branch and when they apparently want to nurse or be carried. The use of this signaling may be related to the efficiency of communication in a noisy environment (Dawkins and Guilford 1997). We believe that vocal rates of rasps, trills, and trill-rasps after weaning especially correlate with food transfers from caretakers to immatures: the rate of these vocalizations drops sharply at

Table 10.3
Vocalization Rate and Received Food Transfers for Sets of Wild Golden Lion Tamarin Twins and Triplets

Golden Lion Tamarin Identification Number	Vocalizations (per Minute)	Rasps (per Minute)	Rasps (Percentage of All Vocalizations)	Food Transfers (per Hour of Focal Observation)
473	9.92	6.24	0.63	2.8
474	6.15	3.17	0.52	0.7
477	4.12	0.82	0.20	1.1
478	7.37	2.15	0.29	0.4
480	7.44	2.54	0.34	1.3
481	5.44	1.50	0.28	1.7
522	1.49	0.77	0.39	1.1
523	3.98	1.44	0.36	2.6
526	2.05	0.40	0.20	3.9
527	13.88	6.57	0.47	5.9
545	8.22	3.59	0.44	0.8
546	3.74	1.87	0.50	0.0
547	8.25	4.67	0.57	3.5

the age when the number of food transfers received decreases (see Feistner and McGrew 1989; Ruiz-Miranda et al. 1999; Chapter 9 this volume for a discussion of the function of food transfers). In four of six sets of golden lion tamarin litters, we observed that the young that vocalized the most received the most food transfers (Table 10.3). The dominance of noisy vocalizations (e.g., rasps and rasp-trills) during solicitation of food transfers is also characteristic of conflict or agonistic situations (August and Anderson 1987; Morton 1977, 1982), and in primates, these calls may serve both to attract caregivers and repel bothersome conspecifics (Todt 1988; Todt et al. 1995). Additionally, the conspicuous vocal behavior of young lion tamarins has several of the characteristics of tonic communication (e.g., serial calling and redundancy) (Hersek and Owings 1993; Schleidt 1973).

CAPTIVE-BORN VERSUS WILD-BORN GOLDEN LION TAMARINS Captive-born golden lion tamarins emit all of the vocalizations that their wild counterparts use, albeit with some differences in use and context. Captive

Figure 10.3. An infant golden lion tamarin vocalizing while separated from adults. (© Ian Yeomans)

adults show lower vocal rates than reintroduced and wild adults due to fewer trills, clucks, and peeps (Figure 10.4). There are significant differences between captive and wild golden lion tamarins in the context in which short calls are emitted (Table 10.4). The short call, or wah-wah (Halloy and Kleiman 1994; Kleiman et al. 1988), is a soft, multisyllabic vocalization used to coordinate group travel (Boinski et al. 1994; Table 10.1). Wild golden lion tamarins emit short calls most often when on the periphery of the group while the group is traveling (or moving/foraging) or during transitions in group activities (e.g., from resting to foraging). In contrast, reintroduced animals emit short calls while the group is foraging in one place, and the caller often remains in place. In further contrast, captive adults emit most of these calls while the group is resting, and the caller tends to remain stationary. Other contextual differences occur with trills and tsicks, which are used while foraging and may also function to coordinate group movements (Boinski et al. 1994; Castro et al. 1998). Captive-born adults at SNZP emitted 50 percent of their short trills while stationary and visible to others, whereas wild animals emit short trills while leaping and foraging and when at the periphery of the group (Boinski et al. 1994). Tsicks by wild animals are emitted in food-related contexts, whereas captive animals also emit them when excited and in a state of alarm (McLanahan and Green 1978).

Table 10.4
Context of Occurrence of Short Calls in Golden Lion Tamarins

Context	Captive (Percent)	Reintroduced (Percent)	Wild (Percent)
Auditory encounter[a]	0	1	4
Visual encounter	0	0	19
Group mobbing[b]	4	5	0
Group vocalizing[c]	11	0	0
Group marking[d]	4	0	0
Group foraging	11	60	18
Group social	4	1	4
Nest box[e]	15	1	—
Group resting	26	0	0
Group traveling[f]	4	20	14
Transition[g]	22	11	42

Note: Numbers represent the percentage of the total number of calls emitted by each rearing type (captive = 27, reintroduced = 75, wild = 86 calls) in diverse contexts. Overall differences among captive, reintroduced, and wild golden lion tamarins were statistically significant ($\chi^2 = 150.6$; df = 18; $p < 0.0001$).

[a]*Encounter* refers to territorial encounters between groups of golden lion tamarins.

[b]*Mobbing* refers to mobbing of predators or non–lion tamarin animals.

[c]*Vocalizing* means other group members were emitting long calls or short calls.

[d]*Marking* refers to scent marking.

[e]*Nest box* refers to animals inside or on top of the artificial nest box.

[f]*Traveling* means the group was moving in a definitive direction, without engaging in other activities.

[g]*Transition* means that 50 percent of the animals were engaged in an activity that differed from the previous one in which the whole group was engaged.

Thus, captive- and wild-born golden lion tamarins seem to use vocalizations in different ways. One explanation for low rates of use of contact calls by captive animals is that group members are rarely out of sight, as they are in the wild. To understand these differences, we must also consider that some of the principal challenges for wild animals—foraging, avoiding predators, looking for shelter, and traveling—are practically nonexistent in captivity. Because the vocalizations of forest primates, in addition to warning about predators and signaling food sources, are likely used to coordinate group movements and maintain group cohesion (Boinski et al. 1994; Boinski 2000), the absence of such contexts would be predicted to result in low rates of vocalizing or the complete absence of some calls. Differences between captive and wild animals suggest that researchers must be cautious when proposing functions for vocalizations based on captive studies.

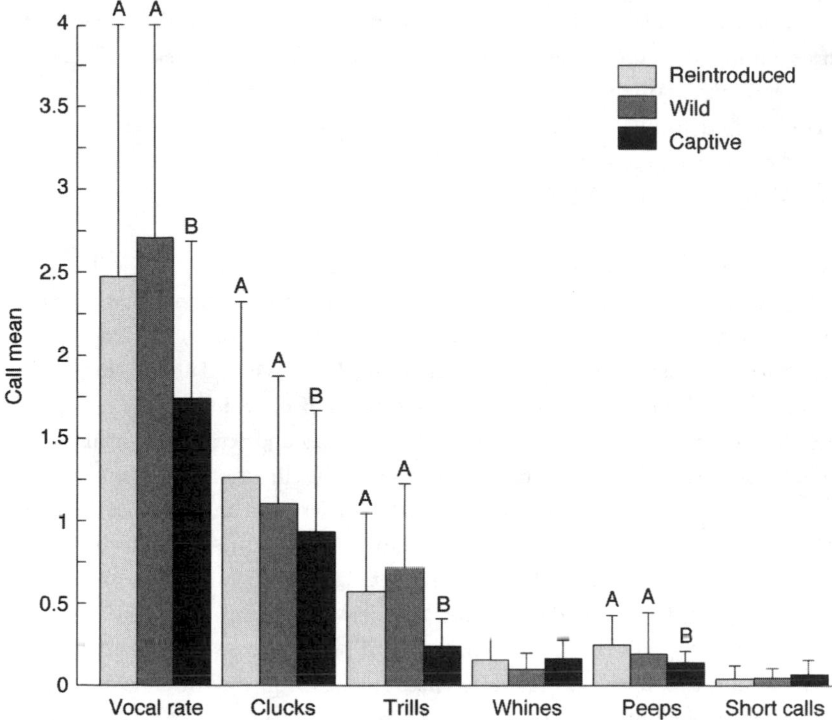

Figure 10.4. Rates of emission of five major categories of vocalizations by captive, reintroduced, and wild golden lion tamarins. The atonal vocalizations were not included because of their low rate of emission by adults. The different letters represent statistically significant differences ($p < 0.05$) obtained using ANOVA. Absence of letters means that no significant differences were found. The Y-bars represent one standard deviation.

Vocalizations Used in Long-Distance Communication

The long calls of lion tamarins can be divided into two types that differ in structure and usage (Table 10.1). These calls have been termed two-phrase long calls and three-phrase (long calls with clucks) long calls (Halloy and Kleiman 1994). The two-phrase and three-phrase long calls appear to function as long-distance communication signals both between and within groups (Table 10.1) and not as a pair-bonding mechanism as was suggested by McLanahan and Green (1978).

SPECIES COMPARISONS At least three of the lion tamarins (*L. rosalia, L. chrysomelas,* and *L. chrysopygus*) have been observed emitting two-phrase long calls. Based on captive observations, these calls differ significantly in acoustic parameters

(Snowdon et al. 1986). The black lion tamarin (*L. chrysopygus*), the largest of the three, emits long calls that are lower in frequency (pitch) measures. Differences in the acoustic structure of the two-phrase long calls of *L. rosalia, L. chrysomelas,* and *L. chrysopygus* parallel differences in cranio-dental morphology and is evidence that they are three distinct species, with *L. chrysomelas* being the most differentiated (Snowdon et al. 1986).

SEX DIFFERENCES IN STRUCTURE AND USAGE The two-phrase long calls of both captive and wild female golden lion tamarins have a higher pitch (i.e., both the highest and lowest frequencies of female long calls are higher) than those of males (Benz et al. 1990; Ruiz-Miranda et al., in preparation). Adults also respond differently to playbacks of male and female two-phrase and three-phrase long calls in the field. Both males and females emit more long calls (two- and three-phrase) and travel more in the direction of the source in response to female playbacks rather than male. Males emit three-phrase long calls in response to male playbacks, but adult females respond to male playbacks with two-phrase long calls only (Ruiz Miranda et al., in preparation). Behavioral observations suggest that wild males and females do not differ in the emission of two-phrase long calls (Ruiz-Miranda et al., in preparation), thus supporting previous studies that found no sex differences in the frequency of emission of long calls by captive animals (McLanahan and Green 1978).

AGE DIFFERENCES Two-phrase long calls appear in the repertoire during the first month of life in wild golden lion tamarins (C. R. Ruiz-Miranda, unpublished data). Immatures (between 4 and 9 months old) emit the two-phrase long calls at considerably lower rates (mean = 0.01 calls/minute, s.d. = 0.03; n = 28 individuals) than adults (mean = 0.12 calls/minute, s.d. = 0.08; n = 19 individuals).

CAPTIVE-BORN VERSUS WILD-BORN GOLDEN LION TAMARINS Captivity may affect the structure, rate, or use of long calls by not providing the appropriate stimulus or context for their expression or by providing an acoustically different rearing environment (Castro et al. 1998). Our studies have found significant differences in the acoustic structure and use of long calls between wild- and captive-born animals (Ruiz-Miranda et al., in preparation). Wild golden lion tamarins emit both types of long calls significantly more often than reintroduced and captive animals, and free-living golden lion tamarins (wild and reintroduced) emit significantly more three-phrase long calls (but not two-phrase long calls) than captives (Ruiz-Miranda et al., in preparation). This likely results from the fact that captive golden lion tamarins are rarely separated from group members and typically lack neighbors. There are also differences among rearing types in the context of emission of both types of long calls (Tables 10.5 and 10.6). Wild animals emit two-phrase long calls mostly during auditory encounters with neighbors and when group members are separated from each other (e.g., during foraging and during transition periods between resting and group travel) (Table 10.5). Reintroduced animals emit the call when group members are dispersed and during transition periods but also while mobbing common marmosets (*Callithrix jacchus*), an introduced species whose natural range is in the northeast of Brazil (Affonso et al. in press; Ruiz-Miranda et al. 2000). Captive animals emit almost half of their two-phase long calls while resting or at the nest box, with an additional percentage emitted when group members are separated from each other.

Table 10.5

Context of Emission of Two-Phrase Long Calls of Golden Lion Tamarins

Context	Captive (Percent)	Reintroduced (Percent)	Wild (Percent)
Auditory encounter[a]	1	2	19
Visual encounter	0	0	12
Group marking[b]	5	1	0
Group vocalizing[c]	8	0	0
Group mobbing[d]	0	11	0
Group social	0	8	5
Group foraging	23	42	26
Group traveling[e]	9	11	8
Transition[f]	12	18	29
Nest box[g]	17	8	0
Resting	27	7	8

Note: Numbers represent the percentage of long calls (captive = 132, reintroduced = 132, wild = 129 calls) emitted while the group was engaged in various activities. Overall differences among captive, reintroduced, and wild golden lion tamarins were statistically significant ($\chi^2 = 240.1$; df = 16; $p < 0.0001$).

[a]*Encounter* refers to territorial encounters between groups of golden lion tamarins.

[b]*Marking* refers to scent marking.

[c]*Vocalizing* means other group members were emitting long calls or short calls.

[d]*Mobbing* refers to mobbing of predators or non–lion tamarin animals.

[e]*Traveling* means the group was moving in a definitive direction, without engaging in other activities.

[f]*Transition* means that 50 percent of the animals were engaged in an activity that differed from the previous one in which the whole group was engaged.

[g]*Nest box* refers to animals inside or on top of the artificial nest box.

Wild and reintroduced animals emit this long call when alone 24 percent and 13 percent of the time, respectively.

There are also evident differences between captive, reintroduced, and wild golden lion tamarins in the three-phrase long call (Table 10.6). Wild animals emit this vocalization mostly during intergroup territorial encounters, whereas reintroduced and captive animals do so mostly within the group. Interestingly, 34 percent of the long calls with clucks emitted by reintroduced animals (Table 10.6) were directed at *C. jacchus*. This call is emitted in proximity to group members by wild (78 percent) and reintroduced animals (100 percent). For the wild golden lion tamarins, 83 percent of the long calls with clucks were emitted after hearing either a two-phrase or a three-phrase long call, whereas only 55 percent of the reintroduced animals' calls were emitted as a response to a long call. It seems that the eliciting stimulus or context for the calls is different for wild- and captive-born animals.

In general, both wild- and captive-born reintroduced golden lion tamarins respond similarly to playbacks of both types of long calls (Kleiman 1990a): piloerection, long calls, and movement toward the source of the playback. However, wild-born golden lion tamarins show a more sustained behavioral response, travel farther, and alter their movement patterns and calling rates more significantly than captive-born golden lion tamarins, suggesting that experience with intergroup encounters likely alters the response to long calls (Kleiman 1990a; Ruiz-Miranda et al., in preparation).

Table 10.6
Context of Emission of Three-Phase Long Calls (Long Calls with Clucks) in Golden Lion Tamarins

Context	Captive (Percent)	Reintroduced (Percent)	Wild (Percent)
Auditory encounter	0	0	22
Visual encounter	0	0	47
Group territorial[a]	0	5	2
Group mobbing	0	34	0
Intragroup[b]	100	61	29

Note: Numbers represent the percentage of long calls (captive = 1, reintroduced = 27, wild = 68 calls) emitted while the group was engaged in various activities. Overall differences between reintroduced and wild were statistically significant ($\chi^2 = 89.3$; df = 4; $p < 0.0001$).

[a] *Territorial* refers to animals engaged in combined scent marking and vocalizing.

[b] *Intragroup* refers to within-group behaviors such as foraging, social, resting, traveling, and transition.

Several authors have suggested that learning seems unnecessary in the ontogeny of vocalization structure in primates (Marler and Mitani 1988; Newman and Symmes 1982; Snowdon 1989b). However, we found significant differences between reintroduced and wild animals in the acoustic structure of the two-phrase and the three-phrase long calls (Ruiz-Miranda et al., in preparation). The two-phrase calls of reintroduced animals have a higher pitch, frequency range, and slope than those of wild animals. The three-phrase long calls of reintroduced and captive animals are shorter in duration than those of wild animals. These differences could be a result of the acoustics of the rearing environment, a consequence of the process of reintroduction, or due to genetic differences between populations (Castro et al. 1998). The first two hypotheses imply that experience can modify call structure: either captive golden lion tamarins tune their calls to the acoustics of the exhibits or reintroduced animals modify their vocalizations to fit a new environment. The differences seen in reintroduced animals do not seem to be a local adaptation because higher frequencies do not travel well in the forest (Morton 1975; Waser and Brown 1986; Wiley and Richards 1982).

THE FUNCTIONS OF LONG CALLS Long calls are probably best considered as advertisement calls with multiple roles or functions. They advertise the presence of a lion tamarin of a particular sex, group membership, or maybe even individuality. For reproductive adults, advertising presence may function to repel intruders, thus defending a territory or access to a mate; for nonreproductive adults, long calls may attract potential mates. Captive, wild, and reintroduced golden lion tamarins all emit the two-phrase long call when separated from group members, but only reintroduced and wild animals emit it in the context of mobbing intruders. The intruders in the case of reintroduced animals are, in many cases, introduced common marmosets (*Callithrix jacchus*), which are believed to compete with golden lion tamarins (Affonso et al. in press; Ruiz-Miranda et al. 2000).

OLFACTORY AND VISUAL COMMUNICATION

There have been a few studies of olfactory communication in captive golden lion tamarins (Kleiman and Mack 1980; Mack and Kleiman 1978) and golden-headed lion tamarins (De Vleeschouwer et al. 2000a; Haazen 1988) but even fewer studies of visual communication (Kleiman et al. 1988). The ethograms for visual and olfactory communication are presented in Kleiman et al. 1988. Reviews of olfactory communication can be found in Epple et al. 1993 and Walraven and Van Elsacker 1992.

Scent Marking

Male and female lion tamarins have well-developed and complex sternal, suprapubic, and anogenital glands that release a variety of different secretions (Epple et al. 1993; Hershkovitz 1977; Kleiman and Mack 1980) whose chemical compositions have not been studied. In scent-marking behavior, the secretions are deposited on a substrate through the rubbing of the glands back and forth along a surface (Figure 10.5). These secretions are chemical signals used for intraspecific communication (Epple et al. 1993; Walraven and Van Elsacker 1992). Among the secretions' functions proposed for callitrichids are signaling of reproductive status and individuality, social regulation, marking territory, parent-offspring interactions (Cebul et al. 1978; Epple et al. 1993; French and Cleveland 1984; Rylands 1990; Walraven and Van Elsacker 1992), and perhaps even navigating within a territory (Mack and Kleiman 1978). In the latter case, an individual may actually be marking arboreal pathways that help it to find its way rapidly through the home range (Eisenberg and Kleiman 1972).

Sternal marking is the most frequent form seen in captive golden lion tamarin adults (Epple et al. 1993; Kleiman and Mack 1980; Kleiman et al. 1988). Golden lion tamarins tend to mark specific sites on their way to food areas and near "ter-

Figure 10.5. An adult golden lion tamarin dragging its sternum and suprapubic region along a branch while carrying an infant. Note the black dye mark on the side of the adult's head: it is used for individual identification. (© Ian Yeomans)

ritory" boundaries (Mack and Kleiman 1978). There are no evident sex differences in marking, but there are age and social status differences among captive golden lion tamarins: reproductive males and females do most of the scent marking. Although females scent mark considerably less at the time of ovulation and after parturition, the frequency of marking by males remains unchanged with female reproductive state (Kleiman and Mack 1980).

Individuals housed alone or in same-sex pairs scent mark infrequently (Kleiman and Mack 1980; Mack and Kleiman 1978). Scent marking by juveniles of both sexes and by nonreproductive individuals is also infrequently seen. Most juveniles do not begin to scent mark as long as they remain with their natal family. In captive animals, circumgenital marking appears earlier (4–5 months) than sternal marking (8–9 months) in both sexes (Hoage 1982; Kleiman and Mack 1980). Females typically begin scent marking at a later age than males, but the behavior will appear rapidly in a nonreproductive female (even a subadult) if the dominant breeding female is removed from a group (D. G. Kleiman, unpublished data).

Observations suggest that females may mark to communicate reproductive status because there are no overt signs of estrus (Stribley et al. 1987), and both female and male marking may be related to territoriality or within-group social status. Both Kleiman (1978b) and De Vleeschouwer et al. (2000a) noted that female golden lion tamarins and golden-headed lion tamarins exhibit lower scent-marking rates around the time of ovulation, which suggests that they do not advertise their presence widely when conception is possible.

Visual Communication

Lion tamarins differ in their visual conspicuousness, from the red-orange *L. rosalia* to the mostly black *L. chrysopygus,* yet visual signals appear to be similar across species (Kleiman et al. 1988). The use of visual signals has been little studied in lion tamarins. The only visual signals that have been studied systematically are the various forms of the arch display (Rathbun 1979) and the posture used by donors during food transfers (Brown and Mack 1978; Hoage 1982) in captive animals.

The arch display is the most prominent and ritualized of the golden lion tamarin's visual signals. It makes the animal look larger and more conspicuous. It is a graded signal, going from a low-intensity arch sit (head elevated, arched back, tail erect, some piloerection) to a high-intensity arch walk (head down, gaze ahead, rigid limbs, piloerection, stiff walking) (Rathbun 1979). Although the arch posture may be accompanied by vocalizations (clucks and whines), the arch walk is silent. It is observed in captive golden lion tamarins during social interactions, including the introduction of unfamiliar individuals (Rathbun 1979), and it is observed in the field during territorial encounters (C. R. Ruiz-Miranda and D. G.

Kleiman, personal observations) and after inspection of a scent mark (C. R. Ruiz-Miranda, personal observations). Females exhibit increases in the frequency of the behavior during periods of sexual activity and during pair formation and show the lowest frequencies during pregnancy (Rathbun 1979). Males begin to show the behavior around 8 months of age, and females around 16 months of age. In general, females only begin to show the behavior when removed from their natal group or when the breeding female is no longer present in the natal group; the developmental pattern is similar to the ontogeny of scent marking (D. G. Kleiman, unpublished data; Rathbun 1979). During pair formation of captive golden lion tamarins, the display is common during the first few days and then gradually disappears as the pairs "settle down" (Rathbun 1979). These observations suggest that the arch walk is a display used during high-arousal competitive/conflict interactions or under conditions when social and reproductive relations among group members are changing.

Golden lion tamarins with food may exhibit a stereotyped posture prior to transferring the food to another golden lion tamarin (Brown and Mack 1978; Hoage 1982; Ruiz-Miranda et al. 1999; Chapter 9 this volume). This signal, presumed to indicate a willingness to "share" food, occurs together with vocal and other visual signals during food transfers, although the visual signal alone is not a good predictor of the willingness to share by the possessor (Brown and Mack 1978).

There are other forms of visual communication in lion tamarins (Kleiman et al. 1988). Some of them, for example, tongue protrusion, can be considered signals or displays, whereas others, such as postures during long calling or when scent marking, may be considered cues. These potential communicative behaviors are typically accompanied by other modes of communication, for example, vocalizations or scent marks.

CONCLUSIONS

There has been little research on communication in three of the four species of *Leontopithecus*. Because of their close phylogenetic relationships, they probably have considerable similarities, but there are differences among these species in habitat, ranging behavior, and social structure (Rylands 1993, 1996; Chapters 7 and 8 this volume), factors that could affect communication patterns. These species could provide useful comparisons of the relationship between signal structure and function and various environmental and social variables.

Two trends in the communication of lion tamarins are conspicuousness and multiplicity of functions. Lion tamarins are highly vocal and use both graded and discrete signals. Intragroup vocal signals show context-specificity but also appear in combination with other calls in multiple contexts, suggesting that complex signals may have different messages. Infant lion tamarins have evolved a conspicuous signaling system that has characteristics of tonic communication (Hersek and Owings 1993). The combination of signals and cues from different sensory modalities, frequently used by lion tamarins, may reflect communication behaviors more complex and dependent on context than estimated previously.

What are the uses of studying communication of lion tamarins for their conservation? First, we can apply this knowledge to the management of wild populations. For example, due to the fact that lion tamarins will approach a speaker and vocalize in response to the "playback" of long calls (Kleiman 1990a), we have been able to census wild populations successfully using the "playback" technique (Kierulff et al. 1997; Pinto 1994; Pinto and Tavares 1994).

A better understanding of lion tamarin communication can also help to explain behavioral differences between wild-born and captive-born golden lion tamarins, information that can be vital for the success of reintroduction programs. There are four main questions to be asked:

1. Are there differences between captive-born and wild-born animals in how they communicate? While the vocal repertoire of captive-born and wild-born animals is the same, the rearing types differ in rate, use, and form of vocalizations used for long-distance communication and the rate and context of vocalizations used for intragroup communication.
2. What is the source of these differences? Those between wild and captive animals may be genetic and caused by founder effect, drift, or selection under captivity; ontogenetic processes; or immediate causes/contextual differences (Arnold 1995; Carlstead 1996; Price 1984). Since the captive management of the genus *Leontopithecus* maximizes retention of genetic variation (Chapter 4 this volume), genetic differences between the captive and wild populations must stem from original founder effects. The similarities in use and rate of vocalizations between reintroduced and wild golden lion tamarins also suggest that immediate context is an important determinant of vocal behavior, whereas differences in acoustic structure suggest that ontogenetic processes and experience may also be involved.
3. Do the differences in communication compromise survival or reproduction? We do not yet know.
4. Can we change captive management techniques to eliminate differences between the zoo and wild populations and thus improve survival after reintroduction while maintaining high standards of animal welfare

(Beck and Castro 1994)? Certainly, enrichment modifications and releasing animals into naturalistic exhibits produce changes in behavior and well-being (Bronikowski et al. 1989; Carlstead 1996; Carlstead and Shepherdson 1994; Erwin and Deni 1979; Price 1992b; Redshaw and Mallinson 1991a, 1991b; Shepherdson 1994). Although semifree ranging conditions intuitively seem optimal, there is no evidence that enrichment of the captive environment or training has improved the survival of reintroduced golden lion tamarins (Castro et al. 1998; Chapter 13 this volume).

ACKNOWLEDGMENTS

This study would not have been possible without the field and logistic support provided by the Reintroduction, Ecology, and Administration Teams of the Associação Mico-Leão-Dourado (AMLD—Golden Lion Tamarin Association). Luis Fernando Moraes, Denise Rambaldi, Dionízio Pessamílio, and Whitson José da Costa also provided vital logistical and moral support. Numerous students assisted in the collection and analyses of data: Ezequiel Moraes, Adriana Grativol, Sira Palerm, Christine Archer, William York, Danyra Quiñones, and Angelina Thompson. Carlos Ruiz-Miranda was funded by the Smithsonian Institution, by Friends of the National Zoo (FONZ) (a grant to David Jenkins), and by the Director's Circle funds of the SNZP. The research was funded by the Smithsonian Institution (Scholarly Studies 1235S30C and 1235S40A, the Nelson Fund, Women's Committee, and Latino Initiative Program). TransBrasil airlines kindly helped with travel. We are also grateful to the National Science Foundation (Grants #BNS-8616480, 8941939, and 9008161 to James Dietz and Andrew Baker), the International Environmental Sciences Program of the Smithsonian Institution (to Devra Kleiman), the Friends of the National Zoo (FONZ) (to Benjamin Beck and Devra Kleiman), the Instituto Brasileiro do Meio Ambiente e dos Recursos Naturais Renováveis (IBAMA—Brazilian Institute of the Environment and Renewable Natural Resources), the Conselho Nacional de Desenvolvimento Científico e Tecnológico (CNPq—Brazilian National Science Council, Grant 463136/2000-4 to CRRM), the Fundação de Amparo à Pesquisa do Rio de Janeiro (FAPERJ—Research Support Foundation of Rio de Janeiro, Grant E-26/171.941/2000 to CRRM), the Frankfurt Zoological Society Fund for Threatened Species (to Benjamin Beck), the World Wide Fund for Nature (to Devra Kleiman and Denise Rambaldi), and William McClure. Additional administrative support and use of the Sound and Video Analyses Laboratory came from the Department of Zoological Research of the SNZP and from the Laboratório de Ciências Ambientais of the Universidade Estadual do Norte Fluminense (Laboratory of Environmental Sciences of the North Fluminense State University), Rio de Janeiro.

ALCIDES PISSINATTI, RICHARD J. MONTALI,
AND FAIÇAL SIMON

11
DISEASES OF LION TAMARINS

Large-scale captive breeding of callitrichids was not initiated until the early 1960s, when their usefulness as laboratory models was discovered, particularly for virus-induced malignant diseases. Besides their value in drug testing (Hiddleston 1976), their importance as research models has been notable in the studies of infectious hepatitis A (Deinhardt et al. 1976; Provost et al. 1977), brain diseases (Gibbs and Gajdusek 1976), cancer (Wolfe et al. 1972), and radiation biology (Gengozian et al. 1978). Prior to the 1960s, they were used mainly for anatomical, developmental, and behavioral studies, and their use in biomedicine was limited to screening in the wild for such diseases as yellow fever (Laemmert et al. 1946; Waddell and Taylor 1945) and malaria (Deane 1976).

It was also in the 1960s that Adelmar F. Coimbra-Filho and Alceo Magnanini first drew attention to the serious consequences of the then widespread capture and commerce of golden lion tamarins for zoos and the pet trade and established a breeding program for the genus in the Rio de Janeiro Zoo (Coimbra-Filho 1965, 1969, 1970a, 1970b; Coimbra-Filho and Magnanini 1962, 1968). In 1972, Coimbra-Filho set up the Biological Bank, in the Tijuca National Park, Rio de Janeiro, specifically for breeding lion tamarins; in 1979, the lion tamarins were moved to a new locality, the Centro de Primatologia do Rio de Janeiro (CPRJ—Rio de Janeiro Primate Center), under the Fundação Estadual de Engenharia do Meio Ambiente (FEEMA—State Foundation for Environmental Engineering), in the municipality of Magé (Coimbra-Filho et al. 1986b; Figure 11.1).

One of the key results of the landmark 1972 conference "Saving the Lion Marmoset" (Bridgwater 1972a; see also Chapter 4 this volume) was the recognition of the importance of a properly managed captive breeding program for the genus

Figure 11.1. Tamarin enclosures nestled at the foot of the mountains in the Centro de Primatologia do Rio de Janeiro (CPRJ—Rio de Janeiro Primate Center). (Photo by Russell A. Mittermeier/Conservation International)

Leontopithecus, and a working group was formed specifically to assess the situation of the captive colonies of lion tamarins and make recommendations concerning their husbandry and management (DuMond 1972). At that time only the Tijuca Biological Bank held *L. chrysomelas* and *L. chrysopygus,* and for the next decade the majority of the research and breeding efforts were concentrated on *L. rosalia.* Mallinson (1996) provided an excellent review of the captive breeding program for the conservation of this species (see also Kleiman and Mallinson 1998; Chapter 4 this volume).

The poor performance of breeding colonies was initially believed to be principally due to poor husbandry and management (Kleiman 1977a). The earliest concern, therefore, regarding the maintenance of ex situ populations was the development of research on their nutrition, social behavior, and reproductive biology (Kleiman 1976, 1977a, 1977b; Kleiman et al. 1982), and despite the wealth of literature now available on the biology of the genus *Leontopithecus,* relatively few studies have been carried out specifically concerning its pathology, which is ultimately important for the management of healthy colonies and for the conservation of the species (DuMond 1972; Hunt and Desroisiers 1994; Valerio et al.

1969). There have been, however, significant advances in our understanding of the pathology and veterinary care of captive lion tamarins and other callitrichids, due especially to research at the Smithsonian National Zoological Park (SNZP) (Bush et al. 1993) and CPRJ/FEEMA.

Here we review some current findings concerning pathological disorders in lion tamarins: (1) trauma, (2) dental disease, (3) reproductive pathology, (4) congenital disorders, (5) nutrition and metabolic problems, (6) stress, (7) viral diseases, (8) bacterial diseases, (9) mycotic infections, (10) parasitic infections, (11) tumors, and (12) toxic and miscellaneous disorders. The research reviewed in this chapter focuses on the three *Leontopithecus* species in captivity, mainly from CPRJ/FEEMA: *L. rosalia, L. chrysomelas,* and *L. chrysopygus.* A more comprehensive coverage of diseases of Callitrichidae includes other conditions to which lion tamarins might also be susceptible (Montali and Bush 1999).

TRAUMA

Fractures and traumatic mutilations at the time of capture are unfortunately quite common and due mainly to the use of inadequate traps and cages and lion tamarin maltreatment during transport. Sometimes they result from fights between unfamiliar and stressed animals confined in small boxes and cages. Golden-headed lion tamarins (*L. chrysomelas*) illegally imported to Belgium (see Konstant 1986; Mallinson 1984) suffered seriously from the lack of proper housing, with fights resulting in serious mutilations of their hands and feet. There are also a number of inhumane practices used in order to tame animals in the illegal animal trade.

Personnel who handle the animals may be ill prepared to do so. Injuries, sometimes not immediately evident, are caused by rough and incorrect handling during removal of animals from cages and when taking measurements, giving injections, or tattooing. Some of the injuries may lead to problems later. Occasionally, infants bang their heads on the edge of the entrance to the nest box as stressed adults carry them to safety; this may result in hemorrhaging of the base of infants' skulls.

Captive lion tamarins may also injure and even kill one another when housed in inappropriate social groups, when groups are too close to one another, and even when housed in natural family groups (Kleiman 1979). In captive colonies, it is important to pay special attention to the social relations of group members. A fight may occur, possibly associated with the assertion of dominance/subordinance relations, which can result in mutilation or death. This has been observed at the SNZP (Kleiman 1979, personal communication) and in the CPRJ/FEEMA colonies of *L. rosalia* and *L. chrysomelas,* but only rarely in those of *L. chrysopygus* (A. Pissinatti, unpublished data).

DENTAL DISEASE

We have investigated oral pathological entities in lion tamarins through the examination of 60 *Leontopithecus* skulls, all from the museum of the CPRJ/FEEMA. Overall, dental disease is relatively common in lion tamarins and appears to be enhanced in captivity. This can lead to secondary infections, thus complicating their maintenance in captivity (Burity et al. 1997c, 1997d). The three species maintained in captivity show significant differences in the frequency of dental problems such as caries, crazing (minute cracks), diseases of the pulp, and occlusion. Canine and incisor teeth are more commonly subject to problems than the cheek teeth. *Leontopithecus chrysopygus* is the most susceptible to dental problems. A tendency for malocclusion in lion tamarins has been associated with tooth loss from caries (McDonald 1974) or with the loss of the vertical dimension and subsequent cross biting. In lion tamarins, the most common dental alterations result from periodontal diseases and tartar formation.

REPRODUCTIVE PATHOLOGY

Significant aspects of the reproductive biology of the genus *Leontopithecus* that have been the subject of research include mating and birth seasonality (Coimbra-Filho and Maia 1979a; see also Dietz et al. 1994a), physiology (French and Stribley 1985; French et al. 1989, 1992; Kleiman 1978b; Kleiman and Mack 1977; Kleiman et al. 1978, 1982; Stribley et al. 1987; Chapter 6 this volume), and caesarian operations (Pissinatti et al. 1984b, 1992). The demographic patterns (fecundity, infant survival, etc.) of the captive colonies worldwide have been monitored since the early 1970s (Ballou 1983–1996; Ballou and Sherr 1997; Kleiman 1977a, 1977b, 1977d; Kleiman et al. 1982; Chapter 4 this volume). French et al. (1996b) provided information on the reproductive biology of the genus *Leontopithecus* at CPRJ/FEEMA, which included a demographic analysis of records over 19 years.

Normal births are usually nocturnal. For this reason, it is important to detect the onset of labor in order to follow the process of parturition and intervene if problems become apparent. Surgical intervention, when necessary, may be problematic (Rothe 1974). Practically nothing is known concerning the occurrence of dystocias in *Leontopithecus* species in the wild. In captivity they can be attributed to a number of causes (e.g., traumatic, genetic, and nutritional due to inadequate management).

Birth problems have been found to be more common in *L. chrysomelas* than in the other species. A comparative study of sexual dimorphism in the pelvic bones indicated marked dimorphism in *L. chrysopygus,* less marked dimorphism in *L. rosalia,* and minimal dimorphism in *L. chrysomelas,* perhaps explaining the higher incidence of problems in this last species (Pissinatti et al. 1992). Individual fetuses that are disproportionately large in relation to the pelvis (i.e., above 10 percent of the mother's weight; see Eisenberg 1978) can cause dystocias. Despite this, many females have, with difficulty and accompanying hemorrhaging, given birth to litters with total weights equivalent to 30 percent the mothers' weights. The infants have never survived in these circumstances (Pissinatti et al. 1984b). Only a few cases of abortions have been observed in lion tamarins in CPRJ/FEEMA.

The prostate is an accessory gland of the reproductive system found in most species of mammals. The prostate glands of *L. rosalia, L. chrysomelas,* and *L. chrysopygus* of various ages have been studied microscopically at CPRJ/FEEMA (Ferreira et al. 1995; A. Pissinatti, unpublished data). Histopathology revealed a high frequency of benign prostatic hyperplasia, as well as other lesions such as fibromuscular hyperplasia, acute and chronic prostatites, and corpora amylacia.

CONGENITAL DISORDERS

Cases of congenital malformation have been registered for a number of New World primates (Colver 1938, Schultz 1960, 1972). Congenital disorders and reduced fecundity are the principal threats to the captive populations of the three species (Feldman and Christiansen 1984; Orzack 1985; Wilcox and Murphy 1985), most especially *L. chrysopygus,* which has very few founders (Chapter 4 this volume). Careful husbandry and management in captivity have maintained high levels of heterozygosity in *L. rosalia* and *L. chrysomelas,* but the genetic management of *L. chrysopygus* remains problematic (Mansour and Ballou 1994).

A low percentage of congenital anomalies, including facial and cephalic malformations, cleft palate, and hydrocephaly, have been noted in neonatal golden lion tamarins from the Species Survival Plan (SSP©) mortality records (Ballou 1983–1996). A relatively high frequency of retrosternal diaphragmatic defects leading to herniation occurred in the early ex situ propagation of the golden lion tamarin, possibly associated with the use of overrepresented founders occultly affected by the condition (Bush et al. 1980, 1996; Montali 1993b; Montali et al. 1980). Subsequently, diaphragmatic defects were reported anecdotally in golden-headed lion tamarins and also observed in a black lion tamarin originating from the wild (F. Simon 1988, personal observations).

An inverted radiographic contrast peritonealogram was developed to score variations in diaphragm contour in golden lion tamarins that were not apparent by conventional methods (Phillips et al. 1996). Individual captive-born golden lion tamarins targeted for reintroduction have been screened, and those with a score above 3 (range 0–5), indicative of an abnormal contour or protrusions, have been excluded from the program. Using this technique, there is now also evidence that the condition occurs in wild golden lion tamarins (Bush et al. 1996). Methods for the surgical correction of this defect are described by Randolph et al. (1981).

A true genetic component has never been established for these diaphragmatic abnormalities, nor have there been shown to be any apparent clinical effects. The prevalance of overt defects decreased markedly after breeding management changed. However, a familial basis for such defects is still possible.

Cecal agenesis was studied in a captive-born golden lion tamarin that died with chronic, severe, multifocal, ulcerative enteritis (Pissinatti et al. 1984a). This animal showed coprophagy, possibly resulting in the ingestion of toxic fecal products as well as potentially pathogenic microorganisms.

Pinder and Pissinatti (1991) reported on the abnormal development of the teeth, claws, and feet in a female and her offspring in a group of eight golden lion tamarins found in a small isolated wild population. These malformations were congenital and possibly the result of inbreeding because the total population of the now destroyed 160 ha forest was estimated at only 25 animals in four groups.

NUTRITION AND METABOLIC PROBLEMS

Increasing attention has been paid to the nutritional requirements, energetics, and metabolism of callitrichids (Allen and Montali 1995; Coimbra-Filho and Maia 1977; Coimbra-Filho and Rocha 1973; Coimbra-Filho et al. 1981, 1984a; DuMond 1972; King 1975; Morris 1976; Ratcliffe 1966; Thompson et al. 1994), which has resulted in significant improvements in the survival rates and well-being of captive colonies (e.g., an early finding was that captive animals must have vitamin D_3 added to their diets or access to sunlight) (Cicmanec 1978). However, much research is still required, especially concerning possible associations between diet (deficiencies) and a number of diseases and metabolic problems (e.g., see Coimbra-Filho and Rocha 1978). Greater attention should be paid to lion tamarin diets in the wild, most especially regarding animal prey (see Coimbra-Filho 1981; Chapter 7 this volume), in order to understand better their metabolic and energetic requirements.

Cholelithiasis (formation of gallstones) is an uncommon pathological finding in nonhuman primates, despite its high prevalence in humans (Anver et al. 1972; Glenn and McSherry 1970). Spontaneous cases have been reported in the genus *Aotus* (see Anver et al. 1972), in *Callithrix jacchus* (see Tucker 1984), and experimentally in the genera *Saimiri* (see Portman et al. 1980) and *Pongo* (see Ruch 1959). Our observations have revealed gallstones in association with a multiseptate gall bladder in all the three species of lion tamarins maintained in captivity. Biochemical analyses of these gallstones have revealed high levels of cystine, as well as high concentrations of calcium oxalate (whewellite). To prevent and treat gallstones, nutritional, morphological, and genetic studies (of the possibility of congenital tendencies) are needed (Pissinatti et al. 1993).

Diabetes mellitus has been studied in two lion tamarins, a male *L. chrysomelas* (CPRJ 021) and a male *L. chrysopygus* (CPRJ 029), both wild born and maintained in CPRJ (Cruz et al. 1993). They died suddenly without any clinical sign of the disease. Microscopically, both cases showed severe vacuolar diffuse degeneration of the Islets of Langerhans, affecting the b cells. The male CPRJ 029 also had aortic atherosclerosis and a nephropathy possibly associated with the diabetes. High glucose levels (above 140 mg/dl) have been recorded in five *L. chrysomelas* and two *C. kuhlii* at CPRJ/FEEMA. Metabolic bone diseases have not been detected in *Leontopithecus* species (Cruz et al. 1997).

STRESS

The following conditions attributed to physiological stress have been recorded and are due to inadequate management: (1) mutilation and death, when individuals from different groups are put together during shipment, (2) alopecia, consisting of abnormal hair loss and mutual hair pulling, which was observed in a privately owned male *L. rosalia* in a very small cage before the animal was confiscated by the authorities, and (3) self-mutilation (aggressive, nervous self-biting). The self-mutilation by an *L. chrysomelas* housed with two other individuals in a small cage led to a severe laceration and eventually death, even when the three animals were transferred to a larger cage.

VIRAL DISEASES

Lymphocytic choriomeningitis virus (LCMV), the etiologic agent of callitrichid hepatitis (CH), is characterized by an acute onset of lethargy, anorexia, elevated liver enzymes, and occasionally jaundice and grand mal seizures. The case fatality rate is high and has a direct association with the feeding of mice that are latently infected with LCMV. Historically, supplementing callitrichids with neonatal mice (pinkies) resulted in sporadic outbreaks of this disease throughout North American zoos in the 1980s and early 1990s, with a high incidence in golden lion tamarins (Montali et al. 1993). In addition, cases of CH may also occur from the callitrichids catching wild mice in their exhibits and eating and sharing them. The clinical signs and characteristic findings of a viral type of hepatitis with acidophilic bodies and immunohistochemical evidence of LCMV antigens are diagnostic for this disease (Montali et al. 1995a).

No cases or evidence of previous exposure to LCMV have been reported from research colonies of callitrichids or identified in a comprehensive serum survey of wild golden lion tamarins in their natural habitat in Brazil (Scanga et al. 1993). Seroconversion to LCMV, a zoonotic disease, has occurred in several callitrichid caretakers, but without evidence of clinical disease. Control of CH can be achieved by avoiding the feeding of mice to callitrichids and by vigilant rodent extermination. Complete eradication of the disease, however, requires zero-level contact of the callitrichids with wild mice (Montali et al. 1995b). Other viral diseases are discussed in Montali and Bush (1999) and Cicmanec (1978).

BACTERIAL DISEASES

Streptococcus zooepidemicus septicemia outbreaks in callitrichids may occur from exposure to contaminated uncooked horsemeat fed to carnivorous species kept in mixed exhibits or from cross contamination during food preparation. This has occurred in golden lion tamarins, as well as red-bellied tamarins (*Saguinus labiatus*) and Goeldi's monkeys (*Callimico goeldii*) (Schiller et al. 1989). The animals developed cervical suppurative lymphadenitis, splenitis, and enteritis, which usually terminated in fatal sepsis; just one animal treated with fluids and antibiotics survived. Raw meat products used in mixed exhibits with susceptible species should be cooked or not fed; sanitary precautions should be taken to prevent cross contamination by food pans or other fomites during food preparation.

Yersinia pseudotuberculosis and *Y. enterocolitica* harbored by rodents and birds may cause sporadic or high mortalities in callitrichids. Animals are often found dead and show severe suppurative enteric and hepatic lesions containing massive colonies of organisms, a feature that is nearly diagnostic. The virulence of these *Yersinia* species varies and has been found to be related to certain plasmids in callitrichids (Brack and Hosefelder 1992). Rodent and avian pest management is important in the control of yersiniosis; some zoos, particularly in Europe, have resorted to autologous vaccines with some anecdotally reported success.

Pasteurella species cause diseases in callitrichids, which include pneumonia, hepatitis, tooth infections, and septicemia. Chronic tooth root abscesses, mainly of the canine teeth, have occurred in black-tailed marmosets (*Callithrix melanura*) and golden lion tamarins presenting signs of lethargy and anorexia, and in advanced cases they have had obvious facial fistulas from infected tooth roots.

An important disease observed in lion tamarins at CPRJ/FEEMA is actinomycosis, a systemic infection characterized by suppurative granulomas of the cervical, thoracic, or abdominal region, which contain gram-positive filamentous bacilli (Al-Doory 1972). Signs observed in *L. chrysopygus* between 6 and 16 years old were weakness, fever, and apathy, with facial or mandibular fistulous abscesses, from which *Actinomyces odontolyticus,* a-hemolytic *Streptococcus* species, *Staphylococcus* species, and *Escherichia coli* were isolated (Gonçalves et al. 1997).

Other bacterial infections at CPRJ/FEEMA include hemorrhagic enteritis caused by *E. coli*, pneumonia caused by *Klebsiella pneumoniae, Staphylococcus aureus,* and b-hemolytic *Streptococcus,* all causing severe pathological changes.

MYCOTIC INFECTIONS

Candida albicans infection has rarely caused problems in the CPRJ/FEEMA colonies, but weak and undernourished individuals sometimes develop infections. A case in a black tufted-ear marmoset (*Callithrix penicillata*) resulted from parasitemia, aggravated by an acute septicemic bacterial infection (Chagas et al. 1986). A similar *Candida* species infection, with the same signs, was found in a golden-headed lion tamarin (*L. chrysomelas*), which had coprophageous habits. *Histoplasma capsulatum* infection has been found in a *Callithrix geoffroyi* from CPRJ/FEEMA, but never in the genus *Leontopithecus.*

PARASITIC INFECTIONS

Protozoans

Toxoplasmosis gondii occurs sporadically in marmosets and tamarins. Acute enteric and pulmonary forms of toxoplasmosis have occurred in golden lion tamarin groups (Griner 1983; Montali et al. 1995b) at North American and European zoos. Clinical signs may resemble an acute toxicosis, with affected animals showing respiratory distress from pulmonary edema or found dead with the typical pathological findings of acute toxoplasmosis. Sources of the disease are usually through contamination of food or quarters by cat feces and from lion tamarins catching and eating mice containing toxoplasma cysts. The best preventive measure is by rodent and feral cat control. Amebiasis and giardiasis have also been diagnosed. They can be controlled, although with some difficulty.

Trypanosoma cruzi, the cause of Chagas' disease in humans, may be carried by natural callitrichid hosts, although public health risks only reside where triatomid bugs exist in tropical areas (Potkay 1992). Neotropical primates are commonly found to be infected (subclinically), but the monitoring of these natural trypanosomal infections has been inadequate. *Trypanosoma cruzi*—positive golden lion tamarins have been identified from reserves in Brazil, although no reported clinical signs or disease implication for other animals or humans have been documented (Lisboa et al. 2000).

The genus *Sarcocystis* has been recorded most frequently for Old World primates (Dubin and Wilcox 1947; Karr and Wong 1975; Korte and Lond 1905; Mandour 1969). Nelson et al. (1966) were the first to record its occurrence in a New World primate, the black-mantled tamarin (*Saguinus nigricollis*). *Sarcocystis* species are found within the fibers of the skeletal musculature, without any evident pathological effects. A species of *Sarcocystis* was found in the heart of a captive-born *L. chrysopygus* at the CPRJ (Cruz and Pissinatti 1986), although it appeared to be an incidental finding because the cause of death of this animal was probably endotoxicosis resulting from ulcerative enteritis, besides a peritonitis and nematodiasis of the pancreatic duct by *Trichospirura leptostoma* (Pissinatti and Tortelly 1984).

Nematodes

At least two spirurid nematodes carried by cockroaches and coprophagous beetles have been reported as important groups of helminths pathogenic for lion tamarins. *Pterygodermatites nycticeba* (formally a *Rictularia* sp.: Montali and Bush 1981) emerged as a significant intestinal spirurid in captive golden lion tamarins (Montali 1993a; Montali et al. 1983) but has been found also in other tamarin species. *Trichospirura leptostoma* is a commonly found spirurid in the pancreatic duct of lion tamarins. This parasite is usually considered incidental but has been implicated in "wasting marmoset disease" in marmosets (Pfister et al. 1990). In lion tamarins, which appear not to be susceptible to "wasting," this parasite has been observed to cause alteration and inflammation of the duct epithelium in *L. chrysopygus* (Pissinatti et al. 1985).

Diagnosis of these forms of pathogenic spirurids in lion tamarins is made by finding the typical thick-shelled embryonated eggs by fecal flotation, although distinguishing the different nematodes by their eggs is difficult. Fenbendazole orally, at 40–60 mg/kg for five consecutive days, or ivermectin, 200–250 mcg/kg given subcutaneously, appears to be effective against these parasites. Infections with any of these spirurids can be associated with heavy cockroach infestation, and vigilant roach control is therefore essential.

Cerebrospinal nematodiasis ("larva migrans") is associated with the aberrant migration of larvae of *Baylisascaris* species, the natural hosts of which are raccoons, skunks, and badgers. The larvae of the three species are indistinguishable, but *B. procyonis* from raccoons is considered the most common source. Affected animals show a variety of CNS deficits, including ataxia, head tilt, blindness, and circling, weeks to months after direct ingestion or eating food contaminated with feces of the definitive host containing *Baylisascaris* eggs. Immunological diagnostic methods are still experimental and inconclusive, and unfortunately only postmortem analyses are definitive. This condition appears to be on the increase in zoo callitrichids, probably associated with the continuing urbanization of raccoons and their habit of leaving exposed scats in trees and on the ground. An increasing incidence in golden-headed lion tamarins in North American zoos has become evident (Pessier et al. 1997). Prevention includes control of raccoons and other wildlife and the judicious planning of facilities to avoid extraneous fecal contamination (Garlick et al. 1996; Huntress and Spraker 1985).

Filariasis has been diagnosed in *L. chrysomelas* contaminated during captivity, as well as in a wild black-handed tamarin (*Saguinus niger*) captured in the eastern Amazon. These were isolated cases at CPRJ/FEEMA.

Pentastomes

About 85 percent of the Pentastomidae identified in Brazil are found as mature parasites in reptiles. Nonhuman primates may serve as intermediate hosts (Rêgo 1980). Only a few adult forms have been found in mammals, amphibians, and birds, although larvae have been recorded for all vertebrates, including reptiles but especially freshwater fish. The incidence of these parasites in Amazonian tamarins indicates a widespread geographic distribution (Cosgrove et al. 1970). Pissinatti and Tortelly (1984) recorded *Porocephalus crotali* in three *L. rosalia* (adult wild-born female CPRJ 044, adult captive-born male CPRJ 092, and adult captive-born male CPRJ 0112). The larvae were present within cysts in the lungs, liver, spleen, diaphragm, and mesentery and were associated with chronic inflammation and giant cells (Pissinatti and Tortelly 1984).

Acanthocephalids

Prosthenorchis elegans (Acanthocephala), thorny-headed worms, cause enteritis, intestinal perforation, hemorrhages, peritonitis, and death in several primate genera and are considered a serious parasite of neotropical primates (Cicmanec 1978). However, the genus *Leontopithecus* seems less susceptible than *Callithrix* and *Saguinus* (A. Pissinatti, unpublished data).

Ectoparasites

A *Cuterebra* species was found in the subcutis of one *L. chrysopygus* in CPRJ/FEEMA. The infection was associated with a high temperature and irritability. The treatment was surgical removal, antisepsis, and antibiotics. The lion tamarins with the darkest fur and that are housed in cages nearest to the forest appear to be most affected by *Cuterebra* species (Pissinatti et al. 1981). Another species of *Cuterebra* causing dermatobiosis was recorded in *L. chrysomelas* (as well as *Cebus xanthosternos* and *Brachyteles arachnoides*) also in cages close to the forest. Reservoirs for this parasite include wild rodents (*Oryzomys* spp., *Proechimys dimidiatus,* and *Nectomys squamipes*), all of which were collected in the forest near the primate cages and parasitized by *Cuterebra* species.

Lion tamarins appear to be relatively free of ectoparasites in the wild; ticks are the most commonly found. Goff et al. (1986, 1987) described *Microtrombicula brennani* and *Speleocola tamarina* (Trombiculidae) from golden lion tamarins in the Poço das Antas Biological Reserve. Wilson et al. (1989) reviewed the ectoparasitic ticks, chiggers, and mites of wild *L. rosalia*, which include immature forms of four species of *Amblyomma* (Ixodidae), three trombiculids (the two species mentioned and *Euschoengastia* spp.), and *Rhyncoptes anastosi* (Rhyncoptidae). Wilson et al. (1989) failed to find gamasid mites and fleas in 90 *L. rosalia* examined, and they concluded that ectoparasites are not a significant threat to the health of golden lion tamarins, which may be attributable to their frequent social and solitary grooming behavior.

TUMORS

Compared with Old World primates, callitrichids seem to have a higher susceptibility to endocrine tumors, particularly pheochromocytomas in golden lion tamarins but also benign thyroid and pituitary neoplasms (Dias et al. 1996). Biliary adenocarcinomas and lymphosarcomas in golden lion tamarins and an adenoma of the adrenal gland in one *L. chrysopygus* have also been diagnosed in specimens from CPRJ/FEEMA.

TOXIC AND MISCELLANEOUS DISORDERS

Exposure of callitrichids to such toxins as organophosphates is sporadic and uncommon. One significant event occurred at the SNZP with inadvertent exposure to brodifcoum, a third-generation anticoagulant rodenticide. Four golden lion tamarins ingested the substance and died of coagulopathy (R. J. Montali, unpublished data). *Leontopithecus chrysopygus* and *L. rosalia* died at the Tijuca Biological Bank as a result of insecticide spraying nearby. Only lion tamarins housed at the edge of the facility were affected, presumably because they ate contaminated insects (A. Pissinatti, unpublished data).

A reintroduced captive-born golden lion tamarin died, presumably from a toxicosis with gastric dilatation, in the Poço das Antas Biological Reserve after eating a large number of unidentified fruits. Another reintroduced golden lion tamarin in the reserve died from a snakebite to the hand (from a *Bothrops* sp.), and a third died after likely being bitten by, but then eating, a highly venomous coral snake (*Micrurus coralinus*) (A. Pissinatti, unpublished data).

LION TAMARINS AT THE RIO DE JANEIRO PRIMATE CENTER AND SÃO PAULO ZOO

Alcides Pissinatti and Faiçal Simon, respectively, can provide complete listings of the mortality and causes of death (when known) from the collections of *L. rosalia, L. chrysomelas,* and *L. chrysopygus* over 19 years (1977–1997) at the CPRJ/FEEMA and over 10 years (1987–1996) at the São Paulo Zoo. Notable is the high occurrence of pneumoenteritis in *L. rosalia* at CPRJ/FEEMA and enteritis in the golden lion tamarins held in the São Paulo Zoo. The prevalent fatal pathological conditions in *L. chrysomelas* at CPRJ/FEEMA are enteritis and dystocia, whereas pneumonia and enteritis have been the more common causes of death in the São Paulo Zoo. The causes of death of *L. chrysopygus* are quite varied at CPRJ/FEEMA, whereas cases of pneumonia and renal lesions predominate in the São Paulo colony. *Leontopithecus caissara* is not held in captivity, but the few animals examined have all died as a result of asphyxiation through handling at the time of capture.

CONCLUSIONS

Callitrichids are subject to many infectious and noninfectious conditions that may undermine breeding programs and the husbandry of captive populations. One unifying theme for the infectious and parasitic diseases is the association of numerous viral, bacterial, and parasitic diseases with cockroaches, mice, and other vermin, which callitrichids seek out in their exhibits and regularly eat. Zoo veterinary and curatorial staff need, therefore, to be familiar with how these diseases are spread and to participate closely in or have oversight over pest control programs, which should be established at all zoos.

Many of the noninfectious conditions may also be products of ex situ conditions and may be rectified by proper management of these primates. The SSP© and Taxon Advisory Group veterinary advisors for the Callitrichidae in Europe, North America, and Australia are a good source of biomedical knowledge and can be readily consulted for more detailed information about any of these species. Special quarantine procedures for use prior to reintroduction have been developed for lion tamarins by the Golden Lion Tamarin Conservation Program of the SNZP, Washington, D.C. (Montali et al. 1995b) and are available in the form of a management protocol through the international golden lion tamarin studbook (Rettberg Beck 1990).

A critical component of conservation efforts for endangered primates is the establishment of captive breeding facilities within countries of origin. The CPRJ has concentrated on the maintenance and breeding of many Atlantic Forest primates for the last quarter century.

ACKNOWLEDGMENTS

Our sincere thanks to J. D. L. Fedullo, M. A. B. V. Guimarães, and S. H. R. Correa, of the Fundação Parque Zoológico (Zoological Park Foundation), São Paulo, for assistance summarizing the necropsy data from CPRJ/FEEMA and the São Paulo Zoo. We are grateful to Maria Inês Castro, Benjamin Beck, and Abigail Sherr of the SNZP and Andrew Baker of the Zoological Society of Philadelphia for help in writing this chapter; to Michael Seres, Yerkes Primate Center, for his help in obtaining bibliographic references; and to Anthony Rylands and Devra Kleiman for their revision of the text. The Grupo Coffin (Refrigerantes Niterói S/A), Souza Cruz S/A, and Durrell Wildlife Conservation Trust, Jersey, have provided valuable financial support to CPRJ/FEEMA for many years. We also thank Adelmar F. Coimbra-Filho for his pertinent comments in reviewing this chapter and Sônia Maria Eduardo de França for her constant dedication and help. Finally we are grateful to the Instituto Brasileiro do Meio Ambiente e dos Recursos Naturais Renováveis (IBAMA—Brazilian Institute of the Environment and Renewable Natural Resources) and Fundação de Amparo a Pesquisa do Rio de Janeiro (FAPERJ—Research Support Foundation of Rio de Janeiro) (Proc. E-26/171573/00).

Part Three

CONSERVATION AND MANAGEMENT OF LION TAMARINS IN THE WILD

MARIA CECÍLIA M. KIERULFF,
PAULA PROCÓPIO DE OLIVEIRA,
BENJAMIN B. BECK, AND ANDRÉIA MARTINS

12
REINTRODUCTION AND TRANSLOCATION AS CONSERVATION TOOLS FOR GOLDEN LION TAMARINS

Reintroduction has been defined by the World Conservation Union (IUCN) as an "attempt to establish a species in an area which was once part of its historical range, but from which it has been extirpated or become extinct," and translocation as the "deliberate and mediated movement of wild individuals to an existing population of conspecifics" (IUCN 1998). Reintroduction is very broadly defined as the return of animals to the wild, but the source of animals, whether they were born in captivity or in the wild, should be considered because the source determines what strategy is used for reintroduction (Chivers 1991). Many authors have defined reintroduction as the release of captive-born animals back into their natural habitat and translocation as the movement of wild-born animals within the species historical range, with or without an existing population of conspecifics (Beck et al. 1991; Caldecott and Kavanagh 1988; Chivers 1991; Kleiman 1989; Konstant and Mittermeier 1982; Stanley-Price 1991; Strum and Southwick 1986).

In this chapter the origin of the golden lion tamarins released in the forest is distinguished by using the term *reintroduction* for captive-born animals or wild-born animals that spent part of their lives in captivity and were released into the wild. The term *translocation* is used for those wild-born golden lion tamarins that were captured in forests threatened with deforestation and immediately transferred to protected forested habitat uninhabited by golden lion tamarins. By the IUCN (1998) definition, both activities should be considered reintroductions since the golden lion tamarins were released into habitat from which the species had been extirpated.

Translocations and reintroductions have been used for a variety of different purposes, including conservation, commerce, recreation, education, research, animal welfare, and, indirectly, the protection of habitat (Armstrong and McLean 1995; Beck et al. 1991; Caldecott and Kavanagh 1988; Dietz et al.1994b; Griffith et al. 1989; Kleiman et al. 1991; Nielsen 1988). Specific conservation objectives for the golden lion tamarin reintroduction and translocation programs are summarized in Table 12.1. A translocation or a reintroduction is a success if it results in a self-sustaining population (Griffith et al. 1989), measurable in terms of its reproductive success and survival (Saltz and Rubenstein 1995). Here we report on the use of these management techniques for the golden lion tamarin (*Leontopithecus rosalia*) and their success based on this criterion. Selected reintroductions and translocations are also currently being carried out on an experimental basis as part of a metapopulation management program for black lion tamarins (*L. chrysopygus*) (see Chapter 14 this volume).

The golden lion tamarin was almost extinguished in the wild because of habitat destruction and intense hunting pressure. The remaining populations were small and fragmented. In 1969, Coimbra-Filho estimated that a total of 900 km^2 of golden lion tamarin habitat remained, with approximately 600 individuals surviving in small patches of forests (Coimbra-Filho 1969). By 1975, estimates suggested that only 100 to 200 individuals might survive (Coimbra-Filho and Mittermeier 1977; Magnanini 1978). Based on a number of criteria, such as available appropriate captive stock, sufficient information about the species' biology in the wild, available protected and secure habitat in the species' original range, and the need to increase the numbers and genetic diversity of the wild population, Kleiman recommended that the golden lion tamarin would be an appropriate spe-

Table 12.1
Objectives of the Reintroduction and the Translocation Programs for Golden Lion Tamarins

Reintroduction	Translocation
1. Increase the size of the wild population	1. Rescue and translocate threatened groups
2. Increase the genetic diversity of the wild population	2. Maintain genetic diversity of the wild population
3. Expand geographic distribution of the wild population	3. Create a new protected population
4. Protect additional tracts of Atlantic Forest	4. Protect additional tracts of Atlantic Forest
5. Contribute to the science of reintroduction	5. Contribute to the science of translocation
6. Enhance programs of public education	6. Track possible effects of small population size and isolation

cies for a reintroduction program as a conservation strategy (see Kleiman 1989, 1990b).

The emergence of a self-sustaining captive population of golden lion tamarins in the early to mid-1980s (Ballou and Sherr 1997; Chapter 4 this volume), the establishment of the Poço das Antas Biological Reserve for the species in 1975, and the beginning of a long-term study of the behavioral ecology of the wild golden lion tamarin population in the reserve made possible, in 1983, the reintroduction of captive-born golden lion tamarins into native Brazilian forest (Beck et al. 1991; Kleiman et al. 1986; Chapter 13 this volume). Since 1984, 146 captive-born and seven confiscated wild-born animals have been released in the Poço das Antas Biological Reserve and on privately owned ranches around it (Table 12.2). The reintroduction of these golden lion tamarins involved extensive postrelease provisioning ("soft release"), management, and veterinary support to maximize their survival and reproduction.

In addition to the reintroduction program, a translocation of golden lion tamarins into protected habitat devoid of golden lion tamarins was also carried out in response to the results of an extensive survey of wild populations conducted in 1991 and 1992 (Kierulff 1993a, 1993b). This survey covered the golden lion tamarins' known or suspected geographic range, and 60 individuals in 12 groups were found surviving in small isolated secondary forest fragments of between 0.2 and 2 km^2 (Kierulff 1993a; Kierulff and Procópio de Oliveira 1996). The translocation of these groups was begun in 1994. A forest of the Fazenda União (União Ranch), owned by the Rede Ferroviaria Federal S.A. (RFFSA—Brazilian Federal Railway Company), contained adequate habitat for these animals, and there were no resident groups in the area. The threatened golden lion tamarin groups were subsequently captured and released into the new site. They were monitored daily following their translocation, but they were not provisioned, as is the custom for the reintroduced captive-born golden lion tamarins around the Poço das Antas Biological Reserve.

The reintroduction and translocation programs for the golden lion tamarin have both conservation and research objectives, although the rationale and emphases are rather different (Table 12.1). The reintroduction in and around the Poço das Antas Biological Reserve was carefully planned to provide new genetic stock from the captive population, whereas the translocation was an emergency measure for the maintenance of genetic variability already severely depleted in the isolated and severely threatened wild populations. Here we report on the success of these programs based on the survival and reproduction of the golden lion tamarins involved. We also examine estimates of the costs involved in reintroducing or translocating individual golden lion tamarins to determine the cost-effectiveness of each management technique.

THE REINTRODUCTION AND TRANSLOCATION PROGRAMS

After 17 years, including surviving "founders" and their offspring, the reintroduced population in December 2000 numbered 359 animals in 50 groups (Beck and Martins 2001; Table 12.2). Nearly 40 percent of the current estimate of 1,000 golden lion tamarins in the Atlantic Forest are reintroduced captive-born individuals and their descendants (Figure 12.1). To date, six groups have been translocated. The total translocated population of golden lion tamarins and their offspring numbered 120 animals in 16 groups in December 2000.

Both the reintroduction and translocation programs have contributed indirectly to the protection of habitat for the golden lion tamarin (Figure 12.2). Reintroduced groups have been established on 21 private ranches adjacent to the Poço das Antas Biological Reserve, which, combined, total about 32 km^2 of forest. This represents about 17 percent of the total remaining forest that contains golden lion tamarin populations. The translocated groups were moved to a forest at the Fazenda União. The 24 km^2 forest there is one of the largest remaining fragments of Atlantic Forest in the state of Rio de Janeiro and was decreed a federal biolog-

Table 12.2
Golden Lion Tamarin Reintroduction Program, 1984–2000

Year	Number Reintroduced	Cumulative Number Reintroduced	Number Reintroduced Still Alive	Number Wild Born Still Alive	Total Alive
1984	14	14	9	1	10
1985	12	26	15	1	16
1986	0	26	5	3	8
1987	21	47	22	5	27
1988	20	67	37	12	49
1989	7	74	29	22	51
1990	7	81	24	34	58
1991	11	92	27	51	78
1992	34	126	42	62	104
1993	7	133	31	87	118
1994	3	136	24	96	120
1995	5	141	22	147	169
1996	6	147	24	176	200
1997	0	147	20	208	228
1998	0	147	14	265	279
1999	0	147	14	288	302
2000	6	153	18	341	359

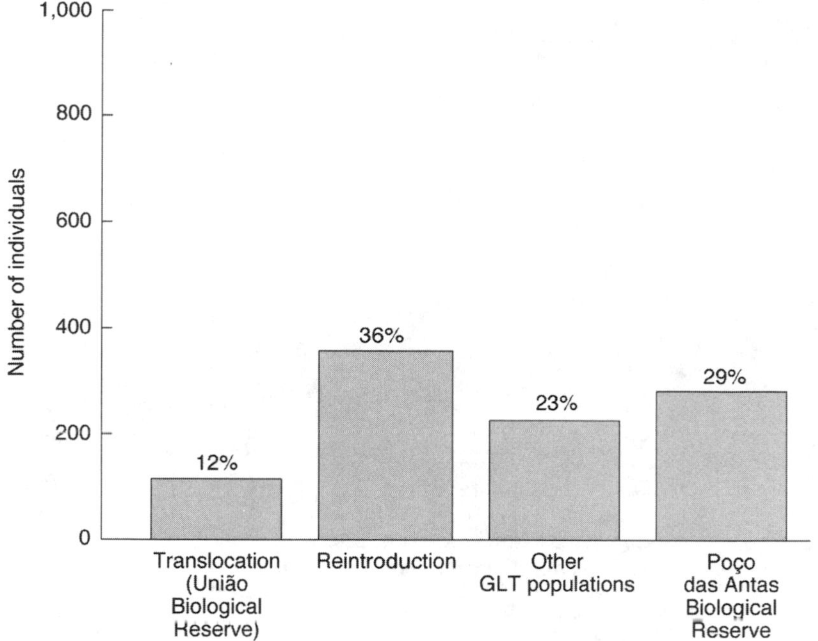

Figure 12.1. Total number of golden lion tamarins (GLTs) surviving in Brazilian forests (total = 1,000 individuals) and percentage of GLTs surviving from the translocation (União Biological Reserve) and reintroduction programs and living within the Poço das Antas Biological Reserve and in unprotected forests.

ical reserve (the União Biological Reserve) in April 1998 (32 km^2). It also represents about 17 percent of the total area currently containing golden lion tamarins.

The principal objective of the translocation was to rescue local wild populations threatened as a result of habitat destruction and uncontrolled hunting (Table 12.1). Since these threatened populations were located at a considerable distance from the main surviving population in the Poço das Antas Biological Reserve, it was assumed that they were genetically different because linear distance and genetic distance have been found to be positively correlated in golden lion tamarins (Grativol et al. 2001). Thus, a further goal of the translocation was to protect this additional genetic diversity.

To date, 6 of the 12 groups have already been translocated to the União Biological Reserve. One group found during the first census in 1991 and 1992 was never seen again, and another group was killed before translocation was possible (Kierulff and Oliveira 1996). These populations, along with those in captivity and the remaining major wild populations, are components of the metapopulation

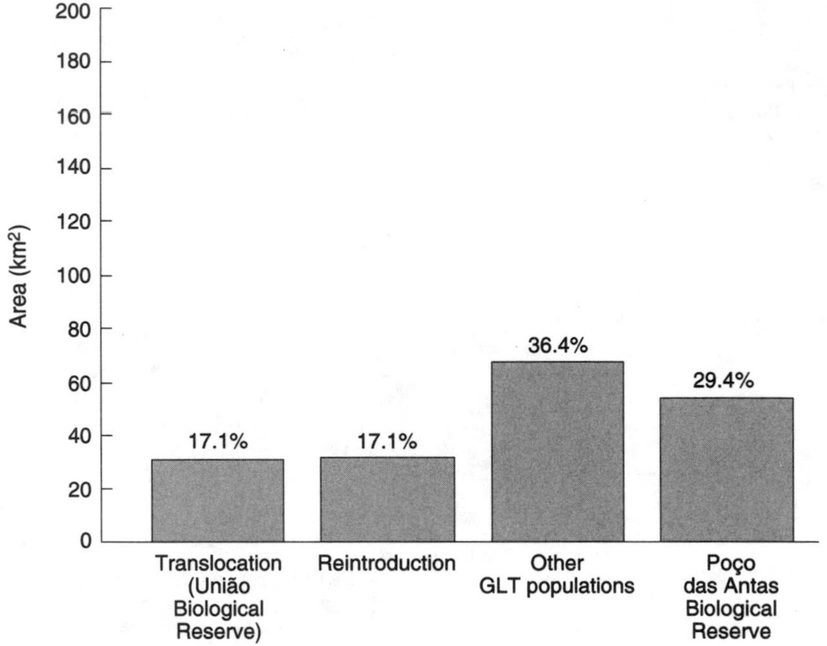

Figure 12.2. Total hectares of forest occupied by golden lion tamarins (GLTs) (total = 18,700 ha) and percentage of area occupied by GLTs surviving from translocation (União Biological Reserve) and reintroduction programs and living within the Poço das Antas Biological Reserve and in unprotected areas.

(Kierulff and Procópio de Oliveira 1996), and in the future both reintroduced and translocated animals can be used to restock areas where genetic variability and continued survival of isolated wild populations are threatened by low heterozygosity and inbreeding (Ballou 1990; Chapter 4 this volume).

EVALUATION OF SUCCESS

In December 2000, the reintroduced population totaled 359. Only 5 percent (18) of these were actually reintroduced, and 95 percent (341) were born in the wild to reintroduced golden lion tamarins or to their wild-born offspring (Table 12.2). The probability of losses of reintroduced golden lion tamarins is high in the first year (70 percent) but then levels off to match the age-specific mortality typical of wild populations. For golden lion tamarins born in the wild to reintroduced parents, the percentage loss is considerably lower in all age classes than for reintroduced captive-born golden lion tamarins. Twenty-eight percent of all losses of golden lion tamarins born in the wild occur in the first month of life, when the infants can not survive independently. Sixty-eight percent survive to 6 months, 45 percent to 1 year, and 45 percent to 2 years of age (*n* = 268) (Beck et al. 1991).

An assessment of the causes of losses in the reintroduced population showed that theft or shooting by humans accounted for 21 percent, natural predation 15 percent, starvation 13 percent, lethargy/diarrhea/anorexia/dehydration 10 percent, hypothermia/exposure 10 percent, wounded in social conflict 8 percent, bee sting 8 percent, eating toxic fruit 5 percent, snakebite 5 percent, hemorrhage following abortion 3 percent, and head injury 3 percent (Beck et al. 1991). The main single cause of loss, therefore, among the reintroduced population is theft and vandalism by humans, but problems with adaptation to the new environment, readily noticeable after the release of captive-born golden lion tamarins (e.g., inability to find food and problems with locomotion and orientation), likely caused the losses in a number of these categories (starvation and injuries, for example). Infants born to reintroduced parents seem to be less affected by these deficits and survive better than the reintroduced captive-born individuals. The best reintroduction strategy for this species is therefore intensive support of the reintroduced captive-born animals to maximize their chances of survival and reproduction, with the aim of seeing their wild-born offspring become truly independent (Beck et al. 1991; Chapter 13 this volume).

The translocated population numbered 120 (16 groups) in December 2000 (Table 12.3). At least 23 of these were translocated, and the other 67 were born at the release site (União Biological Reserve). Survival among the original translocated lion tamarins is high, and the average annual survival rate of monitored adults has been approximately 82 percent. Two of nine confirmed deaths involved animals injured in fights between groups. Six individuals were probably killed by predators, including the only golden lion tamarin that was translocated alone (probably taken by a predator on the first day after release) and another animal that was found injured and later died. The others disappeared and probably dispersed from their original groups.

A total of 107 offspring were born at the translocation release site (União Biological Reserve) from October 1994 to December 2000. Mortality among offspring declined after the first year, and most deaths occurred in the first weeks of life. Eighty-seven offspring survived more than 6 months, 83 offspring survived more than 1 year, and 79 offspring survived to 2 years. The disappearance of individuals after 1 year of age was in many cases through dispersal because many were found traveling alone or in newly formed groups.

It took 12 years for the number of golden lion tamarins in the reintroduced population to exceed the number actually reintroduced, whereas it took only 3 years for the number of golden lion tamarins in the translocated population to exceed the number actually translocated. As reintroduced golden lion tamarins

Table 12.3
Golden Lion Tamarin Translocation Project, 1994–2000

Year	Number Translocated	Cumulative Number Translocated	Number Translocated Still Alive	Number of Individuals Born in the Release Site Still Alive	Total Alive
1994	7	7	7	2	9
1995	17	24	21	5	26
1996	0	24	17	13	30
1997	19	43	33	21	54
1998	0	43	27	36	63
1999	0	43	24	43	67
2000	0	43	23[a]	67[a]	120[a, b]

[a]Data recorded in December 2000 from the 11 groups that were monitored.

[b]Ninety individuals were individually monitored. An additional 30 were in recently formed groups and had not yet been captured and identified. Thus, we cannot determine if these additional individuals were originally translocated or born in the release site.

begin to eat natural foods and to move through their territories, the feeding/observation visits by the field researchers are progressively reduced from daily to 3 days a week, then to once a week, and finally once a month. Provisioning is eventually discontinued except for bait used to trap animals to replace radio collars and dye marks. Twenty-five of the 50 reintroduced groups are now totally independent of provisioning and management, and 8 groups are provisioned only 2 or 3 days a week. The time necessary for a group to become fully independent varied, but all groups were independent 5 years after reintroduction.

Groups from the reintroduced population have begun to meet at territorial boundaries and to display encounter behavior typical of wild lion tamarins. New groups have been naturally established by individuals that dispersed from the original reintroduced groups. Some individuals first thought to have disappeared were later found in other groups in the reintroduced population or as residents of groups in the native wild population.

Social disruption following translocation was common in golden lion tamarin groups, with the replacement of breeding males in established groups by immigrant males. Emigration and immigration and large movements following translocation were more frequent when groups were released between established territories. The amount of unsaturated habitat and the low population density increased the opportunities for the establishment of new groups by individuals dispersing from the original translocated groups. The lower density of groups in the União Biological Reserve allowed them to use home ranges larger than those normally described for the species (Kleiman et al. 1986), but range size decreased with increasing population density (Kierulff 2000).

The translocated population was self-sustaining immediately after release in that it was totally independent of provisioning or additional management. The feeding ecology and ranging patterns of the translocated groups are similar to those of wild golden lion tamarins, and more than 120 plant species have been identified in the diet of two groups that have been studied intensively.

COSTS

Efforts to rehabilitate and reintroduce captive animals are extremely expensive to develop, although the costs will likely be reduced as the techniques are refined (Kleiman et al. 1991). Translocation and reintroduction share many procedures (Figure 12.3), but in general, reintroduction requires more intensive management over a longer period and is therefore more expensive (see Table 12.4). Kleiman et al. (1991a) estimated that after the first 6 years of the reintroduction project each surviving reintroduced golden lion tamarin had cost $22,000 (U.S.), which included zoo management, methodology development, associated studies of ecology and behavior of the wild and reintroduced golden lion tamarins, population monitoring, local community conservation education programs, and administrative infrastructure. After release, however, the procedures and expenses for reintroduction and translocation are quite similar (Table 12.4). Based on only the direct expenses in Brazil, a single reintroduced golden lion tamarin currently costs about $7,000 (U.S.) to monitor and manage after release, compared with $4,600 (U.S.) for a translocated animal, including the cost of capture and postrelease monitoring.

The most expensive stage of the translocation is the capture of a group in an isolated forest fragment. All traditional methods used to capture wild golden lion tamarins (e.g., traps baited with fruits) were tried without success. In some areas it took more than 6 months to capture a single group. The costs of traveling to an area and the per diem for staying there were very high. A new mist-net method was subsequently used, and the time necessary to capture a group decreased to 2 months as a result.

The principal differences in the costs of translocation and reintroduction are the amount and duration of postrelease monitoring and management. The reintroduced captive-born golden lion tamarins need food provisioning to survive. Translocated animals are self-sufficient after release, but for both programs, initial costs are likely to be higher than those ultimately required because the field teams must monitor the individual golden lion tamarins intensively after the initial release.

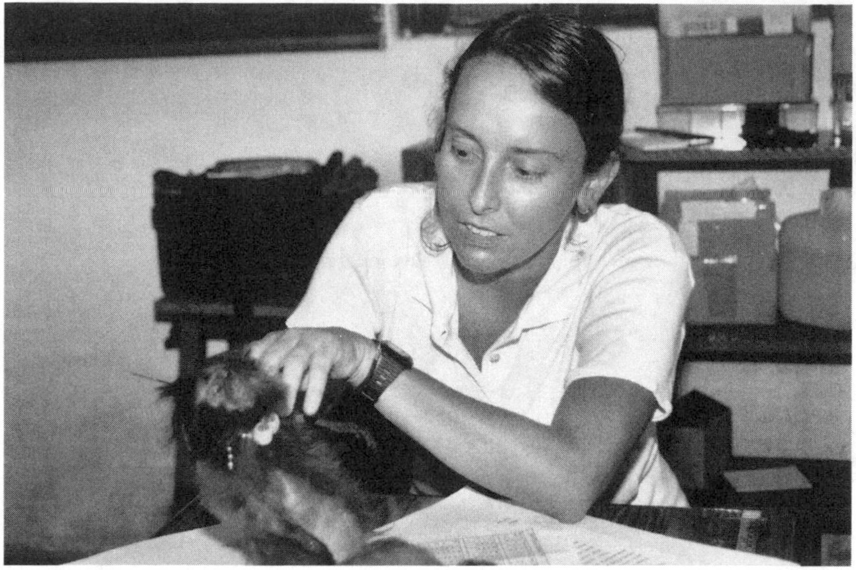

Figure 12.3. Maria Cecília Kierulff checking the tightness of the radio collar on a golden lion tamarin prior to release. (Photo by Devra Kleiman)

Beck and Martins (2001) reported that all of the appropriate golden lion tamarin habitat is at carrying capacity and within practical commuting distance for the seven-person local team involved in the reintroduction program. In the future, adults and subadults will likely be removed from existing reintroduced groups and translocated to unoccupied habitat. Thus, future reintroductions of captive-born animals will be less frequent. However, further reintroduction of captive-born golden lion tamarins may be necessary to increase genetic diversity and improve the demographic status of the entire wild population, to promote conservation education, and to maintain support for the conservation of the species in the zoo community.

CONCLUSIONS

Although each had the overall objective of establishing new populations of golden lion tamarins in otherwise unoccupied but appropriate habitats, reintroduction and translocation programs in the case of golden lion tamarins had different specific objectives. Reintroduction was used to increase the size and genetic diversity of the wild population, whereas translocation was used to rescue threatened wild subpopulations undoubtedly representing significant genetic diversity

Table 12.4
Procedures and Resources for the Reintroduction and Translocation of Golden Lion Tamarins

Reintroduction	Translocation
1. Training before release	1. Capture and transport
2. Transport	2. Postrelease monitoring, requiring
3. Postrelease monitoring, requiring	Observers
Food and nest box	Cars
Observers	Radiotelemetric equipment
Cars	Field and office miscellaneous
Radiotelemetric equipment	
Field and office miscellaneous	

for the species. Both programs have been successful as measured by the golden lion tamarins' survival and reproduction after release. Both techniques appear to have produced growing populations.

Griffith et al. (1989) reported that, in general, translocation projects are more successful than reintroduction projects. The survival (especially in the early stages of the program) and reproduction rates of translocated golden lion tamarins were higher in comparison with those of reintroduced captive-born golden lion tamarins. Population growth was achieved more quickly through translocation than through reintroduction.

Both techniques can be considered expensive, especially in the first year postrelease. Reintroduction is more expensive on a per animal basis because the golden lion tamarins need pre- and postrelease training and intensive management. Reintroduced groups have been provisioned for an average of 5 years before becoming independent. Reintroduction, however, benefits from its connection with the zoo community, resulting in substantial financial and technical contributions and allowing for the zoo community's direct participation in in situ conservation—a benefit for both golden lion tamarins and zoos.

Models of metapopulation management are now being analyzed in order to evaluate the future use of reintroduction and translocation for this species. Under special consideration are the demographic and genetic management of the species and the use of reintroduction and translocation as a means of responding to catastrophes, such as fires or disease epidemics. Despite the positive growth rate of both the reintroduced and translocated populations, golden lion tamarins are still highly endangered because the existing populations are too small for long-term viability. The possibilities are limited for further increasing the in situ populations of the golden lion tamarin through these management techniques. The key for future efforts will be considerably increasing the extent of forest in the region, especially by increasing connections between the remaining isolated populations. Until there are corridors linking the populations, reintroduction and translocation will remain the key tools for the metapopulation management of the species.

ACKNOWLEDGMENTS

Many people, programs, and institutions have provided considerable and invaluable support to the translocation and reintroduction programs. Thanks especially go to the translocation and reintroduction teams, stalwart and dedicated and too many to name individually. Notable for their support of translocation are David Chivers, Alcides Pissinatti, and Ricardo Medeiros of the RFFSA. The Golden Lion Tamarin Conservation Program (GLTCP), the Associação Mico-Leão-Dourado (AMLD—Golden Lion Tamarin Association), and the Centro de Primatologia do Rio de Janeiro (CPRJ—Rio de Janeiro Primate Center) were vital components for the success of these programs. The translocation program was supported by the Brazilian Higher Education Authority (CAPES), the National Geographic Society, the Philadelphia Zoo, the Durrell Wildlife Conservation Trust, the Lion Tamarins of Brazil Fund, the Wildlife Preservation Trust International (now the Wildlife Trust), the Dublin Zoo, the Lincoln Zoo Scott Neotropic Fund, the Margot Marsh Biodiversity Foundation, the Fundação O Boticário de Proteção à Natureza (Boticário Foundation for Nature Protection), the Brookfield Zoo, the Biodiversity Support Program/World Wildlife Fund/Disney Foundation, the Projeto de Conservação e Utilização Sustentável de Diversidade Biológica Brasileira (PROBIO—Project for the Conservation and Sustainable Use of Brazilian Biological Diversity)/Brazil, the Isaac Newton Trust, and Trinity College, Cambridge. The reintroduction program was supported by the Frankfurt Zoological Society Help for Threatened Wildlife, the Smithsonian Institution's Friends of the National Zoo (FONZ), the Smithsonian's International Environmental Studies Program and Nelson Fund, and Zoo Atlanta. The research, reintroductions, and translocations were conducted with due approval of the Conselho Nacional de Desenvolvimento Científico e Tecnológico (CNPq—Brazilian National Science Council), the Instituto Brasiliero do Meio Ambiente e dos Recursos Naturais Renováveis (IBAMA—Brazilian Institute of the Environment and Renewable Natural Resources) through the International Committee for the Conservation and Management (ICCM) of Lion Tamarins, and the Institutional Animal Care and Use Committee of the Smithsonian National Zoological Park, Washington, D.C.

BENJAMIN B. BECK, MARIA INÊS CASTRO,
TARA S. STOINSKI, AND JONATHAN D. BALLOU

13

THE EFFECTS OF PRERELEASE ENVIRONMENTS AND POSTRELEASE MANAGEMENT ON SURVIVORSHIP IN REINTRODUCED GOLDEN LION TAMARINS

Captive-reared animals are thought to require specialized training and preparation to survive and reproduce after reintroduction to the wild, especially mammals and birds because many of their survival-critical behaviors are learned (Box 1991, Kleiman 1989). The specifics would vary by species, but such behaviors might involve selecting appropriate habitat and microhabitat (including shelter and nesting sites); identifying, locating, and processing appropriate food (including hunting and killing prey); recognizing and avoiding predators; interacting appropriately and effectively with conspecifics (e.g., selecting mates, raising young, and defending territory); thermoregulation (including hibernation); pelage/plumage maintenance; locomotion and orientation (including migration); and establishing appropriate interspecific mutualisms. Training is said to be necessary for enough animals to survive and reproduce so as to make the reintroduction successful, to increase the cost-effectiveness of the reintroduction, and to decrease suffering and enhance the welfare of individual reintroduced animals. Indeed, most sets of reintroduction guidelines recommend structured training and preparation (for reviews, see IUCN 1998; Beck 1992). However, despite widespread acceptance that training is effective, and the implementation of this assumption, little actual research has investigated if and when it is effective (Kleiman 1996).

Reintroduction has been a component of the conservation program for golden lion tamarins (*Leontopithecus rosalia*) since 1983. Other components include management of a captive population, study of the behavioral ecology of the wild population in the Poço das Antas Biological Reserve in Brazil, translocation of imperiled wild groups to the União Biological Reserve and other large protected areas, community education, local employment, and reforestation (Kleiman et al. 1990a). More recently the conservation program has grown to include urban planning, agroforestry, and watershed management (see Chapter 3 this volume). Reintroduction links captive and wild lion tamarin populations in a metapopulation management plan for the species by increasing their numbers, genetic diversity, and distribution in the wild (Ballou et al. 1998).

From May 1984 to 31 December 2000, 146 captive-born golden lion tamarins were released into the wild in and around the Poço das Antas Biological Reserve (Beck and Martins 2000). The Golden Lion Tamarin Reintroduction Program has also released 7 wild-born golden lion tamarins that were captured for the pet trade, confiscated, or simply given to it. On 31 December 2000, 18 (17 captive born and 1 wild born) of these 153 were still surviving. By this time there had also been 499 documented live births within the reintroduced population, of which 341 survived. Though not by experimental design, we can now retrospectively examine the influence of different prerelease environments on survivorship. The 7 reintroduced wild-born lion tamarins are not included in these analyses.

PRELEASE ENVIRONMENTS AND POSTRELEASE MANAGEMENT

In this chapter we discuss four general types of prerelease environments for golden lion tamarins. The first is standard zoo or research colony cages (*cages*). These vary widely but in general use a combination of wire mesh, glass, and concrete walls and ceilings and mesh, concrete, or earthen floors to contain the animals within a space of 3 to 125 m^2. Cages may be completely indoors with no natural light, indoors with skylights, or outdoors with exposure to some natural sunlight, wind, and rain. The golden lion tamarins have rarely been exposed to temperatures below 20° or above 35°C and always have a retreat from rain and wind. Cages holding golden lion tamarins are usually separated, thus preventing social and sensory access to conspecifics.

Keepers typically provide food in pans placed at fixed locations in the cage less than 2 m from the floor. The golden lion tamarins are fed at about the same times every day. Food consists of finely cut fruits and vegetables, commercially prepared diets, and mealworms and crickets. The animals may also consume mice, cockroaches, and other small animals that enter their cage opportunistically, usually on the floor of the cage. Water is provided in a bowl or licker.

Cages usually have a three-dimensional climbing structure made of lumber and/or cut natural branches. In some cases this "furniture" has many elements and is quite dense and complex; in others it consists of as little as a single element running diagonally from top to bottom of the cage. Once erected, furniture is rarely rearranged. Some cages have living plants and small trees. Cages usually have few climbing substrates of less than 2 cm in diameter. A small wooden shelter box is provided 2 to 3 m above the floor for sleeping and retreat. There is a continuum of cleaning routines, from daily hosing and disinfection of the cage and shelter box to simple removal of feces and food remains and dry spot cleaning of stains. Other mammals and birds may occupy the same cage, but none is a serious threat to the golden lion tamarins. If properly managed, golden lion tamarins thrive under the *cages* condition, and the captive population became self-sustaining in this type of environment (Chapter 4 this volume).

A second prerelease condition (termed *training*) imposes structured training on the *cages* condition: Golden lion tamarins living in conventional cages are deliberately exposed to changes in their environments that are believed to increase the frequency and efficiency of survival-critical behaviors. For example, caretakers may embed food in wood hollows or wrap it in large leaves to induce the lion tamarins to search for and extract it (Figure 13.1). Fruit may be presented whole rather than cut, requiring the animals to peel or penetrate the rind to access the edible portion (Figure 13.2). Caretakers may scatter crickets in the substrate or place them in a feeder with small holes, forcing the lion tamarins to probe with

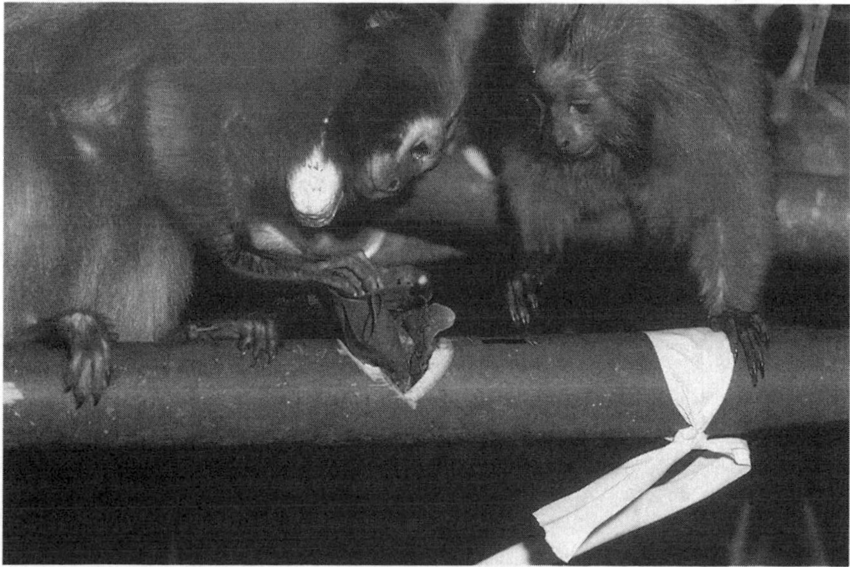

Figure 13.1. Captive golden lion tamarins extracting hidden and wrapped food from a specially designed bamboo "feeder." (Photo by James Dietz)

Figure 13.2. Golden lion tamarins peeling and eating provisioned bananas after reintroduction. (Photo by Devra Kleiman)

their hands to secure them or to wait vigilantly for a cricket to emerge. Animals living in *cages* environments in zoos having extensive enrichment programs may have food-finding challenges similar to those presented in the *training* condition.

Caretakers may rearrange cage furniture periodically to interrupt habitual travel routes and provide branches of less than 2 cm diameter to promote locomotor dexterity. Conspecifics may be allowed to occupy neighboring cages to allow expression of intergroup contact behaviors. Caged or taxidermically mounted predators may be presented to stimulate antipredator behavior. Much of this training resembles environmental enrichment (see Castro et al. 1998), although slightly more risk is tolerated when the golden lion tamarins are being prepared for reintroduction. For golden lion tamarins the *training* environment results in short-term increases in foraging effort and extraction of embedded food (Beck et al. 1986b, 1988).

The third condition, which we call *free range,* allows golden lion tamarins to live at liberty in wooded portions of zoos (see Bronikowski et al. 1989; Castro et al. 1998; Mager and Griede 1986; Stoinski et al. 1997). Free-ranging lion tamarins usually come from the *cages* condition (unless born while their parents are free ranging). Some aspects of the *training* condition (e.g., presentation of cricket feeders or whole fruit) may be used while they are free ranging. In the Northern Hemisphere they remain for 4 to 6 months in the *free-range* condition. Free-ranging golden lion tamarins are given a shelter box 3 to 5 m above the ground, which provides a secure resting and sleeping place and seems also to provide psychological security and a "ground zero" from which they can expand their range and to which they can be returned if they cross a perimeter set by the curator. In a free-ranging golden lion tamarin group most members that weigh more than 450 g wear a radio transmitter on a bead chain collar, allowing them to be tracked and trapped if they get lost or go too far from the box. The areas designated for free-ranging golden lion tamarins in different zoos range from 100 to 2,500 m^2.

Free-ranging golden lion tamarins are given cut up food, but usually in feeders that require manual probing. Feeders and water bowls are raised 3 to 5 m above the ground. They also eat wild insects and spiders, usually 3 to 10 m above the ground. One free-ranging golden lion tamarin took up the habit of catching and eating earthworms on the ground. Potentially toxic foods, such as mushrooms, are present. They drink standing water from tree cavities and lick water droplets from leaves. In this regime, they encounter a variety of natural vegetation, which in some zoos may include 35 m tall trees. There are many different substrates, with horizontal, diagonal, and vertical orientations; diameters less than 2 cm and greater than 15 cm; thorns and spines; and varying bark and leaf textures. The locomotor behavior of golden lion tamarins in the *cages* condition has been found to differ from that of lion tamarins in the *free-range* condition (Stafford et al. 1994).

Potential predators, such as hawks, crows, raccoons, and foxes, occur in *free-range* areas, and there are also frequent competitors for food, water, and the shelter box, such as grey squirrels and starlings. Free-ranging golden lion tamarins have been seen detecting and avoiding predators and fighting with competitors. There are generally more mosquitoes, flies, and stinging insects than in the *cages* or *training* conditions. In the *free-range* condition golden lion tamarins are exposed to heavy rain, high wind, and temperatures from 3° to 40°C, although they always have access to a shelter box.

Golden lion tamarins in the *cages, training,* and *free-range* conditions are given periodic veterinary examinations and are treated for injuries and illnesses. The likelihood of observing injury, illness, or fighting is highest in the *cages* condition and lowest in the *free-range,* but there appear to be no differences in mortality. In all three conditions, if serious fighting breaks out, the curator, caretaker, or veterinarian decides which of the adversaries to remove. None of these conditions allows a choice between potential mates or the opportunity to emigrate or immigrate. Golden lion tamarins maintained in all three conditions are quarantined before shipment to Brazil. The criteria for deciding if an individual is unsuitable for reintroduction are the same for all three prerelease conditions (Beck et al. 1991; Bush et al. 1993, 1996). Golden lion tamarins are kept in acclimation cages at their release sites in Brazil for up to 21 days before they are released.

The fourth condition, *wild born,* is actually the absence of any prerelease environmental condition. The *wild-born* condition includes golden lion tamarins born to reintroduced animals or to their offspring in the forests of the Poço das Antas Biological Reserve or privately owned ranches in the surrounding area. Although they "enter" the reintroduction program with no captive experience, they depend for their social experience on animals previously captive or only a few generations from captivity.

Comparisons among these conditions are confounded by our having used two different general types of postrelease management. The first, which we call *minimal,* was used with captive-born golden lion tamarins mostly in the early years of our reintroduction effort (1984 and 1985), but it is used today with some of the *wild-born* individuals living in self-reliant groups. In the *minimal* condition, food is provided only in the first week after reintroduction and when animals are being habituated to enter live traps for semiannual censuses and changing of radio collars. Monitoring is no more than biweekly. An animal that is ill, injured, hypothermic, or very hungry is rescued and treated and then, when possible, returned it to its group. Because there is little monitoring, there is little opportunity to observe such problems. For captive borns, shelter boxes are initially provided but are not moved if the captive borns leave the release area.

The second postrelease condition, *intensive,* is what is often referred to as a "soft release." Animals are monitored and provided food and water daily for several months after release, and then two or three times a week for up to 2 years (Figure 13.3). A shelter box is offered to each group and moved to the animals if they leave the release area. Lost animals are live-trapped and returned to their group (or to the shelter box if a whole group is lost). Ill, injured, hypothermic, or hungry animals are rescued, treated, and returned. Because monitoring is frequent, there is a high likelihood of detecting such problems.

A small and approximately equal number of captive-born golden lion tamarins in both postrelease conditions were released with mates that were born in the wild or had already lived in the wild for 2 or more years. Most captive borns, however, were released with other captive borns with no experience in the wild.

There are eight possible combinations of prerelease environmental conditions and postrelease management. Table 13.1 shows the sample sizes for each. The data include only the 140 captive-born and 364 *wild-born* golden lion tamarins that had

Figure 13.3 The golden lion tamarin reintroduction team at the start of the day. A team member holds tamarin "feeders," which encourage the reintroduced monkeys to search for hidden food. (Photo by Devra Kleiman)

been at risk for 730 days by 31 December 2000; that is, they had been reintroduced or born in the wild before 31 December 1998. We eliminated 21 *wild-born* golden lion tamarins for which the date of birth (see below) or date of loss was unknown and could not be confidently estimated within 30 days.

This was not a prospectively designed experiment. Because we adjusted pre- and postrelease techniques to maximize survival, cell sample values vary widely (in Table 13.1 two are zero). Survival is expressed as the percentage of the sample in each cell that is alive at 30, 185, 365, and 730 days after reintroduction (after birth for *wild born*). We considered individuals that disappear, or are stolen or rescued, to be lost on the date they were last seen. A few of those that disappeared were observed again after considerable periods. In these cases we erased the first date of loss.

The birth dates had to be estimated for some of the 364 *wild-born* golden lion tamarins because they were first recorded as older infants or juveniles and in a few cases even as adults. This was done by extrapolating from body weights (Dietz et al. 1994a) and supplementing with observations by experienced field observers and/or a knowledge of the reproductive history of the parents.

Table 13.1
Sample Sizes for Each of the Prerelease Environmental Conditions and Postrelease Management Conditions for Reintroduced Golden Lion Tamarins

Prerelease Environment	Minimal Postrelease Management	Intensive Postrelease Management
Cages	4	77
Training	20	0
Free range	0	39
Wild born	77	287

THE EFFECTS OF PRERELEASE ENVIRONMENTS AND POSTRELEASE MANAGEMENT

Cages versus Training versus Free Range

Table 13.2 shows the survival percentages for the three prerelease captive environments, with both postrelease management conditions combined.

There are no significant differences in survival between the three environments (chi-square test, df = 2; column 30 days, $\chi^2 = 0.67$, $p > 0.05$; column 185 days, $\chi^2 = 2.84$, $p > 0.05$; column 365 days, $\chi^2 = 3.37$, $p > 0.05$; column 730 days, $\chi^2 = 2.52$, $p > 0.05$). Captive-born golden lion tamarins reintroduced directly out of cages, with or without training, and those given the opportunity to range freely on zoo grounds before release are all equally likely to survive up to 2 years.

Survivorship is somewhat lower for the *training* condition than for the *cages* or *free-range* condition, but this is probably an artifact: most of the individuals in the prerelease *training* condition received *minimal* postrelease management, whereas most of the golden lion tamarins from *cages* and *free-range* conditions received *intensive* postrelease management. *Intensive* postrelease management is associated with higher survivorship in reintroduced captive-born animals (see below). The percentage surviving for all environmental conditions declines over the first 2 postrelease years.

Captive Born versus Wild Born

Table 13.3 pools the three prerelease environments (*cages, training,* and *free range*) and compares the combined survival data with those for *wild born*.

Golden lion tamarins born in the wild to reintroduced parents or their offspring are significantly more likely to survive at 30 days, 6 months, 1 year, and 2 years

Table 13.2
The Survival of Reintroduced Golden Lion Tamarins from Three Different Prerelease Captive Environments

Prerelease Environment	Surviving to 30 Days (Percent)	Surviving to 185 Days (Percent)	Surviving to 365 Days (Percent)	Surviving to 730 Days (Percent)
Cages	67/81 (83)	48/81 (59)	37/81 (46)	26/81 (32)
Training	15/20 (75)	15/20 (75)	5/20 (25)	3/20 (15)
Free range	31/39 (79)	28/39 (72)	19/39 (49)	13/39 (33)

Note: Rescued animals were counted as lost. Minimal and intensive postrelease management were combined.

after reintroduction than reintroduced golden lion tamarins born in captivity, with all prerelease captive environments combined (chi-square test, df = 1; column 30 days, $\chi^2 = 12.29$, $p < 0.001$; column 185 days, $\chi^2 = 25.29$, $p < 0.001$; column 365 days, $\chi^2 = 69.73$, $p < 0.001$; column 730 days, $\chi^2 = 86.12$, $p < 0.001$).

Of course, *wild-born* golden lion tamarins are all newborns when they enter reintroduction, whereas those from *cages, training,* or *free range* are older (the average age at reintroduction was 3.6, 2.9, and 3.8 years, respectively; the average age for all captive borns combined was 3.34 years). This confounds comparisons since previous analyses show that the amount of time surviving after release is negatively correlated with age (Beck et al. 1991). Differences in survival between *wild-born* and captive-born golden lion tamarins could be due to age differences rather than

Table 13.3
The Survival of Reintroduced Golden Lion Tamarins from Cages, Training, and Free-Range Environments Compared with That of Wild-Born Golden Lion Tamarins

Prerelease Environment	Surviving to 30 Days (Percent)	Surviving to 185 Days (Percent)	Surviving to 365 Days (Percent)	Surviving to 730 Days (Percent)
Cages + training + free range	113/140 (81)	91/140 (65)	61/140 (44)	42/140 (30)
Wild Born	334/364 (92)	310/364 (85)	296/364 (81)	272/364 (75)

Note: Rescued animals were counted as lost. Minimal and intensive postrelease management were combined.

differences in survival-critical behaviors. We can control for age by tracking the survival of *wild-born* and captive-born animals matched by age.

Figure 13.4 shows survival from the age of 2 years for a sample of 140 *wild-born* and 38 captive-born animals. For captive borns, this excludes individuals that were reintroduced when younger than 2 and did not survive to 2 years old, as well as animals that were older than 2 when reintroduced. *Wild-born* golden lion tamarins that did not survive to age 2 were also excluded. *Wild-born* animals showed significantly higher survival from age 2 onwards than their age-matched captive-born cohort ($p < 0.001$, Wilcoxon rank sum test for censored observations). This comparison is unavoidably confounded by time spent in the wild: at the beginning of the period of comparison, the time the 2-year-old *wild-born* golden lion tamarins had spent in the wild was 2 years, whereas the 2-year-old reintroduced captive-born golden lion tamarins had spent, on average, only 9.5 months.

Alternatively, we can compare matched populations according to time spent in the wild, regardless of age. *Wild-born* golden lion tamarins showed significantly higher survival than captive borns ($p < 0.001$, Wilcoxon sum rank test) after having spent 1 month, 3 months, 6 months, 1 year, 2 years, and 3 years in the wild. Comparisons of survival rates between *wild-born* and reintroduced captive-born golden lion tamarins that have spent 4 or more years in the wild showed no significant difference, although the number of animals available for the analysis was

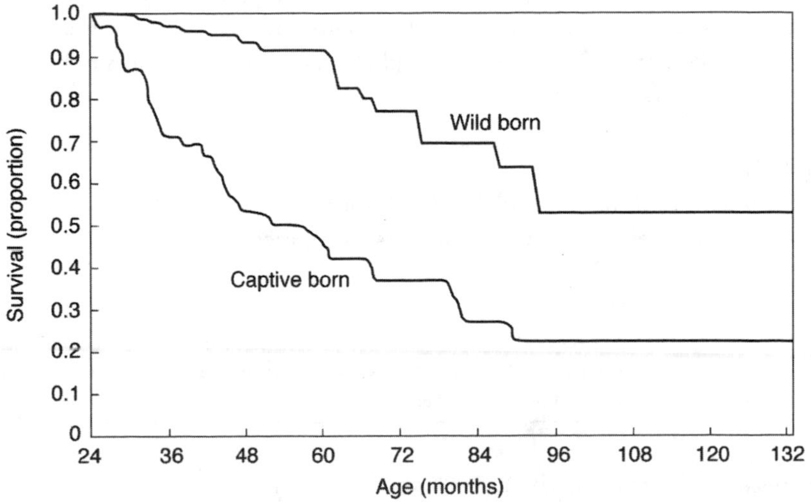

Figure 13.4. Survival in the wild of wild-born (n = 140) and captive-born (n = 38) golden lion tamarins after the age of 2 years.

small (20 or fewer) and comparisons lacked statistical power for definitive conclusions. Because *wild-born* golden lion tamarins survived longer than reintroduced captive-born golden lion tamarins (controlling for age or time spent in the wild), we believe the differences in survivorship reflect differences in some survival-critical behavior(s). This has been investigated by Stoinski (2000) and Stoinski and Beck (2001).

Minimal versus Intensive

Table 13.4 shows the survival percentages at 2 years for the *minimal* and *intensive* postrelease management conditions. The data for *cages, training,* and *free range* are again pooled.

With *intensive* postrelease management captive-born golden lion tamarins are significantly more likely to survive to 2 years ($\chi^2 = 4.22$, df $=1$, $p = 0.04$), whereas *wild-born* golden lion tamarins are significantly *less* likely to survive 2 years ($\chi^2 = 11.12$, $p < 0.001$).

The significantly greater postrelease survivorship for captive borns under the *intensive* condition results in more reproductive opportunities, including captive-born subadults being able to assist their parents in raising *wild-born* offspring. There is also a compelling animal welfare rationale for *intensive* postrelease management.

The lower survivorship of *wild-born* golden lion tamarins under *intensive* post-release management is opposite to what was expected and may actually mean that the survival of these animals is diminished by provisioning and other human activity. However, we think this effect derives from two confounding factors. First, the *wild-born* golden lion tamarins with *minimal* postrelease management lived within the Poço das Antas Biological Reserve and had been born into groups founded by both reintroduced captive-born animals and wild translocated animals in 1983 as part of a rescue effort (Pinder 1986a, 1986b). The exposure to wild golden lion tamarins and their offspring probably conferred a strong survival ad-

Table 13.4
The Survival of Reintroduced and Wild-Born Golden Lion Tamarins under Minimal and Intensive Postrelease Management for 2 Years

Prerelease Environment	Minimal Postrelease Management (Percent)	Intensive Postrelease Management (Percent)
Cages + training + free range	3/24 (13)	39/116 (34)
Wild born	68/77 (88)	204/287 (71)

Note: Rescued animals were counted as lost. Cages, training, and free-range environments were combined.

vantage (although we do not know the mechanism) and obviated the need for postrelease support. In contrast, most of the *wild-born* golden lion tamarins under *intensive* postrelease management had been born into groups founded by only captive-born animals, and thus they might have been less competent in survival-critical behaviors and benefited from postrelease support. Further, less frequent monitoring under the *minimal* condition undoubtedly meant that we were less likely to observe neonates within the first 30 days of life, which is a period of high mortality. Presently, we do not have data to test these alternative explanations.

CONCLUSIONS

The results of the reintroductions of a number of species suggest that captive-bred animals reared in traditional cages and enclosures with standard husbandry routines lack proficiency in many survival-critical behaviors after they are released in the wild. For example, golden lion tamarins reared in cages were hesitant to use natural vegetation and were unable to orient in a novel three-dimensional array (Beck and Castro 1994; Figure 13.5); black-footed ferrets (*Mustela nigripes*) raised in cages with exposure to live hamsters showed little interest in live prairie dogs (*Cynomys* spp.) (Vargas 1994; Vargas and Anderson, in press); caged-reared thick-billed parrots (*Rhynchopsitta pachyrhyncha*) and Hispaniolan parrots (*Amazona ventralis*) showed no tendency to feed on naturally occurring food (Wiley et al. 1992); and cage-reared ring-tailed lemurs (*Lemur catta*) did not initially exploit natural foods and had very small home ranges (Keith-Lucas et al. 1999). This incompetence has resulted in the use of pre- and postrelease skill training in many reintroductions. Beck et al. (1994) found that 35 percent of 145 reintroduction programs used some form of prerelease training while 12 percent used postrelease training.

To examine the efficacy of training captive animals for reintroduction, it is useful to separate programs designed (1) to establish specific skills, such as promoting antipredator behaviors by harassing animals with predators or predator mimics, and (2) to encourage natural behaviors and behavioral flexibility through providing animals with enriched environments that provide physical and/or social challenges, such as our free-ranging environment for golden lion tamarins (see Box 1991).

Intuitively, one would expect such training to enhance postrelease survival, and for this reason many have asserted that training should be used. But the data presented here (Table 13.2) for golden lion tamarins do not support this expectation. Indeed, we conclude that the two different training regimes that we have used confer no short- or long-term survival advantage. Most data available from other reintroduction programs are also consistent with the conclusion that specific skill training does not promote postreintroduction survival. Captive-reared thick-

Figure 13.5. A golden lion tamarin perched on a cage door after being released in Brazil during the first reintroduction in 1984. Reintroduced lion tamarins initially remained on the wood and wire of their forest enclosures and were reluctant to enter the canopy. (Photo by James Dietz)

billed parrots that were formally trained to eat pinecones for 6 months prior to release showed no tendency to feed on cones after reintroduction and had to be recaptured after 2 days (Snyder et al. 1994). Eagle owls (*Bubo b. bubo*) trained to capture live prey showed no greater rate of postrelease success than individuals that did not receive such training (Wayre 1975). Siberian polecats (*Mustela eversmanni*) trained to respond to a stuffed badger paired with an aversive stimulus avoided the badger in captivity but then succumbed to predators within 3 weeks of release (Miller et al. 1991). In all of these studies the behavior of the subjects was changed as a result of training, but the changes did not translate into improved survival upon release.

In some cases the training effect did not persist even in captivity. McLean et al. (1994, 1996) found that captive-reared rufous hare-wallabies (*Lagorchestes hirsutus*) that were trained to recognize a model spent significantly less time feeding and significantly more time watching an out-of-view model of the predator immediately after training. These changes were not observed, however, when the animals were retested 8 months later. In contrast to these negative findings, California condors (*Gymnogyps californianus*) that received an electric shock when perching on the cross-arms of a model power line pole in their prerelease cages tended not to perch on poles after release and thus avoided electrocution and collision with them after reintroduction (Wallace 1997). Cage-reared black-footed ferrets that were exposed to live hamsters and hidden food as juveniles exhibited more effective predatory behavior than ferrets without previous exposure to live prey (Vargas and Anderson 1999). They nonetheless showed lower postrelease survival than ferrets raised in large pens with exposure to live prairie dogs (Biggins et al. 1998).

Preliminary results suggest that prerelease exposure of Houbara bustards (*Chlamydotis undulata macqueenii*) to live foxes resulted in higher postrelease survival (Van Heezik and Maloney 1997). New Zealand takahes (*Porphyrio mantelli*) exposed to sham attacks by stoat (*Mustela erminea*) models spent more time away from the models and showed more vigilance and discomfort in their presence (Holzer et al. 1996). These birds were not reintroduced, and the persistence of the changes over time is not specified. However, reintroduced captive-born takahes that were not specifically trained with predator models showed the same survivorship as wild takahes (Maxwell and Jamieson 1997). New Zealand robins (*Petroica australis*) learned to respond fearfully to predator models after training, but all robins, trained and untrained, disappeared within 6 months of reintroduction (McLean et al. 1999).

The results of training using enriched or free-ranging environments to promote competency and survival are also mixed. As shown here, free-ranging golden lion tamarins showed no advantage in postrelease survival. One reintroduced black lion tamarin (*Leontopithecus chrysopygus*) was believed to have developed effective foraging, motor, and orientation skills after 2 years of free-ranging experience, but data were not presented. The animal was killed by a predator within 4 months of reintroduction (Valladares-Padua et al. 2001). In contrast, black-footed ferrets raised in outdoor pens with live prairie dogs and prairie dog burrows dispersed shorter distances after reintroduction, spent less time above ground, and were more successful hunters than ferrets raised in traditional cages (Vargas et al. 1996). These behavioral differences corresponded with a significant increase in the long- and short-term survival of these animals over their cage-reared counterparts (Biggins et al. 1998; Miller et al. 1995; Reading et al. 1996; Vargas et al. 1996).

Masked bobwhite quail (*Colinus virginianus*) with prerelease experience in an enriched covey-box environment with exposure to dogs, humans, and a trained Harris hawk showed higher postrelease survivorship than untrained controls (Ellis et al. 1977). Providing young wolves (*Canis lupus*) with the opportunity to hunt freely in large fenced areas increased hunting behavior and hunting success after release (Badridze 1994), although the effect on postrelease survival was not examined. Boer (1994) reported that the flight distance of captive-reared European lynx (*Lynx lynx*) from humans increased dramatically after 6 months in a large naturalistic prerelease enclosure, with no specific aversive training. Maxwell and Jamieson (1997) demonstrated that reintroduced captive-reared takahes have the same postrelease survivorship as wild takahes. Before their release the captive-reared birds had foraged in small pens for fern rhizomes, an important winter food (however, there was no suitable control—reintroduced captive borns without exposure to rhizomes—to show that the high survivorship resulted from the enriched environment). These results conform to earlier laboratory evidence that enriched captive environments promote natural behaviors. For example, rats raised in more complex environments escaped predator models more efficiently (Renner 1988), and gerbils reared in tunnels behaved adaptively to threatening stimuli, whereas gerbils reared in cages approached the stimuli (Clark and Galef 1977). However, as with specific skill training, creating general learning opportunities that increase species-typical behaviors may not be sufficient to ensure survival: Common Amakihi (*Hemingnathus virens virens*) raised in aviaries used appropriate foraging strategies upon release significantly more than individuals raised in small cages, although there was no difference in postrelease survival (Kuehler et al. 1996). Overall, the link between enriched environments, behavioral change, and enhanced postrelease survival is not yet solid.

If, in some cases, specific skill training or exposure to enriched prerelease environments does increase postrelease survival, why is this not the case in golden lion tamarins? One possibility is species differences. Kleiman (1989) and McLean et al. (1994) noted that species may vary in the amount of preparation they need for reintroduction. For example, terrestrial species may need less locomotor experience than arboreal species, herbivores may need less training in foraging than carnivores and omnivores, and less social species may need less training in intraspecific social "etiquette" than highly social species. Griffith et al. (1989) found that herbivores were more successfully translocated and reintroduced than omnivores and carnivores. In the reintroduction survey of Beck et al. (1994), all of the successful mammalian reintroductions were of ungulates.

Second, perhaps the prerelease training or free-ranging environment is not sufficiently challenging or realistic (Castro et al. 1998). Rufous hare-wallabies were squirted with a water pistol (McLean et al. 1994) and Siberian polecats were shot with rubber bands (Miller et al. 1991) when exposed to predator models. Despite the "naturalness" of their environment, free-ranging golden lion tamarins get processed food, they have shelter boxes and numerous sturdy substrates for locomotion, and the presence of zoo staff and visitors may deter predators. In addition, the several months of training or exposure to the free-ranging environment may not be long enough. Further, golden lion tamarins may be exposed to the free-ranging environment as juveniles and adults, thereby missing any early sensitive periods for learning some skills (see below). Finally, although the free-ranging environment affords animals the opportunities to travel, navigate, and forage similarly to the way they would in the wild, the individuals may differ in how they take advantage of these opportunities. For example, a pair of golden lion tamarins that ranged free for approximately 5 months at the Smithsonian National Zoological Park (SNZP) spent an average of 75 percent of their time on human-made substrates, were inside the shelter box 61 percent of the time, and were observed using their fingers to reach embedded food (as they would in the wild) only 0.4 percent of the time. Comparable figures for a free-ranging group at Zoo Atlanta were 35 percent, 3 percent, and 4 percent, respectively (T. S. Stoinski and B. B. Beck, unpublished data).

We may have to promote actively any desired behaviors in addition to simply providing exposure to the complexities of a free-ranging environment. For example, food could be reduced, scattered, and hidden to encourage foraging, or frequently used travel routes could be altered to force animals to develop new routes (Menzel and Beck 2000). Beck (1991) and Castro et al. (1998) argued that zoos may have to be less risk-averse to establish captive environments that will be sufficiently realistic to prepare animals adequately for success in the wild.

Third, timing may be crucial. The age of the animals when they are trained or exposed to enriched environments may determine if the experience is effective. Black-footed ferrets have a critical or sensitive period (60 to 90 postnatal days) for learning prey preferences (Vargas 1994). Accordingly, ferrets born into captive prairie dog burrow systems or who begin training in burrows at 60 days of age or earlier have a higher postreintroduction survival rate (Biggins et al. 1998; Reading et al. 1996). Further, the age of animals when they are reintroduced may affect survival and thus the efficacy of training. Younger golden lion tamarins (Beck et al. 1991), black-footed ferrets (Reading et al. 1996), and thick-billed parrots (Snyder et al. 1994) have higher postrelease survival than older individuals, and younger black lion tamarins (Valladares-Padua et al. 2001), sandhill cranes (*Grus canadensis*) (Drewien et al. 1981; Horwich 1989), and European eagle owls (Wayre 1975) are said to be better release candidates.

Even if skill training or exposure to enriched environments does not improve postrelease survival, such experience may be useful because researchers may be able to predict the success of the animals after reintroduction based on the animals' responses. The SNZP group referred to above was "lost" by the second day after reintroduction, while the Zoo Atlanta group is still alive after 2 years in the wild and has reproduced (Beck and Martins 1997, 1999; Stoinski and Beck, in preparation).

Table 13.3 documents that survival in golden lion tamarins that were born in the wild as descendants of reintroduced lion tamarins is higher than in the reintroduced captive-born lion tamarins themselves. This must be due to the wild-born golden lion tamarins' greater proficiency in some as yet unspecified survival-critical behavior(s). Stoinski (2000) compared foraging, locomotion, and social behaviors of reintroduced captive-born golden lion tamarins with those of their first- and second-generation wild-born offspring to specify these differences. Golden lion tamarins are highly social primates presumed to learn many survival-critical behaviors at least in part by observation, so how can a wild-born filial generation be more proficient than the captive-born parents and older siblings from which they presumably learn?

One possibility may involve sensitive periods. The wild-born golden lion tamarins may experience reinforcement contingencies that are critical to survival in the wild early in life when behavioral acquisition may be rapid and lasting. In contrast, the captive-born lion tamarins experience these contingencies later in life, after reintroduction, when there is less plasticity.

Another, not exclusive, explanation is that the captive-born golden lion tamarins become extremely proficient in behaviors that are adaptive for life in captivity and that this proficiency interferes with the later acquisition of behaviors that are adaptive for life in the wild. The wild-born lion tamarins experience no such interference.

A third possibility, again not necessarily exclusive of the first two, is that the wild-born golden lion tamarins may get information about survival-critical behaviors by observing the unsuccessful responses of their captive-born parents and older siblings. Templeton (1998) showed that common starlings (*Sturnus vulgaris*) learn a discrimination task faster when observing demonstrators that make only incorrect choices than when observing demonstrators that make only correct, or both correct and incorrect, choices.

Whatever the mechanism, the data presented here strongly suggest that the key to successful reintroduction of captive-born golden lion tamarins is intensive post-release management. With this they are more likely to give birth to wild-born offspring, the survivorship, and presumably the behavioral proficiency, of which appear to be comparable to those of the truly wild population.

ACKNOWLEDGMENTS

The reintroduction portion of the Golden Lion Tamarin Conservation Program is supported by the Frankfurt Zoological Society Help for Threatened Wildlife, the Friends of the National Zoo, the Smithsonian Institution's International Environmental Studies

Program, the Smithsonian Institution's Nelson Fund, and Zoo Atlanta's Elizabeth Smithgall Watts Fellowship. We are grateful to Andreia Martins, field coordinator for reintroduction, and her eight-member field team that so painstakingly monitors the reintroduced golden lion tamarin population in Brazil. The Associacão Mico-Leão-Dourado (AMLD—Golden Lion Tamarin Association), under the executive directorship of Denise Rambaldi, coordinates all conservation activities for golden lion tamarins in Brazil. This research is conducted with the approval of the Conselho Nacional de Desenvolvimento Científico e Tecnológico (CNPq—Brazilian National Science Council), the Instituto Brasiliero do Meio Ambiente e dos Recursos Naturais Renováveis (IBAMA—Brazilian Institute of the Environment and Renewable Natural Resources), and the Institutional Animal Care and Use Committee of the SNZP. We also thank JoAnne Grumm and Mary Bowman, who coordinated the many dedicated volunteers who monitor free-ranging golden lion tamarins at the SNZP and Zoo Atlanta, respectively.

CLÁUDIO B. VALLADARES-PADUA, JONATHAN D. BALLOU, CRISTIANA SADDY MARTINS, AND LAURY CULLEN, JR.

14
METAPOPULATION MANAGEMENT FOR THE CONSERVATION OF BLACK LION TAMARINS

The black lion tamarin (*Leontopithecus chrysopygus*) is one of the rarest and most endangered primates in the world (Coimbra-Filho 1985b; Murray and Oldfield 1984). The species is endemic to São Paulo, the most developed state of Brazil. Agriculture, cattle ranching, the timber and other industries, and urbanization have reduced the state's original forest cover by about 95 percent, resulting in drastic reductions in the ranges and habitats of the eight primate species occurring there (Ferri 1980; Serra-Filho et al. 1975; Valladares-Padua and Cullen 1994; Victor 1975).

The black lion tamarin was considered extinct from the beginning of the twentieth century until 1970, when it was rediscovered by Brazilian researcher Adelmar F. Coimbra-Filho in the Morro do Diabo State Park (Coimbra-Filho 1970a, 1970b; Chapter 1 this volume). During the 1970s and beginning of the 1980s, Coimbra-Filho and Russell Mittermeier conducted preliminary surveys and concluded that only 100 individuals existed in the wild, distributed between the Morro do Diabo State Park (then 37,156 ha) and the Caetetus State Ecological Station (2,179 ha), both administered by the Instituto Florestal de São Paulo (IF/SP—São Paulo Forestry Institute) (Coimbra-Filho and Mittermeier 1977; Mittermeier et al. 1982).

Protection in these sites was, however, somewhat tenuous. In the early 1980s, the Companhia Energética de São Paulo (CESP—São Paulo Electricity Company) built three hydroelectric plants near the Morro do Diabo State Park. The smallest, Rosana, flooded around 3 percent of the park, including some the best lion tamarin habitat, thus reducing its size to 36,000 ha. Concern for this loss resulted in CESP promoting and implementing a rescue operation for the black lion tamarins inhabiting the area to be flooded (Valle and Rylands 1986; Chapter 1 this volume). This was the starting point for a major long-term research and conservation effort for the black lion tamarin (Mittermeier et al. 1985). Begun in 1983, the conservation actions extended from the Morro do Diabo State Park to cover eventually the entire geographic range of the species. By the 1990s, enough data had been obtained to evaluate the species' conservation status and to conclude that it was still possible to save the black lion tamarin by using a metapopulation management strategy for the few isolated remaining subpopulations.

This chapter is not about the theoretical or quantitative aspects of metapopulation theory. It focuses on the practical aspects of the use of this theory to save a critically endangered species. Here we briefly review the results of almost two decades of conservation research with the black lion tamarin, both in captivity and in the wild, and discuss some applied management techniques to create a minimum viable metapopulation for the species. We also present the results of the initial stages of the current IPÊ plan, compare this approach with some other endangered species conservation initiatives, and summarize some of the lessons learned.

THE CURRENT SITUATION

The estimated number of black lion tamarins in the wild is 1,000 individuals, located in nine geographically separated and isolated subpopulations in the state of São Paulo, Brazil (Table 14.1).

Ranging in size from 4 to 820 individuals, these nine wild subpopulations form the core of the conservation program for this species. However, because of their fragmented nature and their generally small size, these isolated subpopulations by themselves lack the ecological, demographic, and genetic potential to ensure the survival of the species (Valladares-Padua 1993). In addition to the wild population (formed by all the different subpopulations), there is a captive population of 112 individuals in 11 institutions (see Chapter 4 this volume). Most of the animals are at the Centro de Primatologia do Rio de Janeiro (CPRJ—Rio de Janeiro Primate Center), São Paulo Zoo, and Jersey Zoo, Durrell Wildlife Conservation Trust (DWCT). Other zoos involved are in South America (a total of 4 zoos, including CPRJ and São Paulo), North America (1), Europe (a total of 5, including Jersey Zoo), and Australia (1). Based on the 1999 studbook, the captive population is estimated to have 92.3 percent of the wild subpopulations' gene diversity. This is equivalent to the genetic diversity "captured" by 6.5 wild-born individuals (Valladares-Padua 2000) and equal to an average level of relatedness of about 8 percent (higher than first cousins).

The genetic goal of most captive breeding programs is to establish a captive population size sufficient to retain 90 percent of gene diversity over 100 years

Table 14.1
Estimated Number of Black Lion Tamarins in the Wild by Subpopulation

Location	Public/Private	Area (ha)	Subpopulation Size
Morro do Diabo State Park	Public	34,000	820
Caetetus State Ecological Station	Public	2,000	40
Fazenda Rio Claro	Private	1,600	70
Fazenda Ponte Branca	Private	1,200	10
Fazendas Tucano and Rosanella	Private	2,000	10
Buri	Private	100	20
Fazenda Santa Maria I	Private	400	4
Fazenda Santa Mônica	Private	500	10
Fazenda Mosquito	Private	2,000	6
Total		43,800	990

without the addition of new founders (Foose et al. 1995; Chapter 4 this volume). Under this strategy, the captive population would retain a substantial amount of the species' total genetic variation over the long term, regardless of what might happen to any existing wild population. The captive black lion tamarin population as it now stands cannot achieve these standard objectives because of its small size and the high degree of relatedness among individuals. Although over 90 percent of gene diversity is currently retained, this will erode over a relatively short time frame if the captive population is kept at its current size (Ballou and Valladares-Padua 1990). Based purely on biological grounds, under the currently accepted captive breeding standards, the captive population of black lion tamarins is not contributing effectively to the conservation of the species.

LONG-TERM RESEARCH

The black lion tamarin conservation project of the Instituto de Pesquisas Ecológicas (IPÊ—Institute for Ecological Research) began with a long-term research and management plan that included surveys and censuses, evaluation of the species' genetic and demographic status, broader research on its ecology and behavior, conservation management of the captive population, an integrated environmental education program, habitat protection and restoration, and a final goal of restocking the wild population with captive-bred animals to promote black lion tamarins' sustainability.

The first phase of the plan began in the 1980s with a survey and census conducted in the forest fragments located in the Pontal do Paranapanema region, in the far west of the state of São Paulo. These surveys gave no indication of the existence of any further subpopulations. All rumors and information IPÊ received on possible new subpopulations of black lion tamarins were investigated, but these led to no additional findings. In the 1990s, however, IPÊ started a new project to evaluate the conservation situation of all forest fragments larger than 400 ha in the interior of the state and also initiated a survey to locate new subpopulations of the black-faced lion tamarin (*L. caissara*) in the southwestern part of São Paulo, where it had been reported to occur close to the eastern limits of the former range of the black lion tamarin (Prado and Valladares-Padua 2000; Valladares-Padua 1993; Valladares-Padua and Cullen 1994; Valladares-Padua et al. 1999). Seven new subpopulations of *L. chrysopygus* were found, increasing the total wild population by 15 percent.

Genetic studies were initiated in 1985 in order to delineate the variability characteristic of the species. Twenty-five allozyme loci were analyzed in both captive and wild populations using gel electrophoresis. The genetic assessment, based on 25 blood enzymes, showed black lion tamarins to be genetically monomorphic for the 25 loci studied ($P = 0.00$; $H = 0.00$) (Valladares-Padua 1987). These values are the lowest reported to date for any primate species and confirm earlier findings by Forman et al. (1986) that black lion tamarins have substantially reduced genetic variation. They are similar in degree of homozygosity to the cheetah (O'Brien et al. 1983, 1985, 1986) and the northern elephant seal (Bonnell and Selander 1974; see also Chapter 5 this volume).

Based on quantitative analyses, the demography of black lion tamarins in captivity was also monitored (Valladares-Padua and Mamede 1996a). The entire captive population is managed through an international studbook. This database allowed for the implementation of a breeding plan for the species that maximizes retention of genetic diversity, minimizes inbreeding, and permits the calculation of the demographic rates and, consequently, general trends in captive population growth (Valladares-Padua and Ballou 1998). While the captive population of black lion tamarins began with six original animals, which was supplemented with an additional 20 in 1985, by 1999 the population comprised only 112 animals. Its annual rate of increase has been negative in many years, population growth in captivity has been slow when compared with that of captive *L. rosalia* and *L. chrysomelas,* and it only has the equivalent of six founders (Ballou et al. 1998; Chapter 4 this volume).

A comparative study of the ecology and behavior of several groups of *L. chrysopygus* and an environmental education program at the Morro do Diabo State Park and surrounding areas were begun in 1987. The comparative study, conducted between 1987 and 1990, examined whether or not the behavior and ecology of black lion tamarin groups differed significantly between different habitats and, if so, the implications for conservation strategies for the species. Four groups were studied, each in a different forest type. Sixteen habitat variables were compared using multivariate statistical tests. The effects of differences found in the feeding and foraging behaviors of the four groups were investigated by using quantitative and qualitative information on the variation in resource abundance and on the lion tamarins' use of time and space both as individuals and as groups (see Chapter 7 this volume). Even with minimal genetic variability, the species showed considerable ecological and behavioral plasticity (Valladares-Padua 1993, 1997; Valladares-Padua and Cullen 1994), which was believed would prove favorable for its survival in managed conservation transfers such as reintroductions, translocation, or artificial dispersals.

The support of local communities was clearly a vital element for the conservation of the species. Therefore, an environmental education program was established that increased conservation awareness and changed the attitudes and behavior of the rural and urban communities around the Morro do Diabo State Park, which eventually became a regional landmark (Padua and Valladares-Padua 1997; Chapter 15 this volume).

The Vortex model (Lacy 1993) has been used to develop predictions of the long-term survival of the black lion tamarin metapopulation (Ballou et al. 1998; Lacy 1993; Seal et al. 1990). Analyses using this model have shown that in the next 100 years if the subpopulations are considered individually, most if not all of the subpopulations of *L. chrysopygus* will go extinct, leaving only the large Morro do Diabo State Park population. However, if the wild population (with all its subpopulations) is managed as a metapopulation with high rates of dispersal and gene flow among fragments, most if not all of the subpopulations are likely to survive. After 100 years the metapopulation will have a maximum of 831 animals and will retain 97 percent of its genetic diversity (Ballou et al. 1998)

CONSERVATION STRATEGIES

The behavioral and ecological plasticity of the black lion tamarins is an indication that they will be well suited for almost any type, shape, and size of forest within their original range. This plasticity became even more evident when the recent broad survey for the species was conducted, and it obtained a higher number for the total metapopulation size of *L. chrysopygus* than in previous estimates. As stated earlier, new subpopulations of the black lion tamarins were found living in a number of forest fragments, and it was estimated the total number of black lion tamarins in the wild was around 1,000 individuals living in nine structurally and dimensionally different "islands" of forest.

The combination of isolated subpopulations of black lion tamarins living in a mosaic of fragmented forest in a matrix of pastureland constitutes a good example of what Hanski and Gilpin (1991) called a finite metapopulation, in contrast to the original definition of metapopulation presented by Levins (1969). Levins assumed the existence of an infinite number of habitat patches of the same size and quality, whereas Hanski and Gilpin assumed a more realistic finite number of subpopulations. Small subpopulations of a finite metapopulation of an exclusively arboreal primate such as the black lion tamarin are very susceptible to extinction for two main reasons: (1) They occupy discontinuous habitats that may inhibit natural migration and recolonization among subpopulations (Gilpin 1987), and (2) when the number of subpopulations is small, there is a real possibility of all subpopulations going extinct simultaneously, thus causing the total extinction of the metapopulation (Nisbet and Gurney 1982). However, this susceptibility can be reduced if the primates are managed through the systematic transfer of animals among habitat fragments as if they were dispersing naturally. In our case we use the working definition created by McCullough (1996), where the two key proofs of the existence of a metapopulation are a spatially discrete distribution and a high probability of extinction in at least one or more of the local patches. The midterm goal is then to prevent the extinction of black lion tamarins by reducing the subpopulations' genetic and demographic deterioration and, if necessary, by using the ultimate backup of recolonization. The long-term goal, however, is to restore the landscape so the wild population can survive through natural processes.

The role of the captive population in the conservation of black lion tamarins is then clearly defined. Rather than just providing exhibit animals to a small number of zoos, appropriate management and husbandry provide the conditions for the captive lion tamarins to be included as an additional component of the metapopulation. With the vast majority of black lion tamarins in the wild in the Morro do Diabo State Park, prudent conservation strategies dictate that additional subpopulations be established as insurance against any catastrophic threats (biological or otherwise). The captive population has been managed to serve as a significant demographic and genetic reserve for the wild black lion tamarin population while being a symbol for conservation education, public relations, and fundraising for the black lion tamarin conservation program (see Chapter 4 this volume).

Genetic and demographic goals for captive populations are typically developed under the assumption that the captive populations should be of sufficient size, stability, and genetic constitution to preserve the survival of the species if the wild population was to go extinct. Under the metapopulation strategy, the conservation of black lion tamarins depends on all the subpopulations within the metapopulation. The captive population therefore need not be self-sufficient but provide genetic and demographic input to and receive input from other parts or sectors of the metapopulation. This strategy allows for the maintenance of a smaller captive population (150–200 individuals) that continuously retains a higher level of gene diversity (approximately 95 percent) than standard strategies call for. This has been termed the nucleus population strategy and is designed so that the captive population always contains a fairly high proportion of the wild gene diversity. This way, if needed, a fully self-sufficient wild population can be developed from the nucleus population at any time (Foose et al. 1995).

Maintaining a small captive population with high levels of gene diversity requires frequent gene flow into this population. The rate of gene flow required depends on how much gene diversity we wish to retain in the population continuously, the size of the population, and how frequently we move animals and the effects of removing animals from the wild population.

The number of initial imports (the number of animals it takes to bring gene diversity to the target level in one generation) and the number of "maintenance" imports (the number needed to keep gene diversity at the desired level) per generation (5 years) for 98 percent and 95 percent gene diversity for different captive population sizes are shown in Table 14.2. For example, to maintain 98 percent of the wild population's gene diversity in a captive population of 100 individuals requires an initial import of about 60 animals and 44 imports every 5 years thereafter (Valladares-Padua and Ballou 1998). These calculations are based on the assumption that each new founder imported contributes a 0.4 founder genome equivalent to the population (Lacy 1989; Mansour and Ballou 1994).

Table 14.2
Number of Imports Needed to Maintain Different Levels of Wild Gene Diversity in Subpopulations of Different Sizes

		Captive Population Size		
Percentage of Gene Diversity	Number of Imports	100	150	200
98	Initial imports	59.0	56.0	56.0
	Every 5 years	44.0	30.0	23.0
95	Initial imports	13.9	13.4	13.4
	Every 5 years	7.3	4.9	3.7

Maintenance of 98 percent gene diversity would require an unreasonable number of imports each generation and would result in the population rapidly exceeding its desired size unless substantial numbers were also reintroduced. Maintaining 95 percent gene diversity, on the other hand, would require a modest number of initial imports (which could be spread over several years), followed by a much more reasonable number of maintenance imports.

In 1997 the *L. chrysopygus* International Recovery and Management Committee (IRMC) approved the "95 percent gene diversity/150 captive animals" management strategy. This allowed for the inclusion of the entire captive population as one of the populations of the black lion tamarin metapopulation. However, for this approach to work, reintroduction and translocation must be routine and successful components of the plan. Although reintroduction of captive-bred animals is difficult and expensive (Beck et al. 1994), reintroduction has been a successful component of the golden lion tamarin conservation program (Beck et al. 1988; Castro et al. 1998; Chapter 12 this volume).

Implementation of the metapopulation management plan for black lion tamarins began with the first experimental translocation of the species in 1995. Since 1997, a series of experimental "conservation shifts" have been conducted to begin the introduction of new founders into the captive population and to develop successful and low-cost ways of promoting gene flow among the many wild subpopulations.

PRELIMINARY RESULTS OF BLACK LION TAMARIN CONSERVATION SHIFTS

Between 1995 and 2000, three experimental "conservation shifts" were implemented on different occasions:

- *Translocations.* In 1995 and 1999 two black lion tamarin groups with four and six animals respectively were translocated from the Fazenda Rio Claro (Rio Claro Ranch) to the Fazenda Mosquito.
- *Reintroductions.* (1) In 1999 a group of five black lion tamarins was captured at Morro do Diabo State Park. Two females together with a male from the Jersey Zoo, DWCT, were incorporated in the first mixed group reintroduced into the wild. (2) In 2000, after the death of the one reintroduced male and female, one of these two wild females was then mixed with two males from the CPRJ and subsequently reintroduced again into the wild at the Morro do Diabo State Park.

- *Managed dispersal.* In 1999 two males from the group captured at Morro do Diabo were used for a managed dispersal project, which involved physically moving the animals from one section of the reserve to another at an age when they would naturally disperse. One male of the group was transferred to captivity as the first of an initial import of founders to the captive population.

We summarize these conservation shifts in more detail below.

Translocations

TRANSLOCATION OF GROUP A The first group of black lion tamarins, group A, was translocated in 1995, from the Fazenda Rio Claro, in Lençóis Paulista, to the Fazenda Mosquito, in Narandiba. The forest fragment in the Fazenda Mosquito had no black lion tamarins despite being within the species' range. This first group was monitored for 3 years before its translocation and has been monitored since its translocation in 1995. As was found with translocated golden lion tamarin groups (Chapter 12 this volume), the animals showed no difficulties in adapting to the new area and found new food items with ease. Five births occurred in the group between 1995 and 2000 (four males and one female). In May 2001, the group was composed of one adult male and one adult female, along with a young subadult male that entered group A from another translocated group (group B) in September 2000.

TRANSLOCATION OF GROUP B The translocated group B was composed of six animals, four males and two females. Following capture, the group was transported and released at the Fazenda Mosquito in July 1999. Two females were radio tagged, and the group was released at the same site as group A (released in 1995). An infant was observed in December 1999 but did not survive. In February 2000, the oldest female was found dead, but the cause was not determined. Systematic behavioral data on group B have been collected since August 2000. In May 2001 it was composed of three males and one female surviving from the original four males and two females.

Reintroductions

The reintroduction part of the program began in July of 1999. Two reintroductions of mixed wild-captive groups were conducted. The decision to reintroduce mixed groups was made on the assumption that there would be a higher probability of success if the captive-born members could learn the adaptive behaviors necessary to survive in the wild from their wild group mates (see also Chapter 13 this volume). Previous research demonstrated that black lion tamarins follow the strategy of an animal that is familiar with its area (Valladares-Padua 1993; Waser and Wiley 1979). This approach differs from the primary approach used in the golden lion tamarin and other programs in which single animals or whole groups of captive-born animals have been reintroduced (Beck et al. 1991; Chapter 13 this volume).

REINTRODUCTION OF MIXED GROUP A Following its arrival at Morro do Diabo State Park, group A (one captive male and two wild females) was kept for 21 days in an enclosure inside the park headquarters. The group was given native fruits, live insects, and small frogs in an attempt to train the captive male for his new life in the forest. In the early stages the captive male showed no interest in the native food, but after some time he began to mimic the wild females and experiment with the new food. This period was very important for increasing the group cohesion, and after 10 days in the enclosure a sexual response by the male toward the females (an erection) was observed, followed by the females starting to groom the male.

The group was released from the enclosure during the night (7 August 1999). The animals were put inside a tree hole previously used by the wild group and observation began the next day at dawn. Initially, observers avoided close proximity in order to minimize the chance that the females would flee. Food was supplied by attaching palm leaves with fruits to the tree holes used by the group.

The captive male remained with the females, although he spent 3 nights apart from them. After 7 days, however, he was very weak (dehydrated) and was recaptured. He recovered after 1 week and was released again, rejoining the females, which had been followed in his absence. Subsequently, the reintroduced mixed group remained in good condition until November 1999. Observers continued supplying extra food, but in October, with the beginning of the rainy season, some native fruits became available, and the entire group ate them, including the captive male. The male stayed with the females but did not show any signs of sexual behavior. The group was observed for some time every day until October, when systematic data collection was begun (at least 3 full days each month). In November 1999, the male was found to have been killed. The sternal region had been opened; the body was not in rigor mortis. Some tracks of a small feline nearby were observed, and it was believed that the male was killed by an ocelot (*Leopardus pardalis*). The females were close by, above the dead male.

Observations of the females were suspended after the predation event but began again in March 2000. The females were followed until September 2000, when they were captured because the subadult showed signs of illness. It died 3 days later, and the other female was maintained in captivity in order to pair it with two captive CPRJ males, forming the group used for the second reintroduction.

REINTRODUCTION OF MIXED GROUP B Two captive-bred males (2.5 and 1.5 years old) were transported to an enclosure at Morro do Diabo State Park in October 2000. There they were paired with the remaining wild female of group A. Systematic behavior observations determined that the animals showed no agonistic behavior. The older male stayed closer to the female. Native food items such as fruits and insects were provided, and branches of different sizes were put in the enclosure to provide the animals with a more diverse environment (Castro et al. 1998). The young male in particular showed considerable ability in foraging for beetles and butterflies. After 2 weeks the older male was observed mounting the female over a period of 3 days.

Following the same protocol used with the first group, group B was released on 5 December 2000 in the territory of the original female. The two captive males remained in the same tree hole for a month, traveling a maximum of 10 m just to reach the provisioned food. The wild female left during daylight and returned later in the day to sleep with the captive males in the same tree hole. Subsequently, however, they began following the wild female, with the younger male of the two showing a greater ability to move in the canopy.

The older male was found dead in April 2001, close to a decapitated capuchin monkey; they might have been attacked almost at the same time by a predator (possibly a jaguar) that fled when researchers approached. The surviving young male and the wild female continued to be monitored on a daily basis. They were provisioned because of a serious drought in 2000, which resulted in many trees failing to fruit. After 5 months in the wild the young male no longer had problems following the female and was fully competent in foraging for available fruit and invertebrate prey.

Managed Dispersal

In June of 1999 two males were moved in a managed dispersal experiment from their original home range inside the Morro do Diabo State Park to another area 6 km away but still inside the park. The animals were monitored by radiotransmitters every other day. They stayed together until August of the same year, when another adult male spontaneously joined the duo. The youngest male traveled more than 7 km during the period of our observations. It remained a floater and crossed the territories of three other groups without joining any of them. The other male lost the transmitter early on, neither it nor the wild male that joined the dispersed duo of males were located.

CONCLUSIONS

There are intraspecific comparative studies on the ecology and behavior of primates that show the adaptive importance of their behavioral plasticity (Dawson 1976; Richard 1978; Rudran 1978; Kinnaird 1990). Our research on ecology and behavior confirmed that black lion tamarins are flexible and capable of adapting to significantly different habitats. This increases the chances that they will successfully occupy uninhabited forests and that appropriate in situ management will increase their long-term probability of survival. It seems clear that the chance of survival of black lion tamarins is dim if we only rely on reintroduction programs using captive-bred animals. Indeed, Kleiman (1996) specifically recommended not reintroducing captive-born black lion tamarins, in part because the captive population was not secure.

However, as we have shown in the case of *L. chrysopygus,* it is still possible to consider management strategies other than captive breeding and reintroduction, as was originally envisioned. The existence of forest fragments in the original range of the species, both with and without black lion tamarins, allows for the development of long-term in situ metapopulation management strategies for the species.

As such, all wild and captive subpopulations are currently being managed as a metapopulation that includes the shifting (reintroduction, translocation, and managed dispersal) of individuals among the many subpopulations (Figure 14.1). The captive population is being maintained as the nucleus population of the metapopulation, with a maximum number of 150 animals and a genetic goal of maintaining 95 percent of the wild gene diversity (see Chapter 4 this volume). This approach is in accordance with the results of the Vortex population viability analyses, which indicate that if all subpopulations are incorporated into a metapopulation with frequent migration there will be a high probability of maintaining many of the subpopulations (Ballou et al. 1998).

This metapopulation plan, however, depends on the establishment of new reserves and the enlargement of the existing ones. New areas should be forest fragments that are similar to those inhabited by *L. chrysopygus* and outside already protected areas. Another way to increase and improve habitat conditions is to create, whenever possible, forest corridors between neighboring but noncontiguous subpopulations. Despite the controversy about the costs and benefits of forest corri-

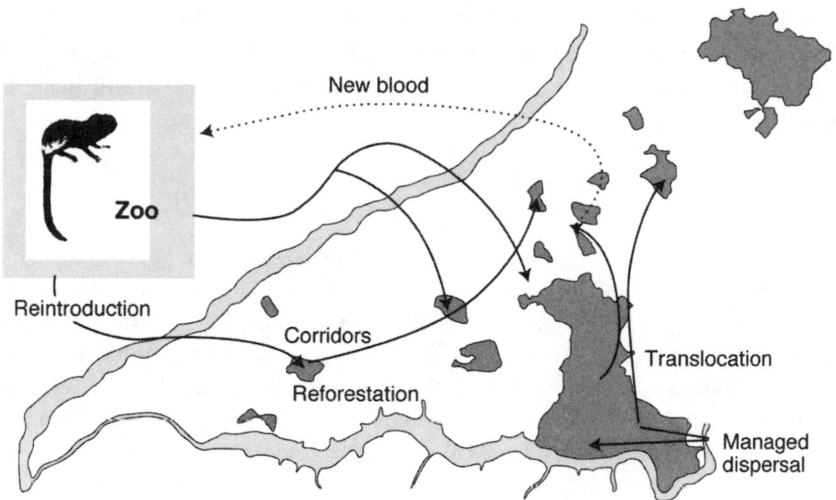

Figure 14.1. Schematic view of the metapopulation management of black lion tamarins in the Pontal do Paranapanema region.

dors (Simberloff et al. 1992), there is no doubt for *L. chrysopygus* that they can play a major role in reducing human intervention in the managed dispersal among the connected subpopulations (Cullen and Valladares-Padua 1999). For this reason, IPÊ has carried out studies of hunting by humans (Cullen et al. 2000, 2001) and begun reforestation programs in collaboration with local landowners to form corridors to connect protected areas to forest fragments (Figure 14.1; see also Chapter 3 this volume).

Finally, the continuity of the environmental education and community involvement program is crucial for the success of *L. chrysopygus* conservation. A program established for the Morro do Diabo State Park region helped the community (1) become aware of the importance of the park as a conservation site and (2) contribute to its protection (Padua 1991; Padua and Valladares-Padua 1997; Chapter 15 this volume). The appreciation and involvement of local communities and landowners can enhance the protection of the species itself as well as the remaining forest habitats. In the black lion tamarin conservation scenario, each and every remaining habitat fragment is an important piece of its metapopulation survival puzzle.

The long-term conservation effort discussed in this chapter shows that even a highly endangered species such as the black lion tamarin has survival chances in the wild if the appropriate management measures are taken. Despite the fact that metapopulation management is in its infancy and there is still a lot of research to be done before we can come to definite conclusions, the 15-year study of black lion tamarins leads us to conclude that the future of the species will likely be through the metapopulation management of its wild and captive populations.

ACKNOWLEDGMENTS

The successes of the black lion tamarin conservation program would not have been possible without the support of the local communities, the enthusiasm of the research team, especially Srs. Jose, Homero, Zezinho, Wanderley, Cicinho, and Alemao, and the staff of the Morro do Diabo State Park, as well as its directors, Marco Antonio Garrido, Giselda Durigan, and Helder H. de Faria. The support from the following institutions was also of crucial importance: the Conselho Nacional de Desenvolvimento Científico e Tecnológico (CNPq—Brazilian National Science Council), the Center for Environmental Research and Conservation (CERC) at Columbia University, Duratex S.A., Fazenda Mosquito, Fundação de Amparo à Pesquisa do Estado de São Paulo (FAPESP—Research Support Foundation of the State of São Paulo), IF/SP, the Instituto Brasileiro do Meio Ambiente e dos Recursos Naturais Renováveis (IBAMA—Brazilian Institute of the Environment and Renewable Natural Resources), the International Committee for Conservation and Management of the Lion Tamarins, the Lion Tamarins of Brazil Fund, the Fundo Nacional do Meio Ambiente (FNMA—National Environment Fund), the National Geographic Society, the Margot Marsh Biodiversity Foundation, the Programa Natureza & Sociedade (Nature and Society Program)—State University of New York/WWF, CPRJ, the Smithsonian Institution, and.the Whitley Animal Protection Trust. Dominic Wormell and the DWCT, Jersey, are very important partners of the project. The authors would like to acknowledge in particular the Wildlife Trust and Virginia Mars for their understanding and continuing support for the conservation of the black lion tamarins.

SUZANA M. PADUA, LOU ANN DIETZ,
DENISE MARÇAL RAMBALDI,
MARIA DAS GRAÇAS DE SOUZA,
AND GABRIEL RODRIGUES DOS SANTOS

15
IN SITU CONSERVATION EDUCATION AND THE LION TAMARINS

Some species are especially effective in stimulating human interest. The Brazilian lion tamarins (genus *Leontopithecus*) are undoubtedly among these, and their attractiveness has been used to support their conservation. In the past two decades, a variety of in situ and ex situ conservation education programs have been designed for the four lion tamarin species, in some cases with extensive impacts. Outside of Brazil, zoos have successfully used the lion tamarins as part of conservation-oriented campaigns and as a method for informing the public about environmental issues.

In Brazil, each in situ education program for the genus *Leontopithecus* has focused on one of the four species, but each has also gradually raised regional awareness about conservation in general. Despite Brazil's rich biodiversity, there are not many conservation programs that take advantage of natural areas for education purposes. With the expansion of urban centers and limited opportunities for direct contact with wildlife, people feel little connection to nature and are often unaware of its importance. This overall lack of awareness and pride in natural resources has contributed to the dramatic loss of habitat in many of Brazil's biomes. Conservation education deals with raising people's overall awareness of and pride about local nature, so it will hopefully help reverse the degradation that has been common throughout the country.

The conservation of ecosystems can often be more easily understood through a focused perspective on a single species (the "flagship" species approach, Dietz et al. 1994b). There are examples worldwide of the success of education programs utilizing individual species for raising awareness about conservation issues in general (Jacobson 1999). Butler (1995) developed strategies using parrot species for promoting broader conservation measures, thus increasing the value attributed to local resources and cultures throughout the Caribbean. In Canada, Blanchard (1995) was able to shift people's attitudes through education, and seabirds that had been overhunted increased in numbers and became as plentiful as at the beginning of the last century. Turtles have also been effective flagship species for education. In Brazil, one of the best examples is the Projeto Tamar, a conservation education program for marine turtles. Data confirm that turtle populations have grown because of the decrease of hunting, and the marine turtle has become a regional symbol of pride and a theme of many local cultural events (Castilhos et al. 1997).

All four lion tamarins, following educational and public awareness programs, have become conservation symbols. Within Brazil, they have promoted pride and enhanced protection for the ecosystems they inhabit. Methods used for these in situ education programs have become models for others, not only at the national level but also internationally.

A characteristic that helped these in situ education programs acquire credibility was their adoption of a research framework from the very beginning, still not a common practice in the conservation education field. The advantages of incorporating research methodologies, such as conducting surveys before and after the implementation of an education program to determine the program's impact, are many. Educators are able (1) to improve strategies without losing time and resources, (2) to assess their overall effectiveness in relation to the proposed objectives, and (3) to disseminate information on what works and what does not, thus allowing other educators to benefit from their experiences (Dietz and Nagagata 1986, 1995, 1997; Jacobson and Padua 1992, 1995; Padua 1991, 1994a, 1994b, 1997; Padua and Jacobson 1993; Padua and Valladares-Padua 1997, 1998).

This chapter discusses the four in situ conservation education programs for the lion tamarins in Brazil. After the features common to all four programs are summarized, each program is discussed in detail.

THE CONTEXT FOR IN SITU EDUCATION PROGRAMS FOR LION TAMARINS

The four lion tamarins are restricted to small and separate regions of the Atlantic Forest of Brazil (see Chapter 2 this volume). The in situ education programs designed for each have played important conservation roles, especially because habitat destruction is the main reason for the lion tamarins' endangered status. All remaining habitats suitable for *Leontopithecus* species deserve special attention. As these habitats are protected, innumerable other species also benefit.

Habitat loss has occurred at exponential rates, especially during the last 50 years.

According to Dietz and Nagagata (1995), public support to reverse the trend can be achieved when there are economic alternatives that support nature conservation and when people are reached through education programs. The latter programs include opportunities for cognitive and affective gains, for giving people skills, and for empowering them to act and behave more ethically toward other living species. Through a variety of educational approaches, people who live around protected areas can be encouraged to participate in conservation instead of being banned from these areas, the usual approach in most regions of Brazil. Padua and Tabanez (1997b) have described examples of people acquiring ownership of the initiatives being implemented and then feeling motivated enough to embrace a cause or help protect natural areas.

The education programs for the lion tamarins vary in their contexts and specifics, but the involvement of local communities has been a common characteristic of all four programs. One of the means to achieve public support has been the dissemination of scientific findings in a simple and direct language that can be understood by all. Activities that are entertaining and educational have helped attract attention and have been effective in involving the general public.

The environmental problems in each of the regions differ, but in general, human pressure on natural environments has increased over the years despite the existing and sometimes isolated conservation efforts and federal legislation. Deforestation, hunting, logging, and capturing animals for the pet trade are all still common, and new social pressures on natural environments have emerged in the last few years. An organized movement in support of agrarian reform has succeeded in settling thousands of families that were previously "landless"; many of these settlements are in or near protected areas, which represents an additional and considerable threat. In the past 5 years, some lion tamarin education programs have had to add this audience to the target groups on which they focus attention. Information has been shared on how to improve output from agricultural plots and at the same time protect forest fragments, and hands-on practices have been promoted that can be good for everyone: humans and other living species.

Similarities of the Four Lion Tamarin Education Programs

During the planning stages of the conservation education programs, threats to lion tamarins were identified. The most common are deforestation, hunting inside the protected areas, fires around forest fragments, invasion by exotic nonnative species, cattle destroying the forest understory, and lion tamarins being captured for the pet trade. The conservation education programs have shared information about these threats and how they can be minimized or prevented. In addition, the programs have shared data gathered from ecological and behavioral research on the different species, which are used when involving the local communities in the conservation of lion tamarins and their habitats. The goals and objectives have thus included increasing the local people's knowledge about lion tamarins and the threats facing them and increasing the value these people place on conservation so that public participation for conservation can be effective. Local support has been critical, especially because the lion tamarins are located in remote areas where there are few guards or other means to protect the natural areas formally.

As the objectives of an education program are developed and a profile of the local populations emerges, it is possible to design appropriate educational strategies in the regional context. Dietz and Nagagata (1995) have equated this with developing a market strategy. In order to sell a product, you must know your public and your context well. In all cases, when surveys were conducted with different local communities, the results indicated that there was little knowledge about lion tamarins but also no evidence of negative attitudes toward these primates. This served as one of the first indications that they might be used as conservation symbols to raise public awareness and pride about the natural ecosystems of each region where they occur.

Target Groups, Designed Activities, and Products

The in situ education programs have reached broad audiences, among which are employees of the protected areas where the lion tamarins are found, community members at large, students of all levels, decision makers, landowners who still have forests on their properties, and, in more recent years, members of the agrarian reform ("landless") movement. The programs have thus been designed to respond to a large variety of contexts, expectations, and needs.

Communities have been involved at all times. With the public participating in the organization of activities and in selecting priorities, the projects have shared ownership, increasing their chances of success. To involve local people, communication has to be open, and the appropriateness of the language used in educational materials continuously tested. Regionalism can be strong in Brazil, so words can be understood differently, even in locations that are not too far apart. Therefore, the clarity of the language used has to be assessed in a number of ways before the final material is produced.

Program implementation has varied according to the realities of the situation—to the number of educators, to the support acquired from the institutions involved or from other organizations, and to the funds gathered to implement the different strategies. Some of the strategies have included offering people different ways of being exposed to the natural world. Nature trails have thus been common, and their effects have been assessed in different studies (Padua et al. 1997, 1998; Tabanez et al. 1997). Some programs have built visitor centers to receive local people, especially students, in an organized way (Figure 15.1). Common activities organized for students have been lectures at schools about the lion tamarins and their forests, courses for teachers and other school employees, educational fieldtrips, contests, plays, art exhibits on ecological themes, and the creation of local clubs and thematic groups.

To reach community members in general, educators have designed activities to attract attention to and to raise awareness about the importance of the local envi-

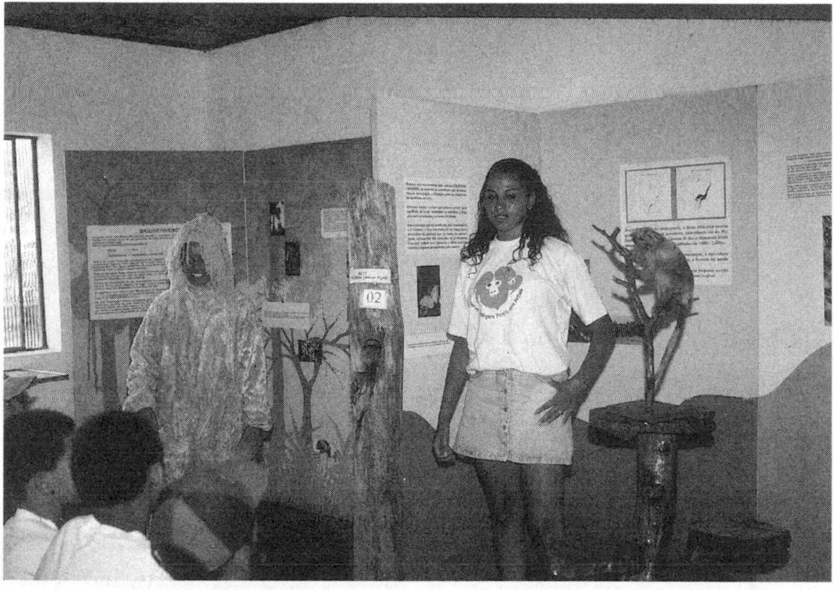

Figure 15.1. At the visitor center in the Poço das Antas Biological Reserve, staff of the Associação Mico-Leão-Dourado (AMLD—Golden Lion Tamarin Association) present information about the biology of golden lion tamarins and their forest habitat to local students. (Photo by Devra Kleiman).

ronment and nature. Examples of these activities are ecological music festivals, marathons, plays, and parades.

For guards and other personnel directly linked to the protected areas, educational efforts have focused on talks, classes, workshops, and participatory meetings that aim at obtaining greater understanding of and involvement in conservation issues. The education programs have frequently been helpful in raising employees' self-esteem, which is crucial for empowering people to become involved in the conservation cause (Padua and Tabanez 1997).

Considerable effort has been made to reach the landowners and farmers. Personal visits to explain a program's objectives have aimed at encouraging them to protect their remaining native forests, to plant native trees in watersheds and other fragile areas, to increase their pride in the lion tamarins and the natural resources on their property, and to turn their properties into permanent reserves.

To reach regional as well as broader audiences, the media have played a critical role. Local radio stations are easily accessed in Brazil and have helped broadcast ecological information through talks and interviews with local people involved in protecting the lion tamarins, field researchers, and other conservationists. Radios have also helped to promote educational events and cultural festivities.

The media have also been an important vehicle for changing values. If forests and lion tamarins are important enough to be discussed on the radio and appear on television and in magazines, then people reason that they must have more value than was previously attributed to them. Generally, the local lion tamarin education programs have not given priority to printed information because a great percentage of the people living in the remote areas where the lion tamarins are found are illiterate. However, respected Brazilian and international newspapers and magazines have published many articles on the conservation of the four lion tamarins and have helped to make them more popular among the general public, which is always important when support for further protection is needed.

All the programs have created products that aim at making the lion tamarins more visible or at educating certain target groups. Some of these products include videos, posters and pamphlets, school notebooks, slide collections for lectures, ecological games, T-shirts, hats, stickers, buttons, traveling exhibits, and instructional materials for the local teachers. Inside the protected areas educational trails have been designed and constructed, as well as many hands-on activities. Nearly all materials and activities have been pretested so adjustments can be made before they are produced on a large scale.

Evaluation has been critical for assessing the effectiveness of the different program stages and their effects in general. From the start, adopting evaluation procedures has made the lion tamarin education programs models to others and a learning process where activities and different approaches have been tested for their efficacy. This has undoubtedly contributed to the overall programs' effectiveness and has helped save human and material resources in the long run.

Interestingly, the individual conservation education programs for the lion tamarins have developed methodologies of their own, and these have served as examples to other programs in Brazil and abroad. The programs have been highly effective in popularizing these species within the country and internationally (Dietz and Nagagata 1986, 1995, 1997; Dietz et al. 1994b; Padua 1991, 1994a, 1994b, 1995, 1997; Padua and Valladares-Padua 1997, 1998; Padua et al. 1990). Indeed, zoo conservation education programs about the lion tamarins worldwide are highly dependent upon the in situ programs in Brazil and frequently use material produced by them.

THE EDUCATION PROGRAMS

The Golden Lion Tamarin Education Program

The first lion tamarin education program was started in 1983 and focused on the golden lion tamarin (*L. rosalia*) at the Poço das Antas Biological Reserve in the state of Rio de Janeiro (Dietz and Nagagata 1986, 1995, 1997). It was one of the first conservation education programs in Brazil with the aim of sensitizing local communities to the importance of wildlife conservation.

The only protected area for the species at the time was the Poço das Antas Biological Reserve. This reserve is administered by the Instituto Brasileiro do Meio Ambiente e dos Recursos Naturais Renováveis (IBAMA—Brazilian Institute of the Environment and Renewable Natural Resources) and is only 5,000 ha, which is insufficient to sustain a viable population. So it became clear that further protected forests were needed.

From the beginning, a research and evaluation plan was designed to assess the knowledge and attitudes of the local community members and the effectiveness of the educational strategies adopted (Dietz and Nagagata 1995, 1997). Initial results of the pre-evaluation survey indicated that local community members had little knowledge about the reserve or the golden lion tamarin. Survey responses also indicated that the community did not understand the link between a species and its habitat. One of the challenges, therefore, was to demonstrate interdependencies among living things and their environments. Additionally, there was a need to build pride among local communities in regard to the remaining natural habitats. Answers showed that there was little notion of the value of forests (Dietz and Nagagata 1995, 1997). In order to change this situation, information needed to be transmitted about the importance of forests, the degree to which they are being destroyed, and the effect of that forest destruction on local communities.

The pre-evaluation also yielded information as to what kinds of activities could be most effective in reaching each of the target audiences (consisting of the public at large as well as landowners) identified among the three municipalities near the Poço das Antas Biological Reserve. The education program developed many activities, with special emphasis on activities that interested local leaders and that appeared most likely to achieve the greatest results for the least cost. Some of these activities included school presentations about the golden lion tamarin, training classes for reserve guards and local teachers, and fieldtrips to the Poço das Antas Biological Reserve.

A one-on-one visit to each ranch was the key activity in garnering initial landowner support. During the visits, the objectives of the golden lion tamarin project were explained, and landowners were presented with the idea of possibly receiving reintroduced golden lion tamarins on their ranches. Special materials were created to show landowners the tax benefits and other advantages of protecting forests and encouraging conservation in general. Once the first ranches received golden lion tamarins, the news spread quickly, and other farmers and landowners quickly became interested in having them. Also, the golden lion tamarin education program created awards to recognize those landowners who participated in conservation initiatives. Thus, the opportunity to receive reintroduced golden lion tamarins on their properties was a key element triggering the increased participation of landowners in habitat protection (Kleiman and Mallinson 1998; Figure 15.2). Twenty-one ranches now have approximately 360 golden lion tamarins in 50 groups, deriving from the reintroduction of about 140 animals between 1984 and 2000, with the resulting protection of an additional 32 km^2 of forest (Chapters 12 and 13 this volume). Other landowners have applied and are waiting to receive golden lion tamarins.

Over the years there have been many instances when the impact of the education program was evident. On several occasions, people from the surrounding villages helped extinguish forest fires, a significant indication of their level of involvement. Another significant result was the return of more than 20 illegal pet golden lion tamarins (representing about 4 percent of the known wild population) to the research and education teams. Other wild animals, many also endangered, were likewise returned: 52 between 1993 and 1995 alone (Pinto et al. 1993).

A formal evaluation was conducted 2 years after the golden lion tamarin education activities began. Some results showed a significant change in the knowledge and attitudes of the local people. These changes can be attributed directly to the

Figure 15.2. The child of a landowner listens to the signals from a radiotransmitter on a golden lion tamarin reintroduced onto his parents' ranch. (Photo by Michael L. Power)

education program because no other activities or media events promoting conservation occurred in the area.

Much was achieved in partnership with IBAMA, the Brazilian federal environmental agency responsible for issues concerning research, management, and conservation for the four species. The program's popularity stimulated greater support, and funds allocated to the Poço das Antas Biological Reserve from federal sources increased over the years. While the laws that regulate biological reserves do not allow for visitation, the golden lion tamarin education program created such a strong demand that IBAMA responded by agreeing to support educational tours and even by allowing the education program to design and implement an environmental education center at the reserve, the first such center ever constructed on a biological reserve in Brazil. These were considerable achievements and have served as precedents for similar situations elsewhere. A second reserve, União, was created in 1998, again as a result of the education program's lobbying efforts.

To continue to build on the many accomplishments of the early years of the golden lion tamarin education program, the Associação Mico-Leão-Dourado (AMLD—Golden Lion Tamarin Association) was created in 1992. This organization currently works in close collaboration with IBAMA, the forest police, and the state and municipal secretariats for education, agriculture, and the environment, as well as the relevant community associations. Periodically, AMLD offers training courses for local teachers and community leaders as well as information and technical assistance to the ranch owners so they can register their land with IBAMA as permanent private reserves termed Reservas Particulares de Patrimônio Natural (RPPNs—Private Natural Heritage Reserves), which have tax-exempt status. As of December 2000, nine RPPNs have been created, protecting around 1,200 ha of forest.

Also, AMLD works to build local community capacity through both formal and informal educational processes. Systematic evaluation of the programs has resulted in the improvement of techniques used to target different audiences, be they community leaders, teachers, the media, ranch owners, or public authorities. A Rural Participatory Diagnosis (DRP) is another important recent initiative the AMLD has undertaken in the municipalities of Silva Jardim and Casimiro de Abreu in order to understand recent changes in the rural communities and the land uses surrounding the Poso das Antas Reserve.

In 1992, the education team first ventured into ecotourism as a way of generating income for the local ranch owners participating in the reintroduction program, as well as the local communities and the AMLD. Local guides, trained by the AMLD, are employed to take the visitors to see one of the reintroduced groups, followed by a visit to the education center and to the Bom Retiro Private Reserve, where they are able to enjoy beautiful scenery, bathe in waterfalls, and walk the forest trails.

The golden lion tamarin education team also promotes a training program for students and professionals that offers the opportunity for participation in the projects through internships of varying duration. By December 2000, 19 students had carried out field research for graduate degrees, and 107 interns had received training.

The success of the golden lion education program is evident from the fact that the golden lion tamarin is now recognized worldwide as a symbol for the conservation of Brazil's Atlantic Forest, local ranch owners are protecting their remaining forested lands, often with reintroduced golden lion tamarins, and considerable national and international support has been received through multiple grants and even the declaration of a new biological reserve, União (see Chapter 1 this volume). Over its history, the achievements of this program opened several doors for educators in Brazil, being a vanguard in its success in involving local communities while at the same time reaching broader audiences.

The Black Lion Tamarin Education Program

The education program for the black lion tamarin (*L. chrysopygus*) was begun in 1988 at Morro do Diabo State Park, which is administered by the Instituto Florestal de São Paulo (IF/SP—São Paulo Forestry Institute). In 1993, a similar program was begun for the second protected area where black lion tamarins are found, the Caetetus State Ecological Station. The black lion tamarin education program has followed a systematic process that has been helpful in assessing the evolution of the strategies applied. The model used was designed by Jacobson (1991) and is known as "Planning/Process/Product" (PPP) (Jacobson and Padua 1992, 1995; Padua 1991, 1994a, 1995, 1997; Padua and Jacobson 1993). By evaluating all phases, from planning to final product, the model helps ensure effectiveness at each step, from conception through completion.

The local community had very little environmental knowledge but showed great interest in becoming more informed (Padua 1991, 1994a, 1997). The need to disseminate information and to broaden understanding about the importance of conservation became crucial to help reverse the ongoing forest devastation. Much of the forest of the entire region had been cut during the last 50 years. Today less than 2 percent remains. Recent effects on Morro do Diabo State Park include selective logging, an airstrip, a railway line, a highway dividing the park resulting in countless road kills, and a hydroelectric dam that flooded 5 percent of the best forests in the region (Valladares-Padua 1987, 1993). Caetetus State Ecological Station is entirely surrounded by coffee plantations and cattle ranches whose activities cause serious environmental damage through the use of pesticides, the invasion by nonnative exotic species, the cattle's effect on the undergrowth, and the domestic animals that compete with wildlife, as well as through hunting, poaching, and other human invasive activities.

The black lion tamarin education program's main goal is the conservation of the remaining forest fragments. The objectives are to foster among local people an appreciation of the black lion tamarin and the remaining natural areas, including both the protected areas and other forest remnants in the region.

The first public targeted was the student population, but because students alone would be unlikely to alter the rapid destructive process underway, public participation and involvement has also been continuously emphasized. The programs have thus sponsored several outreach activities targeting all community members, from local authorities and business people to laborers and members of the agrarian reform, or the "landless" movement.

The western portion of the state of São Paulo, also known as the Pontal do Paranapanema, where Morro do Diabo State Park is located, has been one of the major regions of conflict over land ownership in Brazil. A creative new initiative has joined researchers and educators in a conservation extension program where all are benefiting. Workshops have introduced conservation principles, such as the advantages of planting trees and the ways to improve land plots and protect forest fragments. A tree nursery "school" was established at Morro do Diabo State Park, which has resulted in the creation of 11 community nurseries that are now furnishing seedlings for forest corridors, buffer zones to protect forest fragments, and "stepping stones" to enhance the connections among the remaining forest fragments. In addition, the planting of trees has clearly demonstrated that it is possible to integrate human and environmental needs. The aim of this conservation extension program is to develop means for people to improve their agricultural plots and their quality of life and at the same time protect nature more effectively. The program is a symbol for how to integrate the activities of what are seen as conflicting groups: conservationists and members of the agrarian reform movement. The results are being shared with many groups, including the Golden Lion Tamarin Conservation Program because the "landless" are also present in its region.

Park employees of Morro do Diabo State Park and Caetetus State Ecological Station have always been encouraged to cooperate, and as a consequence, trail signs and other materials for activities have been designed and mostly executed locally with very few extra resources. The IF/SP has been very supportive as well as several other institutions interested in the protection of black lion tamarins, allowing for a more rapid and less bureaucratic implementation of the different program stages.

At Morro do Diabo State Park, nature trails have been established and a visitor center was built with specially designed activities to stimulate curiosity and learning. At Caetetus State Ecological Station, the education program was begun after a teacher's course had been offered, which created a demand and raised expectations regarding its continuity. The program was established in a participatory way, involving local people who have subsequently supported the implementation of many of its strategies. The IF/SP has improved the infrastructure for student visits and allocated guards and personnel to perform different tasks. The Instituto de Pesquisas Ecológicas (IPÊ—Institute for Ecological Research) and IF/SP have succeeded in obtaining funds to build an education center at the Caetetus State Ecological Station.

Among other research areas, the education team for black lion tamarins conducted a study to assess how much information parents learn from their children (Padua et al. 1994, 2000); the study was carried out at the Fazenda Rio Claro (Rio Claro Ranch), a private forest owned by the Duratex Company. Results indicated that parents need to be addressed directly and with specially designed strategies if they are to learn via the activities in which their children participate.

Indications of the education program's effectiveness at different sites include the increase of families visiting Morro do Diabo State Park, university students spending time at the park, local teachers requesting environmental education courses, and the nature guides showing increasing interest in acquiring skills to improve their performance. Several events have demonstrated that the community is involved in the park's conservation: the community's request for a garbage location that would not threaten the park's flora and fauna, members of the community helping to extinguish two forest fires alongside the park's employees, public pressure against illegal logging at a nearby farm, stakeholders forming a group to implement sustainable alternatives that could absorb the local unskilled population rather than importing solutions from outside the community, and finally the community demanding the continuity of the program itself, which had lapsed during a change in leadership (Padua and Tabanez 1997b; Padua and Valladares-Padua 1997, 1998). Among the "landless" the best example of involvement was in May 2001 when a jaguar died after being run over by a car on the road that crosses Morro do Diabo State Park. Members of the agrarian reform movement blocked the road and for more than 5 hours impeded traffic, explaining that the animals are landless like themselves but with even fewer possibilities of defending themselves because they cannot speak. They felt that they had to represent the animals that are continuously killed on the road and demanded a solution from the highway authorities.

During all stages of the black lion tamarin conservation program, students and young conservationists have applied for training in diverse areas, resulting in the creation, in 1992, of IPÊ, a nongovernmental organization (NGO) enabling the lessons learned from the black lion tamarin program to be applied to programs for other species and ecosystems. This organization now has a training facility, the Brazilian Center for Conservation Biology, to help increase the skills of conservationists from all over the country. Thus, lessons learned from the conservation of the black lion tamarin have been applied to other fields, and the process itself as well as specific components essential for conservation to occur are now shared through education.

The Golden-Headed Lion Tamarin Education Program

The endangered golden-headed lion tamarin (*L. chrysomelas*) occurs in the Atlantic Forest of the southeastern part of the state of Bahia, Brazil. Biologically, this is one of the richest regions of the planet, but it has also been strongly affected by human activities. The Una Biological Reserve was created in 1980 to protect the golden-headed lion tamarin (Chapter 1 this volume). Although 11,400 ha of forest are included within the reserve's boundaries, only 7,022 ha have been formally acquired, insufficient for the maintenance of a genetically viable lion tamarin population in the long term (Santos and Blanes 1997, 1999). The conservation of remaining forest fragments is vital for the protection of the golden-headed lion tamarin (Alves 1990; Nagagata 1994a, 1994b; Pinto 1994; Santos 1995).

One of the principal crops in southern Bahia is cocoa (*Theobroma cacao*). Once thriving, the cocoa plantations began to decline about 15 years ago due to a drop in international prices and the spread of a highly infectious fungus (witch's broom disease, *Crinipellis perniciosa*) (Chapter 1 this volume). The decline of the cocoa plantations resulted in unemployment and invasion of forested land by unemployed workers who cleared land for subsistence agriculture. In addition, landowners began cutting their forests to sell the wood to sawmills, to diversify their crops, and for cattle ranching (Alves 1990; Argolo 1992; Pinto and Rylands 1997). These activities increased the rate of forest fragmentation in and around the Una Biological Reserve.

Cristina Alves coordinated one of the first conservation education programs for *L. chrysomelas,* Projeto Mico-Leão-Baiano, from 1990 to 1997. Prior to initiating the conservation education program, Alves interviewed more than 600 adults from the cities of Ilhéus and Itabuna in order to assess their knowledge and attitudes toward the golden-headed lion tamarin and forest conservation. Few people were aware of the negative effects of deforestation, and few knew that the golden-headed lion tamarin was endemic to the local forest. The project involved numerous educational activities, with an initial emphasis on local schools and the establishment of a nature education center near Itabuna. School-related activities included a portable exhibit about golden-headed lion tamarins and forest conservation, development of an environmental education manual for elementary and secondary school teachers, a slide show presentation for schools ("The Golden-Headed Lion Tamarin Goes to School"), and an environmental education training course for teachers.

Nagagata (1994a, 1994b) evaluated the education activities of the Projecto Mico-Leão-Baiano and their effectiveness in changing the knowledge, attitudes, and behavior of the target audience, community dwellers, as well as a nontarget audience, farmers, in the region. She concluded that both groups had good knowledge of the golden-headed lion tamarin but that community dwellers tended to have more positive attitudes toward forest conservation than farmers did. She recommended that a conservation education program be developed and implemented specifically for farmers and that the program be delivered on a personal (one-on-one) level and modified for different kinds of farmers (e.g., small farms versus large farms).

As a result of Nagagata's work, a new conservation education project for the golden-headed lion tamarin was initiated in 1995. Coordinated by Gabriel dos Santos through the Instituto de Estudos Sócio-Ambientais do Sul da Bahia (IESB—Institute for Social and Environmental Studies of Southern Bahia), the goal of this current project has been to raise public awareness of the importance of protecting the Una Biological Reserve and the remaining nearby forests (Blanes and Mallinson 1997; Santos 1995; Santos and Blanes 1997, 1999). The target audiences have, therefore, been landowners and community members at large, especially those living near and around the reserve. Project coordinators and staff have been from the local community, which has made it easier to define priorities and gain local acceptance. This has also facilitated the identification of local resources and the establishment of partnerships.

Based on local problems and local information needs, the main messages of the current education program have focused on forest protection, property rights, environmental laws, land use alternatives, and information about the Una Biological Reserve. A series of signs were placed around the edges of the reserve, indicating its existence, the importance of protecting wild animals, and the prohibition of hunting. The education program has subsequently identified specific audiences and designed strategies for each accordingly. The audiences are as follows:

- *Landowners.* Landowners decide the fate of forests on their land and, therefore, are the most important target group for conservation purposes. Project staff have visited them personally and shared information of the importance of conservation, fire control, environmental legislation, alternative land uses, and other related topics. Much of this information has also been transmitted to the farm workers through lectures and talks.
- *Una Biological Reserve guards.* Forest guards often know a great deal about the forest and its wildlife and the people living in and around the reserve. They are potentially key people for transmitting the education program's conservation message. A rapid participatory diagnosis was made to understand the problems faced by forest guards and the pressures on the reserve. Monthly meetings between project staff and forest guards have been organized so that information can be shared on a continuous basis.
- *Teachers and rural school personnel.* The conservation education team contacted the Una Education Department and head teachers to recommend the inclusion of environmental education in the regional curricula. The team also conducted a survey of rural schools and made recommendations as to where further schools were needed.

- *Hunters and extractors.* Two groups of hunters have been identified in the region: those who hunt for subsistence and those who hunt for sport. In both cases, whenever they are found hunting, they receive information about the importance of conservation and the laws that forbid hunting. *Piaçava* palm (*Attalea funifera*) fibers are used locally for thatching and for brooms. *Piaçava* collectors are contracted by landowners; therefore, it has been necessary for the education program to inform landowners who hire their services that collection within the reserve is forbidden.
- *Ecotourism program staff.* In addition to the conservation education program, IESB owns and operates a local ecotourism attraction "Ecoparque de Una." This attraction helps promote ecologically sustainable tourism in the region.

Partnerships have been important for the success of the conservation education program. The education team has developed a cooperative link with its counterpart in the AMLD, which works for the conservation of the golden lion tamarin in Rio de Janeiro. Despite regional differences, the exchanges of information and lessons learned have been helpful and have served as an inspiration in many instances. This partnership has resulted in the production of a video that stresses the importance of the involvement of landowners in the conservation of RPPNs. As a result, landowners have been inspired to support conservation initiatives.

Seventy percent of the landowners contacted have adopted some kind of conservation practice, and 10 percent are implementing sustainable agricultural activities under the guidance of IESB rural extension specialists (Santos and Blanes 1997). To broaden the scope of activities in the region, IESB has developed a Public Policy Program that interacts with IBAMA and the Instituto Nacional de Colonização e Reforma Agrária (INCRA—National Institute for Colonization and Agrarian Reform). The idea is to stop logging and forest clearance and participate in planning the agrarian reform settlements occurring around the Una Biological Reserve.

Since the education program was established, there have been no problems from fire inside the reserve (Santos and Blanes 1997), and hunting has decreased by 50 percent since 1993 and 1994 (based on IBAMA records). The acceptance of the education program by the landowners was confirmed when five of them offered their properties for research, opening new doors to reach rural laborers and their families as well as to broaden the studies being conducted in the area.

Although results indicate advances in many conservation areas, as in all education processes, continuity is the key to achieve broader effectiveness. Areas around the Una Biological Reserve need to be better protected, and the landowners must always be encouraged to preserve their forests. Education can also play an important role in resolving the tenure problems of squatters inside the reserve, especially in dealing with INCRA.

The Black-Faced Lion Tamarin Education Program

The black-faced lion tamarin (*L. caissara*) occurs in the state of Paraná in the south of the Atlantic Forest and is one of the rarest and most endangered primates. Its stronghold is the island of Superagüi, most of which is a national park of 21,400 ha. Described only in 1990, the black-faced lion tamarin's distribution, behavior, and ecology are poorly known (Lorini and Persson 1994b; Prado 1999). Based on its experience of working with black lion tamarins, IPÊ was invited to begin an ecological study on black-faced lion tamarins in 1996, and it was quickly evident that an environmental education program was needed (Prado 1999; Prado et al. 2000). The education program follows the PPP continuous evaluation model, with activities being created to transmit information gathered through field studies to the local communities.

Because Superagüi is an island and part of the Superagüi National Park, the approaches for black-faced lion tamarin education have been quite distinct from those for the other three species. The village of Superagüi is small and has only a very basic infrastructure. Fishing is the most important activity of the community, but during spawning, fishing is not allowed, and many families lack even their basic necessities. Ecologically sound economic alternatives are evidently of great importance to encourage the villagers to stay on the island and not sell their properties to outsiders who usually have no attachment to the land.

Ecotourism has been a viable alternative, but the local population lacks the means to take advantage of the ever-increasing flux of people "invading" its land. Tourists leave behind a great deal of garbage—one of the main problems facing the villagers. With no effective garbage collection, waste accumulates and frequently ends up in the ocean. Transportation constraints result in high prices at the few local markets and the unavailability of materials and skilled professionals. The school system is very poor, and most teachers themselves have reached only the fourth-grade level. The result is a very low literacy rate.

The education program began by focusing on the Superagüi students and introducing them to concepts and activities designed to raise their interest in environmental matters (Figure 15.3). The black-faced lion tamarin has become the conservation symbol for the education program on the island of Superagüi, with the aim of building up pride for the species and the local forest as a whole.

Games, guided walks, and nature tours help students experience the natural environment in a new way. Artwork is posted in local commonly visited places to

Figure 15.3. Children on the island of Superagüi dressed up in black-faced lion tamarin costumes. (Photo by Sandra Navas)

raise curiosity, awareness, and pride in the environment among adults. Videos are shown whenever possible, especially some produced by Brazilian TV networks that have focused on the black-faced lion tamarin and the red-tailed Amazon (*Amazona brasiliensis*). As people see themselves, their friends, and their relatives on the screen, they become proud of belonging to a place that is of interest to the entire country.

The education program also encourages valuing traditional culture. Local artisans have been encouraged to prepare exhibits when tourists are present. Workshops for women, teenagers, and artisans are now occurring on a continuing basis, and the education team is introducing motifs that focus on regional wildlife, which has a twofold benefit: to raise awareness among local people about the importance of regional nature and to disseminate to broader audiences what is found in Superagüi National Park. In addition, the production of artifacts that focus on local nature can improve family revenues, which in turn may perhaps increase its overall intrinsic value to the local community.

A newsletter, *Supernotícias* (Supernews—because of Superagüi), is produced periodically to inform people of IPÊ's research in the island, regional environmental curiosities, and interviews with local people about the Atlantic Forest. Because many people are illiterate, the newsletter increases comprehension by using a layout that includes a black-faced lion tamarin and a parrot as part of the letterhead, as well as other drawings.

The education program for the black-faced lion tamarin has involved all sectors of the Superagüi community in conservation. The effort of being inclusive has produced positive results, with local people supporting and participating in the education team's initiatives. Raising people's self-esteem has probably been one of the most important aspects of the education program on the island of Superagüi. Local people have discovered a new pride in their work and in their environment. The importance of this strategy is based on the idea that those who have self-confidence are more likely to participate in what they believe to be right (Padua and Tabanez 1997a).

The introduction of community-based ecotourism has inspired the organization of a group of women who are now producing artifacts and preparing meals for tourists. Through educational workshops the black-faced lion tamarin and the red-tailed parrot are now pictured on embroidered towels, puppets, and others items that are being sold to tourists, who also learn in the process. This has been a win-win situation where people and nature together benefit.

Partnerships have also emerged from the education program. An NGO, Sociedade de Pesquisa em Vida Selvagem e Educação Ambiental (SPVS—Society for Wildlife Research and Environmental Education), has been working with IPÊ, mainly on the education program for the red-tailed Amazon. Since SPVS is based in Paraná and has been working in the region for many years, its team has helped IPÊ reach leaders and decision makers to influence policies that can enhance conservation.

The black-faced lion tamarin education program is the most recent of the four, and it still has a great deal to accomplish. However, the acceptance and support of the local people and of IBAMA's park director, Guadalupe Vivekananda, are strong indications that the seeds are being sown appropriately. Continuous dedication to and evaluation and improvement of specific strategies are required so that conservation principles can flourish and become an integral part of the region's life and culture. (See Vivekananda 2001 for a review of the history of the human impact on the environment in and around Superagüi National Park.)

CONCLUSIONS

Conservation education programs have been an important component of multidisciplinary efforts to conserve the four lion tamarin species and the forest habitats on which they depend. Although the four species are by no means safe from the threat of extinction, these integrated programs have begun to yield positive conservation results. However, only sustained conservation action, including education, over the long term will ensure the survival of these unique creatures and their diverse and biologically rich forest habitats.

The conservation education programs for the four lion tamarin species, together with other conservation measures, serve as models of effective and integrated efforts toward species and habitat protection. Through public awareness programs, people are encouraged to value and take greater pride in their local natural resources. The "empowerment" and action of local people have shown that individuals can contribute greatly to the conservation of natural areas. This is especially important in Brazil because of the richness of the country's biological diversity and the lack of resources for its protection.

It may be extreme to say that environmental education alone can be responsible for effective conservation. Nevertheless, because people's behavior is the main cause of destruction of natural habitats worldwide, it is of critical importance for individuals to be reached in ways that can change their attitudes toward nature. Over time it has become clear that conservation education, although demanding time and continuous inputs, serves as an important tool to promote the changes that are so evidently needed. The lion tamarin programs are certainly examples of what can be achieved when education processes are implemented with commitment.

Cooperation among the teams working for the conservation of the lion tamarins through education has been outstanding. The golden lion tamarin community education program was a pioneer in Brazil and as such served as inspiration to other programs. However, all four programs now share their experiences with each other and with other conservation programs in Brazil and abroad. In this way all have benefited from the lessons learned and have been able to progress more effectively.

Another highlight of the education programs for lion tamarins has been the partnerships established among NGOs and among NGOs and governmental agencies (see Chapter 3 this volume). Thus, the education programs for lion tamarins have also served as models that integrate individuals, institutions, and communities for species and habitat conservation.

ACKNOWLEDGMENTS

Acknowledgments for the Golden Lion Tamarin Conservation Program (GLTCP) and AMLD can be found in Chapter 3 (this volume). The AMLD is grateful to all the institutions that have supported it with resources and especially thank the local communities in Silva Jardim and Casimiro de Abreu for working with it as its educational and public policy programs evolved.

The successes of the black lion tamarin environmental education program would not have been possible without the support of the local communities; enthusiasm of the education team, especially Sr. Arnaldo and family and Marlene Tabanez; IF/SP—Secretaria do Meio Ambiente (SMA—Secretary of the Environment) de São Paulo; Apenheul Zoo, Holland; Canadian Embassy in Brazil; Fanwood Foundation; IdeaWild; U.S. Fish and Wildlife Service; National Geographic Society; Whitley Animal Protection Trust; World Wildlife Fund; Conselho Nacional de Desenvolvimento Científico e Tecnológico (CNPq—Brazilian National Science Council); and, at the University of Florida, Center for the Latin American Studies, the Program for Studies in Tropical Conservation, and the Tropical Conservation and Development Program.

Acknowledgments for staff of the IESB and the golden-headed lion tamarin education program can be found in the Acknowledgments section of Chapter 3 (this volume).

The black-faced lion tamarin education program is grateful to Sandra Navas and Fabiana Prado for their enthusiasm and dedication, IBAMA, and Superagüi National Park's director, Guadalupe Vivekananda. The black-faced lion tamarin receives support from Fundação O Boticário de Proteção à Natureza (Boticário Foundation for Nature Protection), Fundo Nacional do Meio Ambiente (National Environment Fund), IdeaWild, Lincoln Park Zoo Scott Neotropic Fund, Lion Tamarins of Brazil Fund, Margot Marsh Biodiversity Foundation, Programa Natureza & Sociedade (Nature and Society Program)—State University of New York/WWF, and Wildinvest-UK.

DEVRA G. KLEIMAN AND ANTHONY B. RYLANDS

16

LION TAMARIN BIOLOGY AND CONSERVATION: A SYNTHESIS AND CHALLENGES FOR THE FUTURE

The research, education, and conservation efforts for the lion tamarins are exceptional in the world of conservation biology, primatology, and biodiversity-focused research. The programs for the four species have been at the forefront of work in any number of disciplines during the past quarter century, but they have especially shown how a multidisciplinary team approach to basic and applied research can significantly improve the prospects for the recovery of endangered species (and the recovery of the native habitats they occupy). At a time when the need for basic biological research to support conservation efforts is under some discussion (e.g., Kinnaird and O'Brien 2001; Sheil 2001; Whitten et al. 2001), the lion tamarin programs are a powerful reminder that behind any successful conservation effort must be a research base. An experimental scientific approach to conservation is still the best paradigm we have available for making decisions and elaborating, and changing, strategies.

Any conservation strategy is based on expected outcomes and cost-benefit analyses, however unsophisticated they may be. Prediction through modeling not only refines a strategy but can also create new ones. However, the modeling requires data. The Population Viability Analysis (PVA) Workshop for the lion tamarins in 1990 was riddled with unknowns and uncertainties, and one of the major results was the identification of areas that required more research (Rylands 1993/1994; Seal et al. 1990). By contrast, the recommendations derived from a second, similar workshop in 1997, by then termed Population and Habitat Viability Analysis (Lacy 1993), were of specific actions and strategies for in situ conservation (Ballou et al. 1998). They were based on major distribution surveys for the four species carried out during the 1990s (Kierulff 1993a, 1993b; Lorini and

Persson 1994b; Martuscelli and Rodrigues 1992; Pinto 1994; Pinto and Rylands 1997; Pinto and Tavares 1994; Valladares-Padua and Cullen 1994), on field studies that had provided the needed demographic data (in 1990, based almost entirely on data available for the golden lion tamarin) (Baker and Dietz 1996; Dietz et al. 1994a, 1994c, 1996), and on a considerably improved database concerning forest remnants, landownership, and the possibilities for creating protected areas (Santos 1995; Valladares-Padua and Martins 1996b; Valladares-Padua et al. 1999).

Ongoing research continued to improve the husbandry and management of the captive populations (Ballou 1992, 1996; Ballou and Foose 1996; Ballou and Lacy 1995); resulted in the golden lion tamarin translocation program, which created a new population now accounting for more than 10 percent of the entire wild population (Chapter 12 this volume); and provided the understanding necessary for the reintroduction and translocation program of the black lion tamarin (Chapter 14 this volume).

The lion tamarin programs are also the best and most long-lasting example of the integration of field and captive conservation efforts on a global scale. The zoo community has long insisted that its main function is conservation, but many detractors believe that zoo professionals basically pay lip service to the global environmental crisis and use conservation as a marketing strategy to bring in visitors and raise funds for exhibits. The lion tamarin conservation efforts are at least one example where they are wrong. Through a variety of venues, the zoo community has probably supported well over 50 percent of the research and recovery work on lion tamarins. Zoos and their publics support lion tamarins because they are a "flagship" species (Dietz et al. 1994b), and they do so with monies that would not be available for the protection and preservation of habitat alone. However, habitats are being protected as a result of these species recovery efforts.

Zoo behavioral research and the data-based species management for golden lion tamarins set the global standard for captive endangered species management programs (Chapter 4 this volume). The golden lion tamarin program is probably the only zoo-based program of its kind in which the ownership of the vast majority of the captive population, owned for decades by zoos, has been returned to the range country—a symbolic but significant message to the citizens of Brazil.

The decades of field research on the lion tamarins have contributed significantly to our knowledge base concerning their habitat and critical resource needs, which in turn has been the basis of decisions concerning the type and size of protected areas that we need to create and manage for their survival. The protracted nature of the research has meant that decisions concerning protection are informed by natural events, such as long-term fluctuations in environmental parameters and population size that can only be seen and measured by extensive and prolonged observations. Only recently, a population decline of golden lion tamarins studied in the Poço das Antas Biological Reserve, apparently due to a significant change in predation pressure (Franklin and Dietz 2001), has resulted in significant changes in group sizes and compositions that would have been considered the norm if the study had only been in effect for the last few years.

The long-term research has also informed biological arguments about the evolution of mating systems, especially in defining what constitutes genetic and social monogamy (Chapter 8 this volume). Following known individuals and groups of golden lion tamarins for more than a decade, Andrew Baker, James Dietz, and colleagues can now speak with authority about the percentage of groups in which polyandry and polygyny occur and the specific contexts that lead to a deviation from the norm of genetic and social monogamy. The field research on the muriqui (*Brachyteles arachnoides*) at the Caratinga Biological Station in Minas Gerais begun by Karen Strier in 1983 (Strier 1987, 1992) and that on the golden lion tamarins (Chapter 8 this volume) are the longest continuous field studies of any of the New World primates and stand next to classic long-term studies such as those by Jane Goodall and colleagues on chimpanzees (*Pan troglodytes*) in Tanzania's Gombe Reserve, by Wolfgang Dittus on toque macaques (*Macaca sinica*) in Sri Lanka, and by Jeanne and Stuart Altmann on yellow baboons (*Papio cynocephalus*) in Kenya.

The conservation needs of the lion tamarins have led to the creation of a number of national nongovernmental organizations (NGOs) in Brazil that are having a powerful effect on attitudes about the natural environment in a country struggling with the need for development while trying to preserve its natural heritage (Chapter 3 this volume). Lion tamarins and muriquis are *the* symbols for the conservation of the Atlantic Forest and, indeed, for this entire country. Both have been the theme of special stamp issues and the focus of considerable national media attention. Most recently, golden lion tamarins were voted nationwide to be put on a new Brazilian R$50 banknote.

The education and outreach programs (Chapter 15 this volume) for the lion tamarins can take full credit for the popularity of these species in Brazil; the creative use of the media to bring the conservation message to the public has been the major factor driving the engine of public opinion. Twenty years ago, the Brazilian public neither knew nor cared about these unique primates, nor for that matter most other environmental issues. Today, there is hardly any Brazilian with access to TV, radio, and print media who is ignorant of the lion tamarins, and most understand and are proud that these species are unique and special to Brazil.

The lion tamarin education programs are also distinctive because they are themselves research based. Additionally, they have had conservation rather than development as their primary focus at a time when Integrated Conservation and Development Programs (ICDPs) have dominated much international conservation work (Oates 1999). Finally, the lion tamarin education programs are classic examples of adaptive management at work: strategies have changed as (1) the threats have changed and (2) research results have provided data suggesting improved methodologies for reaching different target groups.

Lion tamarin conservation emerged as an issue in Brazil at a time when fledgling democratic institutions were being put in place and successive governments tried to control inflation while supporting the legitimate needs of millions of people at the edge of or in poverty. Even today, the lion tamarin education programs are challenged by the activities of the "landless" movement ("Sem Terra"), a political force that is the most recent to threaten the degradation and destruction of surviving forest fragments. In this climate, the partnership between the Instituto Brasileiro do Meio Ambiente e dos Recursos Naturais Renováveis (IBAMA—Brazilian Institute of the Environment and Renewable Natural Resources) and the lion tamarin programs has been extraordinary. The staff of IBAMA, most recently Maria Iolita Bampi and her colleagues, have been incredibly supportive of the lion tamarin conservation programs, often creatively responding to changing political pressures, government strikes, and new legislation with suggestions and solutions that facilitate efforts to protect the species and their habitats. Many of the IBAMA staff have devoted their careers to ensuring that environmental issues are considered "up front" and have done a balancing act as Brazil climbs its dizzying developmental ladder.

The lion tamarin conservation programs are also at the forefront of metapopulation management efforts, last-ditch attempts to preserve genetically viable populations of endangered species existing in unimaginably fragmented habitats. The maps of the current distributions of the black, golden, and black-faced lion tamarins are a shocking reminder of how much of the Atlantic Forest, one of the world's richest in endemism and biodiversity, has been lost to us and future generations (Chapter 2 this volume). These metapopulation management programs, which approximate the management of endangered species in captivity, are the only way we can ensure survival of both lion tamarins and their habitats in the foreseeable future. Our ultimate goal must be to re-create continuous forest to maintain lion tamarin populations of sufficient size and genetic diversity to permit the continuation of natural evolutionary processes (Myers and Knoll 2001). What we humans have done in the name of development, we now need to undo in the name of a sustainable future for our descendants.

BIOLOGICAL CHALLENGES FOR THE FUTURE

The lion tamarins have a disjunct distribution and are spread over thousands of kilometers; they are also morphologically distinct. Yet they lack significant genetic variation (Chapter 5 this volume); indeed, they probably have the least amount of genetic variation of any primate genus. Understanding better how the lion tamarin mating system and the four species' isolation during the Pleistocene (when the Atlantic Forest divided into multiple refugia) contributed to their lack of genetic variation (Kinzey 1982; Rylands et al. 1996) will add significantly to our knowledge of the role of bottlenecks in evolution and the ability of species to adapt to low levels of genetic variation.

Cláudio Valladares-Padua and colleagues (Chapter 14 this volume) examined the ability of black lion tamarins to exist in different forest habitat types and found them to be highly flexible and adaptable. Given the low genetic variation in the genus and the degree to which the Atlantic Forest has been destroyed and degraded time and time again for centuries (Dean 1995), it is a wonder that lion tamarins have survived at all! As the conservation community attempts to create "green" corridors and expand the amount of continuous forest available for lion tamarins, it will be significant to determine what limits there are in terms of the species' abilities to adapt to different environments. How far can they expand their ranges and distribution? We can now evaluate a population's degree of adaptation to its environment by measuring reproductive and "stress" hormone levels using noninvasive methods such as Jeffrey French and his colleagues (Chapter 6 this volume) have developed. An analysis of endocrine responses within different habitat types or to changing environmental conditions, including catastrophic events, can answer many important questions about lion tamarin physiology, ecology, and evolution. Among other significant issues, we can determine (1) whether or not lion tamarins in extreme habitats exhibit higher stress levels and reduced reproductive success and even (2) what constitutes a forest habitat type in which lion tamarins cannot survive.

The importance of long-term data on demography and genetics for decision making about the critical requirements for a species cannot be understated. For golden lion tamarins, the population decline in the Poço das Antas Biological Reserve has underscored the value of long-term data for decision making. Additionally, the more data are collected, the more we will understand about the natural variation in home range size and population density in different habitat types. Cecília Kierulff, Becky Raboy, and colleagues (Chapter 7 this volume) demonstrate just how much variability there is in the key characteristics of the natural history (e.g., group size and density, home range size, and daily path length) for these four species. Our challenge is to separate the effects of habitats and ecological conditions from genetic influences on these characteristics.

This brings us to the variability in the social, rearing, and mating systems of lion tamarins. What are the key habitat characteristics that affect the number of breeding females per group and how many offspring a group can rear annually? How do we really define the variation in group size and age-sex structure together with the mating system? Can we look at the typical monogamous mating system of lion tamarins as a series of probabilities that one male and one female will mate exclusively over a series of years and determine how demographic stochasticity affects these probabilities? Again, the recent population changes in the Poço das Antas Biological Reserve have provided an opportunity to look at a relationship among a series of demographic and behavioral characteristics that is rarely available to biologists.

Lion tamarins have some unique characteristics within the Callitrichidae and share further characteristics with marmosets and other tamarins about which we know little. Lion tamarins use tree holes for sleeping at night, the only callitrichids to do so. Their foraging strategies are more complicated than those of most other callitrichids because they feed on fruits, invertebrate and vertebrate prey, nectar, and exudates (Chapter 7 this volume). Some lion tamarins are sympatric with *Callithrix* species; others are not. How all of these ecological differences originated and how they affect behavior and other life history characteristics are still unknown.

The lion tamarins do not exhibit the physiological inhibition that is typically seen in marmosets and tamarins (Chapter 6 this volume), and thus more than one female may occasionally give birth. Another physiological issue relates to the degree to which males and other helpers exhibit physiological changes during the rearing phase and whether or not this might be universal among callitrichids. All of these biological issues remain to be explored and resolved.

SOCIAL, POLITICAL, AND ORGANIZATIONAL CHALLENGES FOR THE FUTURE

The preceding section focuses on the biological challenges for the future lion tamarin programs. There are, however, equally interesting and pertinent socioeconomic and political problems to be solved.

There are currently 5,000 to 10,000 golden-headed lion tamarins still existing in southern Bahia: the largest extant population of any of the lion tamarin species. The reason for this lies in the historical pattern of land ownership in the region: very large cocoa estates where only a small part was given over to plantations. With the change in the regional economy over the last two decades (Alger and Araújo 1996; Alger and Caldas 1994), the once relatively widespread forests are now being rapidly diminished. The Una Biological Reserve, similar in size to the Poço das Antas Biological Reserve for the golden lion tamarins, is too small to retain a genetically viable population of this species. In a decade, the plight of the golden-headed lion tamarin could be comparable to that of the golden lion tamarin 30 years ago. We cannot, yet again, watch and wait as a threatened species reaches numbers so small that the species' survival becomes critically endangered. We need to act now to find new, creative ways of preserving enough habitat, both private and publicly owned, to maintain viable populations of golden-headed lion tamarins throughout a large part of their range. Opportunities still remain to maintain landscape mosaics, with a combination of protected areas, forests in corridors, and other areas with benign land uses (IESB 1996; Saatchi et al. 2001).

As Denise Rambaldi and her colleagues (Chapter 3 this volume) state, many NGOs became established during the period when the international community met at the Earth Summit in Rio de Janeiro in 1992 to discuss the global environmental crisis. Many of them have since disappeared, and a major challenge for the future is to ensure the survival of those that have survived and are truly dedicated to resolving environmental problems. Creating a group of truly self-sufficient NGOs must be an objective of the national and international conservation communities that is tackled with as much innovation and verve as is shown by these NGOs themselves. These NGOs, increasingly recognized for their competence, will make a major contribution to the future of conservation in Brazil (Fonseca and Pinto 1996). Only they can develop locally relevant marketing strategies and tools; gain political access and influence within state and national governments; exert a continuing and pervasive impact on the political process, especially as it relates to environmental issues; and continue to partner with the government to create new and unique methods for increasing the size and security of protected areas—for example, through expanding private ownership of conservation lands, as with the concept of the Reserva Particular de Patrimônio Natural (RPPN—Private Natural Heritage Reserve).

The remaining Atlantic Forest is too small and fragmented to secure the survival of the lion tamarins and the astonishing biodiversity that exists therein. The lion tamarin programs are all exploring and implementing the development of corridors between isolated forest fragments as a means of re-creating a continuous forest block that permits natural dispersal and evolutionary processes. The creation of natural corridors is at the forefront of the conservation movement; overcoming the biological and political obstacles is the challenge faced by the NGOs and partners concerned with lion tamarin conservation.

For the foreseeable future, three of the four lion tamarin species must be managed as metapopulations to secure their survival. Much has been written about what metapopulations are and their significance in a conservation context (Chapter 14 this volume). The lion tamarin programs may be the only endangered species recovery efforts with a clear theoretical and practical approach to metapopulation management.

The biology and policy of conservation are evolving disciplines; over the previous 30-year period, the philosophy and focus of the environmental movement have changed. We can be certain of only one thing: more changes in philosophy and focus will occur as we gain knowledge and experience in solving environmental problems. That is the essence of adaptive management. One of the major mechanisms for incorporating the most cutting edge concepts into a program is through peer evaluation. Program reviews are rare in the conservation community (Kleiman et al. 2000); the Golden Lion Tamarin Conservation Program requested and received a program evaluation in 1997 by five of the top conservationists in Brazil and North America. The importance of this review in re-establishing objectives cannot be underestimated. Our challenge for the future is to welcome constructive reviews of our programs to ensure that they are constantly improved and that we are meeting our research and conservation goals.

Thirty years ago, there were no protected areas for lion tamarins in Brazil. All species were in decline in the wild, and the numbers of golden lion tamarins in captivity (the only species in zoos) were declining precipitously. Today, there are six protected areas in Brazil, three of the four species are managed successfully in zoos, and for the most part the species' decline in the wild has been arrested (Chapters 2 and 4 this volume). Our knowledge of lion tamarin biology has mushroomed as has the political clout of the national and international NGO community in the environmental arena. Yet we still have a distance to go to ensure that these species survive. Hopefully, by 2032, we will be able to review another 30 years of advances in science and conservation to equal the progress we have made since the initial "Saving the Lion Marmoset" conference (Bridgwater 1972a).

REFERENCES

Abbott, D. H. 1984. Behavioral and physiological suppression of fertility in subordinate marmoset monkeys. *American Journal of Primatology* 6: 169–86.

———. 1993. Social conflict and reproductive suppression in marmoset and tamarin monkeys. In W. A. Mason and S. P. Mendoza (eds.), *Primate social conflict,* pp. 331–72. Albany: State University of New York Press.

Affonso, A. G., Ruiz-Miranda, C. R., and Beck, B. B. In press. Interações ecológicas entre mico-leão-dourado (*Leontopithecus rosalia* Linnaeus, 1758) reintroduzido e mico estrela (*Callithrix jacchus* Linnaeus, 1758) introduzido em fragmentos de mata Atlântica, RJ. In S. L. Mendes and A. G. Chiarello (eds.), *A primatologia no Brasil—8.* Santa Teresa, ES: Sociedade Brasileira de Primatologia.

Albernaz, A. L. K. M. 1997. Home range size and habitat use in the black lion tamarin (*Leontopithecus chrysopygus*). *International Journal of Primatology* 18: 877–87.

Albuquerque, F. 1999. Cuidado cooperativo a prole em *Callithrix jacchus:* Dinâmica em ambiente natural. Ph.D. dissertation, Universidade Federal do Rio Grande do Norte, Natal.

Al-Doory, Y. 1972. Systemic processes. In R. N. T. W. Fiennes (ed.), *Pathology of simian primates. Part 2,* pp. 224–42. Basel: S. Karger.

Alger, K. and Araújo, M. 1996. Desmatamento dos últimos remanescentes florestais próximos à Reserva Biológica de Una: Uma ameaça à biodiversidade e à economia local. In *Alternativas econômicas para conservação e desenvolvimento da região de Una, Bahia. Resumos das pesquisas, 1994–1995,* pp. 2–5. Ilhéus, BA: Instituto de Estudos Sócio-Ambientais do Sul da Bahia (IESB).

Alger, K. and Caldas, M. 1994. The declining cocoa economy and the Atlantic forest of southern Bahia, Brazil: Conservation attitudes of cocoa planters. *Environmentalist* 14: 107–19.

Alger, K., Araújo, M., Trevizan, S., and Santos, G. dos. 1996. Dinâmica do uso da terra no entorno da Reserva Biológica de Una. In *Alternativas econômicas para conservação e desenvolvimento da região de Una, Bahia. Resumos das pesquisas, 1994–1995,* pp. 12–16. Ilhéus, BA: Instituto de Estudos Sócio-Ambientais do Sul da Bahia (IESB).

Allen, M. E. and Montali, R. J. 1995. Nutrition and diseases in zoo animals. *Erkrankungen der Zootiere* 37: 215–31.

Allnutt, T. F. 1997. Population viability and reserve expansion in coastal Brazil: Contributions of remote sensing and geographic information systems. Master's thesis, University of Maryland, College Park.

Alonso, C., Leite, J. C. L., and Schiel, N. 1997a. As primeiras sete semanas de coabitação em casais recém-formados de *Leontopithecus chrysomelas*. In S. F. Ferrari and H. Schneider (eds.), *A primatologia no Brasil—5,* pp. 195–204. Belém, PA: Universidade Federal do Pará, Sociedade Brasileira de Primatologia.

Alonso, C., Porfírio, S., and Moura, A. C. A. 1997b. Ocorrência de poliginia, incesto, e expulsão do macho reprodutor em um grupo de *Leontopithecus chrysomelas*. In S. F. Ferrari and H. Schneider (eds.), *A primatologia no Brasil—5,* pp. 279–87. Belém, PA: Universidade Federal do Pará, Sociedade Brasileira de Primatologia.

Alves, M. C. 1988. Tamarins and cocoa don't mix . . . or do they? *On the Edge,* no.36: 3–4, 12.

———. 1990. The role of cacao plantations in the conservation of the Atlantic forest of southern Bahia, Brazil. Master's thesis, University of Florida, Gainesville.

———. 1991. Community conservation education for the Atlantic forests of southern Bahia, Brazil, with emphasis on the golden-headed lion tamarin (*Leontopithecus chrysomelas*). Unpublished report, World Wildlife Fund, Washington, DC.

———. 1992. A community conservation education for the Atlantic forests of southern Bahia, Brazil, with emphasis on the golden-headed lion tamarin (*Leontopithecus chrysomelas*). Unpublished report, World Wildlife Fund, Washington, DC.

Amato, F., Sugamosto, M. L., and Grando, E. T. 2000. Uso do sistema de informações geográficas—SIG para o controle da ocupação human no entorno do Parque Nacional do Superagüi. In M. S. Milano and V. Theulen (eds.), *II Congresso Brasileiro de Unidades de Conservação. Anais. Vol. II. Trabalhos técnicos,* pp. 365–71. Rede Nacional Pró-Unidades de Conservação, Campo Grande, Mato Grosso do Sul, and Fundacão O Boticário de Proteção à Natureza, São José dos Pinhais, Paraná.

Anderson, S., Bankier, A. T., Barrel, B. G., de Bruijn, M. H. L., Coulson, A. R., Drouin, J., Eperon, I. C., Nierlich, D. P., Roe, D. A., Sanger, F., Shereier, T. H., Smith, A. J. H., Staden, R., and Young, I. G. 1981. Sequence organization of the human mitochondrial genome. *Nature, London* 290: 457–65.

Anver, M. R., Hunt, R. D., and Chalifoux, L. V. 1972. Cholesterol gallstones in *Aotus trivirgatus*. *Journal of Medical Primatology* 1: 241–46.

Argolo, A. J. S. 1992. Considerações sobre a ofidiofauna dos cacauais do sudeste da Bahia, Brasil. Unpublished report, Universidade Federal de Santa Cruz, Ilhéus, Bahia.

Armstrong, D. P. and McLean, I. G. 1995. New Zealand translocations: Theory and practice. *Pacific Conservation Biology* 2: 39–54.

Arnason, U., Xu, X., and Gullberg, A. 1996. Comparison between the complete mitochondrial DNA sequences of *Homo* and the common chimpanzee based on nonchimeric sequences. *Journal of Molecular Evolution* 42: 145–52.

Arnold, S. J. 1995. Monitoring quantitative genetic variation and evolution in captive populations. In J. D. Ballou, M. Gilpin, and T. J. Foose (eds.), *Population management for survival and recovery: Analytical methods and strategies in small population conservation,* pp. 295–317. New York: Columbia University Press.

Asa, C. S., Porton, I., Baker, A. M., and Plotka, E. D. (1996). Contraception as a management tool for controlling surplus animals. In D. G. Kleiman, M. E. Allen, K. V. Thompson, and S. Lumpkin (eds.), *Wild mammals in captivity: Principles and techniques,* pp. 451–67. Chicago: University of Chicago Press.

Audi, A. 1986. A contribuição da CESP (Companhia Energética de São Paulo) na implantação do plano de manejo da Reserva Estadual "Morro do Diabo." In M. T. de Mello (ed.), *A primatologia no Brasil—2,* pp. 257–59. Brasília: Sociedade Brasileira de Primatologia.

August, P. V. and Anderson, J. G. T. 1987. Mammal sounds and motivation-structural rules: A test of hypotheses. *Journal of Mammalogy* 68: 1–9.

Badridze, J. L. 1994. Captive-raised wolves become wild in Georgia. *Reintroduction News,* no. 8: 14–15.

Baker, A. J. 1991. Evolution of the social system of the golden lion tamarin (*Leontopithecus rosalia*): Mating system, group dynamics, and cooperative breeding. Ph.D. dissertation, University of Maryland, College Park.

Baker, A. J. and Dietz, J. M. 1996. Immigration in wild groups of golden lion tamarins (*Leontopithecus rosalia*). *American Journal of Primatology* 38: 47–56.

Baker, A. J. and Woods, F. 1992. Reproduction of the emperor tamarin (*Saguinus imperator*) in captivity, with comparisons to cotton-top and golden lion tamarins. *American Journal of Primatology* 26: 1–10.

Baker, A. J., Dietz, J. M., and Kleiman, D. G. 1993. Behavioural evidence for monopolization of paternity in multi-male groups of golden lion tamarins. *Animal Behaviour* 46: 1091–103.

Bales, K. L. 2000. Mammalian monogamy: Dominance, hormones and maternal care in wild golden lion tamarins. Ph.D. dissertation, University of Maryland, College Park.

Bales, K. L., Dietz J. M., Baker, A. J., Miller, K., and Tardif, S. D. 2000. Effects of allocare-givers on fitness of infants and parents in callitrichid primates. *Folia Primatologica* 71: 27–38.

Bales, K. L., O'Herron, M., Baker, A. J., and Dietz, J. M. 2001. Sources of variability in number of live births in wild golden lion tamarins (*Leontopithecus rosalia*). *American Journal of Primatology* 54: 211–21.

Bales, K. L., French, J. A., and Dietz, J. M. 2002. Explaining variation in maternal care in a cooperatively breeding mammal. *Animal Behaviour* 63: 419–35.

Ballou, J. D. 1983–1996. *International studbook for golden lion tamarins.* Published annually, 1983–1996. Washington, DC: National Zoological Park.

———. 1985. *1984 International golden lion tamarin studbook.* Washington, DC: National Zoological Park.

———. 1987. Small populations, genetic diversity and captive carrying capacities. *Proceedings 1987 AAZPA Annual Conference,* pp. 33–47.

———. 1989. Emergence of the captive population of golden-headed lion tamarins *Leontopithecus chrysomelas. Dodo, Journal of the Jersey Wildlife Preservation Trust* 26: 70–77.

———. 1990. Small population overview. In U. S. Seal, J. D. Ballou, and C. Valladares-Padua (eds.), Leontopithecus: *Population Viability Analysis Workshop report,* 56–66. Apple Valley, MN: International Union for Conservation of Nature and Natural Resources/Species Survival Commission (IUCN/SSC) Captive Breeding Specialist Group (CBSG).

———. 1992. Genetic and demographic considerations in endangered species captive breeding and reintroduction programs. In D. McCullough and R. Barrett (eds.), *Wildlife 2001: Populations,* pp. 262–75. Barking, UK: Elsevier Science Publishing.

———. 1996. Small population management: Contraception of golden lion tamarins. In P. N. Cohn, E. D. Plotka, and U. S. Seal (eds.), *Contraception in wildlife: Book 1,* pp. 339–58. Lewiston, NY: Edwin Mellen Press.

——— (ed.). 1997–2001. *Tamarin Tales,* nos. 1–5. Washington, DC: National Zoological Park.

Ballou, J. D. and Foose, T. J. 1996. Demographic and genetic management of captive populations. In D. G. Kleiman, M. E. Allen, K. V. Thompson, and S. Lumpkin (eds.), *Wild mammals in captivity: Principles and techniques,* pp. 263–83. Chicago: University of Chicago Press.

Ballou, J. D. and Houle, C. 1999. *1998 International studbook golden lion tamarin* Leontopithecus rosalia. Washington, DC: National Zoological Park.

———. 2000. *1999 International studbook golden lion tamarin* Leontopithecus rosalia. Washington, DC: National Zoological Park.

Ballou, J. D. and Lacy, R. C. 1995. Identifying genetically important individuals for management of genetic diversity in captive populations. In J. D. Ballou, M. Gilpin, and T. Foose (eds.), *Population management for survival and recovery,* pp. 76–111. New York: Columbia University Press.

Ballou, J. D. and Mickelberg, J. 2001. *2000 International studbook golden lion tamarin* Leontopithecus rosalia. Washington, DC: National Zoological Park.

Ballou, J. D. and Sherr, A. 1997. *1996 International studbook golden lion tamarin* Leontopithecus rosalia. Washington, DC: National Zoological Park.

Ballou, J. D. and Valladares-Padua, C. 1990. Population extinction model of lion tamarins currently in protected areas. In U. S. Seal, J. D. Ballou, and C. Valladares-Padua (eds.), Leontopithecus: *Population Viability Analysis Workshop report,* pp. 79–94. Apple Valley, MN: International Union for Conservation of Nature and Natural Resources/Species Survival Commission (IUCN/SSC) Captive Breeding Specialist Group (CBSG).

Ballou, J. D., Lacy, R. C., Kleiman, D. G., Rylands, A. B., and Ellis, S. (eds.). 1998.

Leontopithecus *II: The second population and habitat viability assessment for lion tamarins* (Leontopithecus). Apple Valley, MN: World Conservation Union/Species Survival Commission (IUCN/SSC) Conservation Breeding Specialist Group (CBSG).

Baptista, L. F. and Gaunt, S. L. L. 1997. Bioacoustics as a tool in conservation studies. In J. R. Clemmons and R. Buchholz (eds.), *Behavioral approaches to conservation in the wild,* pp. 212–42. Cambridge: Cambridge University Press.

Barroso, C. M. L., Schneider, H., Schneider, M. P. C., Sampaio, M. I. C., Harada, M. L., Czelusniak, J., and Goodman, M. 1997. Update on phylogenetic systematics of New World monkeys: Further DNA evidence for placing the pygmy marmoset (*Cebuella*) within the marmoset genus *Callithrix*. *International Journal of Primatology* 18: 651–73.

Bateson, P. 1994. The dynamics of parent-offspring relationships in mammals. *Trends in Ecology and Evolution* 9: 399–403.

Beck, B. B. 1991. Managing zoo environments for reintroduction. *Proceedings 1991 AAZPA Annual Conference,* pp. 436–40.

———. 1992. AAZPA (American Association of Zoological Parks and Aquariums) Reintroduction Advisory Group Guidelines for Reintroduction of Animals Born or Held in Captivity. Unpublished.

Beck, B. B. and Castro, M. I. 1994. Environments for endangered primates. In E. F. Gibbons, E. Wyers, E. Waters, and E. Menzel (eds.), *Naturalistic environments in captivity for animal behavior research,* pp. 209–70. Albany: State University of New York.

Beck, B. B. and Martins, A. F. 1997. Annual report of golden lion tamarin reintroduction. Unpublished. Washington, DC: National Zoological Park.

———. 1999. Update on the golden lion tamarin reintroduction. *Tamarin Tales,* no. 3: 6–7.

———. 2000. Annual report of golden lion tamarin reintroduction. Unpublished. Washington, DC: National Zoological Park.

———. 2001. Update on the golden lion tamarin reintroduction program. *Tamarin Tales,* no. 5: 7–8.

Beck, B. B., Dietz, J. M., Kleiman, D. G., Castro, M. I., Lémos de Sá, R. M., and Luz, V. L. F. 1986a. Projeto Mico-Leão. IV. Reintrodução de micos-leões-dourados (*Leontopithecus rosalia* Linnaeus, 1766) (Callitrichidae, Primates) de cativeiro para seu ambiente natural. In M. T. de Mello (ed.), *A primatologia no Brasil—2,* pp. 243–48. Brasília: Sociedade Brasileira de Primatologia.

Beck, B. B., Kleiman, D. G., Dietz, J. M., Dietz, L. A, Castro, M. I., and Rettberg-Beck, B. 1986b. Preparation of golden lion tamarins for reintroduction to the wild. Paper presented at the XIth Congress of the International Primatological Society, Göttingen, Germany.

Beck, B. B., Castro, M. I., Kleiman, D. G., Dietz, J. M., and Rettberg-Beck, B. 1988. Preparing captive-born primates for reintroduction. Paper presented at the XIIth Congress of the International Primatological Society, Brasília.

Beck, B. B., Kleiman, D. G., Dietz, J. M., Castro, M. I., Carvalho, C., Martins, A., and

Rettberg-Beck, B. 1991. Losses and reproduction in reintroduced golden lion tamarins *Leontopithecus rosalia. Dodo, Journal of the Jersey Wildlife Preservation Trust* 27: 50–61.

Beck, B. B., Rapaport, L. G., Stanley-Price, M. R., and Wilson, A. C. 1994. Reintroduction of captive born animals. In P. J. S. Olney, G. M. Mace, and A. T. C. Feistner (eds.), *Creative conservation: Interactive management of wild and captive animals,* pp. 265–86. London: Chapman and Hall.

Belyaev, D. K. 1979. Destabilizing selection as a factor in domestication. *Journal of Heredity* 70: 301–8.

Benz, J. J. 1993. Food-elicited vocalizations in golden lion tamarins: Design features for representational communication. *Animal Behaviour* 45:443–55.

Benz, J. J., French, J. A., and Leger, D. W. 1990. Sex differences in vocal structure in a callitrichid primate, *Leontopithecus rosalia. American Journal of Primatology* 21: 257–64.

Benz, J. J., Leger, D. W., and French, J. A. 1992. Relation between food preference and food-elicited vocalizations in golden lion tamarins (*Leontopithecus rosalia*). *Journal of Comparative Psychology* 106: 142–49.

Bergallo, H. de G., Rocha, C. F. D. da, Alves, M. A. dos S., and Van Sluys, M. (eds.). 2000. *A fauna ameaçada de extinção do estado do Rio de Janeiro.* Rio de Janeiro, RJ: Editora da Universidade do Estado do Rio de Janeiro.

Bernardes, A. T., Machado, A. B. M., and Rylands, A. B. 1990. *Fauna brasileira ameaçada de extinção.* Belo Horizonte, MG: Fundação Biodiversitas.

Biggins, D. E., Godbey, J. L., Hanebury, L. R., Luce, B., Marinari, P. E., Matchett, M. R., and Vargas, A. 1998. The effect of rearing methods on survival of reintroduced black-footed ferrets. *Journal of Wildlife Management* 62: 643–53.

Blanchard, K. 1995. Reversing population declines in seabirds on the north shore of the Gulf of St. Lawrence, Canada. In S. K. Jacobson (ed.), *Conserving wildlife: International education and communication approaches,* pp. 51–63. New York: Columbia University Press.

Blanes, J. and Mallinson, J. J. C. 1997. Landowner's Environmental Education Program for Una and surrounding areas. *Tamarin Tales,* no. 1: 10–11.

Boer, M. 1994. Reintroduction of the European lynx (*Lynx lynx*) to the Kampinoski National Park/Poland—a field experiment with zooborn individuals. Part I: Selection, adaptation and training. *Der Zoologische Garten* 64: 366–78.

Boinski, S. 1991. The coordination of spatial position: A field study of the vocal behaviour of adult female squirrel monkeys. *Animal Behaviour* 41(1): 89–102.

———. 2000. Social manipulation within and between troops mediates primate group travel. In S. Boinski and P. A. Garber (eds.), *On the move: How and why animals travel in groups,* pp. 421–69. Chicago: University of Chicago Press.

Boinski, S., Moraes, E., Kleiman, D. G., Dietz, J. M., and Baker, A. J. 1994. Intra-group vocal behaviour in wild golden lion tamarins, *Leontopithecus rosalia:* Honest communication of individual activity. *Behaviour* 130: 53–75.

Bonnell, M. and Selander, R. 1974. Elephant seals: Genetic variation and near extinction. *Science* 184: 908–9.

Box, H. O. 1977. Quantitative data on the carrying of young captive monkeys (*Callithrix jacchus*) by other members of their family groups. *Primates* 18: 475–84.

———. 1978. Social interactions in family groups of marmosets (*Callithrix jacchus*). In D. G. Kleiman (ed.), *The biology and conservation of the Callitrichidae*, pp. 239–49. Washington, DC: Smithsonian Institution Press.

———. 1991. Training for life after release: Simian primates as examples. In J. H. W. Gipps (ed.), *Beyond captive breeding: Re-introducing endangered mammals to the wild*, pp. 111–23. Oxford: Clarendon Press.

Brack, M. and Hosefelder, F. 1992. In vitro characteristics of *Yersinia pseudotuberculosis* of nonhuman origin. *Zentralblatt für Bakteriologie Mikrobiologie und Hygiene* 277: 280–87.

Bradbury, J. W. and Vehrencamp, S. L. 1998. *Principles of animal communication*. Sunderland, MA: Sinauer Associates.

Brand, H. M. 1981. Urinary oestrogen excretion in the female cotton-topped tamarin (*Saguinus oedipus oedipus*). *Journal of Reproduction and Fertility* 62: 467–73.

Brazil. 1988. *Constituição. República Federativa do Brasil, 1988*. Brasília, DF: Senado Federal.

Brazil, FEEMA. 1989. Plano Diretor—Estação Ecológica Estadual do Paraíso. Coordenácão de Dinâmica de Ecossistemas, Departamento de Estudos e Projetos, Fundação Estadual de Engenharia do Meio Ambiente (FEEMA), Rio de Janeiro.

Brazil, IBAMA. 1990. Directive (Portaria) numbers 1.203 (18 July 1990); 1.204 (18 July 1990); 1.2342 (28 November 1990). Brasília, DF: Instituto Brasileiro do Meio Ambiente e dos Recursos Naturais Renováveis (IBAMA).

———. 1992. Directive (Portaría) number 106-N (30 September 1992). Brasilia, DF: Instituto Brasileiro do Meio Ambiente e dos Recursos Naturais Renováveis (IBAMA).

———. 1999. Directive (Portaría) number 746 (15 December 1999). Brasilia, DF: Instituto Brasileiro do Meio Ambiente e dos Recursos Naturais Renováveis (IBAMA).

Brazil, MA/IBDF/FBCN. 1981. *Plano de Manejo. Reserva Biológica de Poço das Antas*. Rio de Janeiro, RJ: Ministério da Agricultura (MA), Instituto Brasileiro de Desenvolvimento Florestal (IBDF), and Fundação Brasileira para a Conservação da Natureza (FBCN).

Brazil, Paraná, SEMA. 1995. *Lista vermelha de animais ameaçados de extinção no estado do Paraná*. Curitiba, PR: Secretaria de Estado do Meio Ambiente (SEMA), Deutsche Gesellschaft für Technische Zusammenarbeit (GTZ) (GmbH).

Brazil, São Paulo, SEMA. 1998. *Fauna ameaçada no estado de São Paulo*. São Paulo, SP: Centro de Editoração (CED), Secretaria de Estado do Meio Ambiente (SEMA).

Brazil, Secretaria Municipal de Educação e Cultura de Una, Bahia. 2000. Música, festa e alegria: Festival de música de Una (Bahia) sensibiliza a comunidade para a preservação do meio ambiente. In I. Tamaio and D. Carreira (eds.), *Caminhos e aprendizagens: Educação ambiental, conservação e desenvolvimento*, pp. 23–26. Brasília, DF: WWF-Brasil.

Bridgwater, D. D. (ed.). 1972a. *Saving the lion marmoset*. Wheeling, WV: Wild Animal Propagation Trust.

———. 1972b. Introductory remarks with comments on the history and current status of the golden marmoset. In D. D. Bridgwater (ed.), *Saving the lion marmoset,* pp. 1–6. Wheeling, WV: Wild Animal Propagation Trust.

Brisson, M. J. 1756. *Regnum animale.* Paris. viii + 382 pp.

Bronikowski, A. M. and Altmann, J. 1996. Foraging in a variable environment: Weather patterns and the behavioral ecology of baboons. *Behavioural Ecology and Sociobiology* 39: 11–25.

Bronikowski, E. J., Beck, B. B., and Power, M. L. 1989. Innovation, exhibition and conservation: Free-ranging tamarins at the National Zoological Park. *Proceedings 1989 AAZPA Annual Conference,* pp. 540–46.

Brooks, T. M. and Balmford, A. 1996. Atlantic forest extinctions. *Nature, London* 380: 115.

Brooks, T. M., Pimm, S. L., and Oyugi, J. O. 1999. Time lag between deforestation and bird extinction in tropical forest fragments. *Conservation Biology* 13: 1140–50.

Brown, J. L. 1987. *Helping and communal breeding in birds.* Princeton, NJ: Princeton University Press.

Brown, K. and Mack, D. S. 1978. Food sharing among captive *Leontopithecus rosalia. Folia Primatologica* 29: 268–90.

Brown, R. E. 1985. The rodents I: Effects of odours on reproductive physiology (primer effects). In R. E. Brown and E. W. McDonald (eds.), *Social odours in mammals,* pp. 245–354. Oxford: Clarendon Press.

Brown, W. M. 1985. The mitochondrial genome of animals. In R. J. MacIntyre (ed.), *Molecular evolutionary genetics,* pp. 95–130. New York: Plenum Press.

Brown, W. M., George, M., Jr., and Wilson, A. C. 1979. Rapid evolution of animal mitochondrial DNA. *Proceedings of the National Academy of Sciences* 76:1967–71.

Burity, C. H. de F., Mandarim-de-Lacerda, C. A., and Pissinatti, A. 1997a. Craniometric sexual dimorphism in *Leontopithecus* Lesson, 1840 (Callitrichidae, Primates). *Primates* 38: 101–8.

———. 1997b. Sexual dimorphism in *Leontopithecus* Lesson, 1840 (Callitrichidae, Primates): Multivariate analysis of the cranial measurements. *Revista Brasileira de Biologia* 57: 231–37.

Burity, C. H. de F., Alves, F. M. U., Pissinatti, A., and Cruz, J. B. da. 1997c. A comparative analysis of dental anomalies in three species of *Leontopithecus* Lesson, 1840 in captivity (Callitrichidae, Primates). In S. F. Ferrari and H. Schneider (eds.), *A primatologia no Brasil—5,* pp. 167–73. Belém, PA: Universidade Federal do Pará, Sociedade Brasileira de Primatologia.

Burity, C. H. de F., Alves, F. M. U., and Pissinatti, A. 1997d. Alterações orais em três espécies de *Leontopithecus* mantidas em cativeiro (Callitrichidae—Primates). *Revista Brasileira de Ciências Veterinárias* 4: 9–12.

Burity, C. H. de F., Mandarim-de-Lacerda, C. A., and Pissinatti, A. 1999. Cranial and mandibular morphometry in *Leontopithecus* Lesson, 1840 (Callitrichidae, Primates). *American Journal of Primatology* 48: 186–96.

Bush, M., Montali, R. J., Kleiman, D. G., Randolph, J., Abramowitz, M. D., and Evans,

R. F. 1980. Diagnosis and repair of familial diaphragmatic defects in golden lion tamarins. *Journal of the American Veterinary Medicine Association* 177: 858–62.

Bush, M., Beck, B. B., and Montali, R. J. 1993. Medical considerations of reintroductions. In M. E. Fowler (ed.), *Zoo and wild animal medicine: Current therapy,* vol. 3, pp. 24–26. Denver, CO: W. B. Saunders Co.

Bush, M., Beck, B. B., Dietz, J. M., Baker, A. J., Everitte, J. A., Jr., Pissinatti, A., Phillips, L. G., Jr., and Montali R. J. 1996. Radiographic evaluation of diaphragmatic defects in golden lion tamarins, *Leontopithecus rosalia rosalia:* Implications for reintroduction. *Journal of Zoo and Wildlife Medicine* 27: 346–57.

Butler, P. 1995. Marketing the conservation message: Using parrots to promote protection and pride in the Caribbean. In S. K. Jacobson (ed.), *Conserving wildlife: International education and communication approaches,* pp. 103–18. New York: Columbia University Press.

Butynski, T. M. 1990. Comparative ecology of blue monkeys (*Cercopithecus mitis*) in high- and low-density subpopulations. *Ecological Monographs* 60: 1–26.

Caine, N. G. 1993. Flexibility and co-operation as unifying themes in *Saguinus* social organization and behaviour: The role of predation pressures. In A. B. Rylands (ed.), *Marmosets and tamarins: Systematics, behaviour, and ecology,* pp. 200–219. Oxford: Oxford University Press.

Caldecott, J. O. and Kavanagh, M. 1988. Strategic guidelines for nonhuman primate translocation. In L. Nielsen and R. D. Brown (eds.), *Translocation of wild animals,* pp. 64–75. Milwaukee: Wisconsin Humane Society.

Câmara, I. de G. 1993. Action plan for the black-faced lion tamarin. *Neotropical Primates* 1(3): 10–11.

———. 1994. Conservation status of the black-faced lion tamarin, *Leontopithecus caissara. Neotropical Primates* 2(suppl.): 50–51.

Camargo, J. L. C. 1995. Banco de sementes em diferentes tipos de vegetacao em Poço das Antas. In *Resumos do I Encontro de Pesquisadores da REBIO Poço das Antas/IBAMA,* Casimiro de Abreu, Rio de Janeiro.

Campos, J. C. C. and Heinsdijk, D. 1970. A floresta do Morro do Diabo. *Silvicultura, São Paulo* 7: 43–58.

Canavez, F. C., Alves, G., Fanning, T. G., and Seuánez, H. N. 1996. Comparative karyology and evolution of the Amazonian *Callithrix* (Platyrrhini, Primates). *Chromosoma (Berlin)* 104: 348–57.

Canavez, F. C., Ladasky, J. L., Muniz, J. A. P. C., Seuánez, H. N., and Parham, P. 1998. β2-Microglobulin in neotropical primates (Platyrrhini). *Immunogenetics* 48: 133–40.

Canavez, F. C., Moreira, M. A. M., Ladasky, J. J., Pissinatti, A., Parham, P., and Seuánez, H. N. 1999a. Molecular phylogeny of New World primates (Platyrrhini) based on β2-microglobin DNA sequences. *Molecular Phylogenetics and Evolution* 12: 74–82.

Canavez, F. C., Moreira, M. A. M., Simon, F., Parham, P., and Seuánez, H. N. 1999b. Phylogenetic relationships of the Callitrichinae (Platyrrhini, Primates) based on β2-microglobulin DNA sequences. *American Journal of Primatology* 48: 225–36.

Capobianco, J. P. 1998. Parque Nacional do Superagüi tem limites ampliados. *Parabólicas,* Instituto Sócioambiental, São Paulo 5(36): 4.

Carlstead, K. 1996. Effects of captivity on the behavior of wild mammals. In D. G. Kleiman, M. E. Allen, K. V. Thompson, and S. Lumpkin (eds.), *Wild mammals in captivity: Principles and techniques,* pp. 317–33. Chicago: University of Chicago Press.

Carlstead, K. and Shepherdson, D. 1994. Effects of environmental enrichment on reproduction. *Zoo Biology* 13:447–58.

Carvalho, C. T. de. 1965. Comentários sobre os mamíferos descritos e figurados por Alexandre Rodrigues Ferreira em 1790. *Arquivos de Zoologia,* São Paulo 12: 7–70.

Carvalho, C. T. de and Carvalho, C. F. de. 1989. A organização social dos sauís-pretos (*Leontopithecus chrysopygus* Mikan), na reserva em Teodoro Sampaio, São Paulo (Primates, Callithricidae). *Revista Brasileira de Zoologia* 6: 707–17.

Carvalho, C. T. de, Albernaz, A. K. M., and Lucca, C. A. T. de. 1989. Aspectos da bionomia do mico-leão-preto (*Leontopithecus chrysopygus* Mikan) (Mammalia, Callithricidae). *Revista do Instituto Florestal,* São Paulo 1(1): 67–83.

Carvalho, J. C. de M. 1968. Lista das espécies de animais e plantas ameaçadas de extinção no Brasil. *Boletim da Fundação Brasileira para a Conservação da Natureza (FBCN),* Rio de Janeiro 3: 11–16.

Castilhos, J. C., Dias da Silva, A. C., and Alves, D. A. R. 1997. Resgate cultural e a conservação das tartarugas marinhas. In S. M. Padua and M. F. Tabanez (eds.), *Educação ambiental: Caminhos trilhados no Brasil,* pp. 47–156. Nazaré Paulista, SP: Instituto de Pesquisas Ecológicas (IPÊ).

Castro, M. I. 1990. A comparative study of anti-predator behavior in three species of lion tamarins (*Leontopithecus*) in captivity. Unpublished M.Sc. thesis, University of Maryland, College Park.

Castro, M. I., Beck, B. B., Kleiman, D. G., Ruiz-Miranda, C. R., and Rosenberger, A. L. 1998. Environmental enrichment in a reintroduction program for golden lion tamarins (*Leontopithecus rosalia*). In D. J. Shepherdson, J. D. Mellen, and M. Hutchins (eds.), *Second nature: Environmental enrichment for captive animals,* pp. 97–128. Washington, DC: Smithsonian Institution Press.

Cebul, M. S., Alveario, M. C., and Epple, G. 1978. Odor recognition and attachment in infant marmosets. In H. Rothe, H.-J. Wolters, and J. P. Hearn, *Biology and behaviour of marmosets,* pp. 141–46. Göttingen: Eigenverlag H. Rothe.

Chagas, W. A., Pissinatti, A., Santos, M. A. J., and Pimenta, A. L. P. 1986. Candidiase em *Callithrix penicillata* (Geoffroy, 1812). Callitrichidae—Primates. In *Anais do Congresso Brasileiro de Medicina Veterinária* 20: 127.

Chaoui, N. J. and Hasler-Gallusser, S. 1999. Incomplete sexual suppression in *Leontopithecus chrysomelas:* A behavioural and hormonal study in a semi-natural environment. *Folia Primatologica* 70: 47–54.

Chapman, C. A. 1990. Ecological constraints on group size in three species of neotropical primates. *Folia Primatologica* 55:1–9.

Chaves, R., Sampaio, I., Schneider, M. P., Schneider, H., Page, S. L., and Goodman, M.

1999. The place of *Callimico goeldii* in the callitrichine phylogenetic tree: Evidence from von Willebrand Factor gene intron 11 sequences. *Molecular Phylogenetics and Evolution* 13: 392–404.

Chivers, D. J. 1991. Guidelines for re-introductions: Procedures and problems. In J. H. W. Gipps (ed.), *Beyond captive breeding: Re-introducing endangered mammals to the wild,* pp. 89–99. Oxford: Clarendon Press.

Cicmanec, J. L. 1978. Medical problems encountered in a callitrichid colony. In D. G. Kleiman (ed.), *The biology and conservation of the Callitrichidae,* pp. 331–36. Washington, DC: Smithsonian Institution Press.

Clark, M. M. and Galef, B. G. 1977. The role of the physical rearing environment in the domestication of the Mongolian gerbil (*Meriones unguiculatus*). *Animal Behaviour* 25: 298–316.

Clemmons, J. R. and Buchholz, R. (eds.). 1997. *Behavioral approaches to conservation in the wild.* Cambridge: Cambridge University Press.

Cleveland, J. and Snowdon, C. T. 1982. The complex vocal repertoire of the adult cotton-top tamarin (*Saguinus oedipus oedipus*). *Zeitschrift für Tierpsychologie* 58: 231–70.

———. 1984. Social development during the first twenty weeks in the cotton-top tamarin (*Saguinus o. oedipus*). *Animal Behaviour* 32: 432–44.

Clutton-Brock, T. H. 1975. Feeding behaviour of red colobus and black and white colobus in East Africa. *Folia Primatologica* 23: 165–207.

———. 1977. Some aspects of intraspecific variation in feeding and ranging behaviour in primates. In T. H. Clutton-Brock (ed.), *Primate ecology: Studies of feeding and ranging behaviour in lemurs, monkeys and apes,* pp. 539–56. London: Academic Press.

Clutton-Brock, T. H. and Harvey, P. H. 1977. Species differences in feeding and ranging behaviour in primates. In T. H. Clutton-Brock (ed.), *Primate ecology: Studies of feeding and ranging behaviour in lemurs, monkeys and apes,* pp. 557–84. London: Academic Press.

Coimbra-Filho, A. F. 1965. Breeding lion marmosets *Leontideus rosalia* at Rio de Janeiro Zoo. *International Zoo Yearbook* 4: 109–10.

———. 1969. Mico Leão, *Leontideus rosalia* (Linnaeus, 1766): Situação atual da espécie no Brasil (Callitrichidae—Primates). *Anais da Academia Brasileira de Ciências* 41(suppl.): 29–52.

———. 1970a. Considerações gerais e situação atual dos micos-leões escuros, *Leontideus chrysomelas* (Kuhl, 1820) e *Leontideus chrysopygus* (Mikan, 1823) (Callithricidae, Primates). *Revista Brasileira de Biologia* 30: 249–68.

———. 1970b. Acêrca da redescoberta de *Leontideus chrysopygus* (Mikan, 1823) e apontamentos sôbre sua ecologia (Callithricidae, Primates). *Revista Brasileira de Biologia* 30: 609–15.

———. 1972. Mamíferos ameaçados de extinção no Brasil. In Academia Brasileira de Ciências (ed.), *Espécies da fauna brasileira ameaçadas de extinção,* pp. 13–98. Rio de Janeiro: Academia Brasileira de Ciências.

———. 1976a. Os sagüis do gênero *Leontopithecus* Lesson, 1840 (Callithricidae—Primates). Master's thesis, Universidade Federal do Rio Janeiro, Rio de Janeiro.

———. 1976b. *Leontopithecus rosalia chrysopygus* (Mikan, 1823), o mico-leão do Estado de São Paulo (Callitrichidae—Primates). *Silvicultura, São Paulo* 10: 1–36.

———. 1978. Natural shelters of *Leontopithecus rosalia* and some ecological implications (Callitrichidae, Primates). In D. G. Kleiman (ed.), *The biology and conservation of the Callitrichidae,* pp. 79–98. Washington, DC: Smithsonian Institution Press.

———. 1981. Animais predados ou rejeitados pelo sauí-piranga, *Leontopithecus r. rosalia* (L., 1766) na sua área de ocorrência primitiva (Callitrichidae, Primates). *Revista Brasileira de Biologia* 41: 717–31.

———. 1985a. Sauí-una ou mico-leão-de-cara-dourada *Leontopithecus chrysomelas* (Kuhl, 1820). *FBCN/Informativo,* Rio de Janeiro 9(2): 3.

———. 1985b. Sauí-preto ou mico-leão-preto (*Leontopithecus chrysopygus* Mikan, 1823). *FBCN/Informativo,* Rio de Janeiro 9(3): 3.

Coimbra-Filho, A. F. and Câmara, I. de G. 1996. *Os limites originais do bioma mata Atlântica na região Nordeste do Brasil.* Rio de Janeiro, RJ: Fundação Brasileira para a Conservação da Natureza (FBCN).

Coimbra-Filho, A. F. and Magnanini, A. 1962. *Aves da Restinga.* Rio de Janeiro, RJ: Comissão Permanente da Reserva Biológica de Jacarépagua.

———. 1968. Animais raros ou em vias de desaparecimento no Brasil. *Anuário Brasileiro de Economia Florestal,* no. 19: 149–77.

———. 1972. On the present status of *Leontopithecus,* and some data about new behavioural aspects and management of *L. rosalia.* In D. D. Bridgwater (ed.), *Saving the lion marmoset,* pp. 59–69. Wheeling, WV: Wild Animal Propagation Trust.

Coimbra-Filho, A. F. and Maia, A. de A. 1977. A alimentação de sagüis em cativeiro. *Brasil Florestal,* no. 29: 15–26.

———. 1979a. O processo da muda dos pêlos em *Leontopithecus r. rosalia* (Linnaeus, 1766) (Callitrichidae, Primates). *Revista Brasileira de Biologia* 39: 83–93.

———. 1979b. A sazonalidade do processo reprodutivo em *Leontopithecus rosalia* (Linnaeus, 1766) (Callitrichidae, Primates). *Revista Brasileira de Biologia* 39: 643–51.

Coimbra-Filho, A. F. and Mittermeier, R. A. 1972. Taxonomy of the genus *Leontopithecus* Lesson 1840. In D. D. Bridgwater (ed.), *Saving the lion marmoset,* pp. 7–22. Wheeling, WV: Wild Animal Propagation Trust.

———. 1973. Distribution and ecology of the genus *Leontopithecus* Lesson, 1840 in Brazil. *Primates* 14: 47–66.

———. 1976. Hybridization in the genus *Leontopithecus, L. r. rosalia* (Linnaeus, 1766) × *L. r. chrysomelas* (Kuhl, 1820) (Callitrichidae, Primates). *Revista Brasiileira de Biologia* 36: 129–37.

———. 1977. Conservation of the Brazilian lion tamarins (*Leontopithecus rosalia*). In H. S. H. Prince Rainier III of Monaco and G. H. Bourne (eds.), *Primate conservation,* pp. 59–94. New York: Academic Press.

———. 1978. Reintroduction and translocation of lion tamarins: A realistic appraisal. In H. Rothe, H. J. Wolters, and J. P. Hearn (eds.), *Biology and behaviour of marmosets,* pp. 41–46. Göttingen: Eigenverlag H. Rothe.

———. 1982. Hope for Brazil's golden lion tamarin. *IUCN/SSC Primate Specialist Group Newsletter,* no. 2: 13.

Coimbra-Filho, A. F. and Rocha, N. da C. 1973. Aspectos do processo nutricional de animais selvagens em cativeiro. *Brasil Florestal* 14: 19–35.

———. 1978. Acerca de disfunção pigmentar em *Leontopithecus rosalia chrysomelas* (Kuhl, 1820), seu tratamento e recuperação (Callitrichidae, Primates). *Revista Brasileira de Biologia* 38: 165–70.

Coimbra-Filho, A. F., Magnanini, A., and Mittermeier, R. A. 1975. Vanishing gold: Last chance for Brazil's lion tamarins. *Animal Kingdom* 78(6): 21–27.

Coimbra-Filho, A. F., Rocha, N. da C., and Pissinatti, A. 1980. Morfofisiologia do ceco e sua correlação com o tipo odontológico em Callitrichidae (Platyrrhini, Primates). *Revista Brasileira de Biologia* 40: 177–85.

Coimbra-Filho, A. F., Rocha e Silva, R. da, and Pissinatti, A. 1981. Sobre a dieta de Callitrichidae em cativeiro/The diet of Callitrichidae in captivity. *Revista Biotérios* 1: 83–93.

———. 1984a. Gomas enriquecidas na alimentação de sagüis em cativeiro. In M. T. de Mello (ed.), *A primatologia no Brasil,* pp. 133–36. Brasília, DF: Sociedade Brasileira de Primatologia.

———. 1984b. Cromogenia anomala em *Leontopithecus chrysopygus* (Mikan, 1823). In M. T. de Mello (ed.), *A primatologia no Brasil,* pp. 217–19. Brasília, DF: Sociedade Brasileira de Primatologia.

Coimbra-Filho, A. F., Pissinatti, A., and Rocha e Silva, R. da. 1986a. O acervo do Museu de Primatologia (CPRJ/FEEMA). In M. T. de Mello (ed.), *A primatologia no Brasil—2,* pp. 505–14. Brasília, DF: Sociedade Brasileira de Primatologia.

———. 1986b. *O Centro de Primatologia do Rio de Janeiro.* Rio de Janeiro, RJ: Fundação Estadual de Engenharia do Meio Ambiente (FEEMA).

———. 1991. Hibridismo e duplo hibridismo em *Leontopithecus* (Callitrichidae, Primates). In A. B. Rylands and A. T. Bernardes (eds.), *A primatologia no Brasil—3,* pp. 89–95. Belo Horizonte, MG: Fundação Biodiversitas and Sociedade Brasileira de Primatologia.

Coimbra-Filho, A. F., Rylands, A. B., Pissinatti, A., and Santos, I. B. 1991/1992. The distribution and conservation of the buff-headed capuchin monkey, *Cebus xanthosternos,* in the Atlantic forest region of eastern Brazil. *Primate Conservation,* nos. 12–13: 24–30.

Coimbra-Filho, A. F., Dietz, L. A., Mallinson, J. J. C., and Santos, I. B. 1993. Land purchase for the Una Biological Reserve, refuge of the golden-headed lion tamarin. *Neotropical Primates* 1(3): 7–9.

Colver, F. 1938. *Variations and diseases of the teeth of animals.* London: Bale and Danielson.

Conservation International. 1995. *Prioridades para conservação da biodiversidade da mata Atlântica do Nordeste.* Map. Scale 1:2,500,000. Conservation International, Washington, DC, and Belo Horizonte, Fundação Biodiversitas, Belo Horizonte, and Sociedade Nordestina de Ecologia, Recife.

———. 1997. *The economics of biodiversity conservation in the Brazilian Atlantic forest. Project profile, 1997.* Washington, DC: Conservation International.

Conservation International do Brasil. 2000. *Avaliação e ações prioritárias para a conservação da biodiversidade da mata Atlântica e campos sulinos.* Brasília, DF: Conservation International do Brasil, Fundação SOS Mata Atlântica, Fundação Biodiversitas, Instituto de Pesquisas Ecológicas (IPÊ), Secretaria do Meio Ambiente do Estado de São Paulo, Instituto Estadual de Florestas de Minas Gerais. Secretaria de Biodiversidade e Florestas do Ministério do Meio Ambiente. 49 pp.

Consórcio Mata Atlântica–UNICAMP. 1992. *Reserva da Biosfera da Mata Atlântica. Plano de ação.* Vols. 1 and 2. Consórcio Mata Atlântica, São Paulo, and Universidade Estadual de Campinas, Editora UNICAMP, Campinas.

Cosgrove, G. E., Nelson, B. M., and Self, J. T. 1970. The pathology of pentastomid infection in primates. *Laboratory Animal Care* 20: 354–60.

Costa, L. P., Leite, Y. L. R., Fonseca, G. A. B. da, and Fonseca, M. T. da. 2000. Biogeography of South American forest mammals: Endemism and diversity in the Atlantic forest. *Biotropica* 32(4b): 872–81.

Coutinho, P. E. G. and Corrêa, H. K. M. 1996. Polygyny in a free-ranging group of buffy-tufted-ear marmosets, *Callithrix aurita. Folia Primatologica* 65: 25–29.

Crockett, C. M. and Eisenberg, J. F. 1987. Howlers: Variation in group size and demography. In B. B. Smuts, D. L. Cheney, R. M. Seyfarth, R. W. Wrangham, T. T. Struhsaker (eds.), *Primate societies,* pp. 54–69. Chicago: University of Chicago Press.

Cronin, J. E. and Sarich, V. M. 1975. Molecular systematics of the New World monkeys. *Journal of Human Evolution* 4: 357–75.

———. 1978. Marmoset evolution: The molecular evidence in marmosets. In N. Gengozian and F. Deinhardt (eds.), *Primates in medicine,* vol. 10, pp. 12–19. New York: Karger.

Cruz, J. B. and Pissinatti, A. 1986. *Sarcocystis* sp. em *Leontopithecus chrysopygus* (Mikan, 1823). In M. T. de Mello (ed.), *A primatologia no Brasil—2,* pp. 479–82. Brasília, DF: Sociedade Brasileira de Primatologia.

Cruz, J. B., Pissinatti, A., and Nascimento, M. D. 1993. Spontaneous *Diabetes mellitus* in *Leontopithecus chrysomelas* (Kuhl, 1820) and *Leontopithecus chrysopygus* (Mikan, 1823) Callitrichidae—Primates. In M. E. Yamamoto and M. B. C. de Souza (eds.), *A primatologia no Brasil—4,* pp. 195–204. Natal, RN: Universidade Federal do Rio Grande do Norte, Sociedade Brasileira de Primatologia.

Cruz, J. B., Pissinatti, A., Nascimento, M. D., and Costa, C. H. C. 1997. Osteodistrofia fibrosa em formas híbridas de *Callithrix* (Erxleben, 1777) Callitrichidae—Primates. In M. B. C. de Sousa and A. A. L. Menezes (eds.), *A primatologia no Brasil—6,* pp. 241–48. Natal, RN: Universidade Federal do Rio Grande do Norte, Sociedade Brasileira de Primatologia.

Cullen, L., Jr., and Valladares-Padua, C. 1999. Onças como detetives ecológicos para a mata Atlântica de São Paulo. *Ciência Hoje* 156: 54–57.

Cullen, L., Jr., Bodmer, R. E., and Valladares-Padua, C. 2000. Effects of hunting in habitat fragments of the Atlantic forests, Brazil. *Environmental Conservation* 95: 49–56.

———. 2001. Ecological consequences of hunting in Atlantic forest patches, São Paulo, Brazil. *Oryx* 35(2): 137–44.

Dawkins, M. S. and Guilford, T. 1997. Conspicuousness and diversity in animal signals. In D. H. Owings, M. D. Beecher, and N. S. Thompson (eds.), *Perspectives in ethology,* pp. 55–75. New York: Plenum Press.

Dawson, G. A. 1976. Behavioral ecology of the Panamanian tamarin *Saguinus oedipus* (Callitrichidae: Primates). Ph.D. dissertation, Michigan State University, East Lansing.

———. 1978. Composition and stability of social groups of the tamarin, *Saguinus oedipus geoffroyi,* in Panama: Ecological and behavioral implications. In D. G. Kleiman (ed.), *The biology and conservation of the Callitrichidae,* pp. 23–37. Washington, DC: Smithsonian Institution Press.

———. 1979. The use of time and space by the Panamanian tamarin, *Saguinus oedipus. Folia Primatologica* 31: 253–84.

Dean, W. 1995. *With broadax and firebrand: The destruction of the Brazilian Atlantic forest.* Berkeley: University of California Press.

Deane, L. de M. 1976. Epidemiology of simian malaria in the American continent. In *1st Inter-American Conference on Conservation and Utilization of American Nonhuman Primates in Biomedical Research, PAHO Scientific Publication No. 17,* pp. 144–63. Washington, DC: Pan American Health Organization.

De Bois, H. 1994. Progress report on the captive population of golden-headed lion tamarins, *Leontopithecus chrysomelas*—May 1994. *Neotropical Primates* 2(suppl.): 28–29.

Defler, T. R. 1995. The time budget of a group of wild woolly monkeys (*Lagothrix lagothricha*). *International Journal of Primatology* 16: 107–20.

Deinhardt, F., Wolfe, L., and Ogden, J. 1976. The importance of rearing marmoset monkeys in captivity for conservation of the species and for biomedical research. In *1st Inter-American Conference on Conservation and Utilization of American Nonhuman Primates in Biomedical Research, PAHO Scientific Publication No. 17*, pp. 65–71. Washington, DC: Pan American Health Organization.

Della Serra, O. 1951. Divisão do gênero *Leontocebus* (Macacos Platyrrhina) em dois gêneros sob bases de caracteres dento-morfológicos. *Papéis Avulsos de Zoologia* 10: 147–54.

Deshler, W. O. 1975. Recomendações para o manejo do Morro do Diabo. *Publicações do Instituto Florestal,* São Paulo no. 6: 1–31.

De Vleeschouwer, K. 2000. Social organisation, reproductive biology and parental care: An investigation into the social system of the golden-headed lion tamarin (*Leontopithecus chrysomelas*) in captivity. Ph.D. dissertation, University of Antwerp, Antwerp.

De Vleeschouwer, K., Heistermann, M., Van Elsacker, L., and Verheyen, R. F. 2000a. Signaling of reproductive status in captive female golden-headed lion tamarins (*Leontopithecus chrysomelas*). *International Journal of Primatology* 21: 445–65.

De Vleeschouwer, K., Leus, K., and Van Elsacker, L. 2000b. An evaluation of the suitability of contraceptive methods in golden-headed lion tamarins (*Leontopithecus chrysomelas*), with emphasis on melengestrol acetate (MGA) implants: I. Effectiveness, reversibility and medical side effects. *Animal Welfare* 9: 251–71.

De Vleeschouwer, K., Van Elsacker, L., Leus, K., and Heistermann, M. 2000c. An evaluation of the suitability of contraceptive methods in golden-headed lion tamarins (*Leontopithecus chrysomelas*), with emphasis on melengestrol acetate (MGA) implants: II. Endocrinological and behavioural effects. *Animal Welfare* 9: 385–401.

De Vleeschouwer, K., Van Elsacker, L., and Leus, K. 2001. Multiple breeding females in captive groups of golden-headed lion tamarins (*Leontopithecus chrysomelas*): Causes and consequences. *Folia Primatologica* 72: 1–10.

Dias, J. L. C., Montali, R. J., and Strandberg, J. D. 1996. Endocrine neoplasia in New World primates. *Journal of Medical Primatology* 25: 34–41.

Dietz, J. M. 1993. Population Viability Analysis of golden-headed lion tamarins in Una Biological Reserve, Brazil. Unpublished final report, period 1991–1993. Washington, DC: World Wildlife Fund–U.S.

———. 1997. Conservation of biodiversity in neotropical primates. In M. L. Reaka-Kudla, D. E. Wilson, and E. O. Wilson (eds.), *Biodiversity II: Understanding and protecting our biological resources,* pp. 341–56. Washington, DC: Joseph Henry Press.

Dietz, J. M. and Baker, A. J. 1993. Polygyny and female reproductive success in golden lion tamarins (*Leontopithecus rosalia*). *Animal Behaviour* 46: 1067–78.

Dietz, J. M., Kleiman, D. G., Beck, B. B., Dietz, L. A., and Rettberg, B. 1985. Recommendations for the conservation and management of the golden-headed lion tamarin in captivity and in the wild. Unpublished report. Brasília, DF: Departamento de Parques Nacionais e Reservas Equivalentes, Instituto Brasileiro de Desenvolvimento Florestal (IBDF).

Dietz, J. M., Coimbra-Filho, A. F., and Pessamílio, D. M. 1986. Projeto Mico-Leão. I. Um modelo para a conservação de espécie ameaçada de extinção. In M. T. de Mello (ed.), *A primatologia no Brasil—2,* pp. 217–22. Brasília, DF: Sociedade Brasileira de Primatologia.

Dietz, J. M., Baker, A. J., and Miglioretti, D. 1994a. Seasonal variation in reproduction, juvenile growth and adult body mass in golden lion tamarins (*Leontopithecus rosalia*). *American Journal of Primatology* 34: 115–32.

Dietz, J. M., Dietz, L. A., and Nagagata, E. Y. 1994b. The effective use of flagship species for conservation of biodiversity: The example of lion tamarins in Brazil. In G. M. Mace, P. J. S. Olney, and A. T. C. Feistner (eds.), *Creative conservation: Interactive management of wild and captive animals,* pp. 32–49. London: Chapman and Hall.

Dietz, J. M., Sousa, S. N. de, and Silva, J. O. O. 1994c. Population structure and territory size in golden-headed lion tamarins. *Neotropical Primates* 2(suppl.): 21–23.

Dietz, J. M., Baker, A. J., and Allendorf, T. D. 1995. Correlates of molt in golden lion tamarins (*Leontopithecus rosalia*). *American Journal of Primatology* 36: 277–84.

Dietz, J. M., Sousa, S. N. de, and Billerbeck, R. 1996. Population dynamics of golden-headed lion tamarins *Leontopithecus chrysomelas* in Una Reserve, Brazil. *Dodo, Journal of the Wildlife Preservation Trusts* 32: 115–22.

Dietz, J. M., Peres, C. A., and Pinder, L. 1997. Foraging ecology and use of space in wild golden lion tamarins (*Leontopithecus rosalia*). *American Journal of Primatology* 41: 289–305.

Dietz, J. M., Baker, A. J., and Ballou, J. D. 2000. Demographic evidence of inbreeding depression in golden lion tamarins. In A. G. Young and G. M. Clarke (eds.), *Genetics, demography and population viability,* pp. 203–11. Cambridge: Cambridge University Press.

Dietz, J. M., Baker, A. J., and Di Fiore, A. In preparation. The breeding system of wild golden lion tamarins (*Leontopithecus rosalia*): Evidence for extragroup paternity.

Dietz, L. A. 1985. Captive-born golden lion tamarins released into the wild: A report from the field. *Primate Conservation,* no. 6: 21–27.

———. 1998. Community conservation education project for the golden lion tamarin, Brazil: Building support for habitat conservation. In R. Hoage and K. Moran (eds.), *Culture: The missing element of conservation,* pp. 85–94. Washington, DC: Kendall/Hunt Publishing Co., National Zoological Park, Smithsonian Institution.

Dietz, L. A. and Nagagata, E. Y. 1986. Projeto Mico-Leão. V. Programa de educação comunitária para a conservação do mico-leão-dourado—*Leontopithecus rosalia* (Linnaeus 1766). Desenvolvimento e avaliação de educação como uma tecnologia para a conservação de uma espécie em extinção. In M. T. de Mello (ed.), *A primatologia no Brasil—2,* pp. 249–56. Brasília, DF: Sociedade Brasileira de Primatologia.

———. 1995. Golden Lion Tamarin Conservation Program: A community education effort for forest conservation in Rio de Janeiro state, Brazil. In S. K. Jacobson (ed.), *Conserving wildlife: International education and communication approaches,* pp. 64–86. New York: Columbia University Press.

———. 1997. Programa de conservação do mico-leão-dourado: Atividades de educação comunitária para a conservação da Mata Atlântica no Rio de Janeiro. In S. M. Padua and M. F. Tabanez (eds.), *Educação ambiental: Caminhos trilhados no Brasil,* pp. 133–46. Nazaré Paulista, SP: Fundo Nacional do Meio Ambiente, Instituto de Pesquisas Ecológicas (IPÊ).

Di Fiore, A. and Rodman, P. S. 2001. Time allocation patterns of lowland woolly monkeys (*Lagothrix lagotricha poeppigii*) in a neotropical *terra firma* forest. *International Journal of Primatology* 22: 449–80.

Digby, L. J. 1995. Infant care, infanticide and female reproductive strategies in polygynous groups of common marmosets (*Callithrix jacchus*). *Behavioral Ecology and Sociobiology* 37: 51–61.

Digby, L. J. and Barreto, C. E. 1993. Social organization in a wild population of *Callithrix jacchus.* I. Group composition and dynamics. *Folia Primatologica* 61: 123–34.

———. 1996. Activity and ranging patterns in common marmosets: Implications for reproductive strategies. In M. A. Norconk, A. L. Rosenberger, and P. A. Garber (eds.), *Adaptive radiations of the neotropical Primates,* pp. 173–85. New York: Plenum Press.

Digby, L. J. and Ferrari, S. F. 1994. Multiple breeding females in free-ranging groups of *Callithrix jacchus. International Journal of Primatology* 15: 389–97.

Dover, G. 1982. Molecular drive: A cohesive mode of species evolution. *Nature, London* 299: 111–17.

Drewien, R. C., Derrickson, S. R., and Bizeau, E. G. 1981. Experimental release of captive parent–reared greater sandhill cranes at Grays Lake Refuge, Idaho. In J. C. Lewis (ed.), *Proceedings 1981 Crane Workshop,* pp. 99–111. Tavernier, FL: National Audubon Society.

Dubin, I. N. and Wilcox, A. 1947. *Sarcocystis* in *Macaca mulatta. Journal of Parasitology* 33: 151–53.

DuMond, F. 1972. Recommendations for a basic husbandry program for lion marmosets. In D. D. Bridgwater (ed.), *Saving the lion marmoset,* pp. 120–36. Wheeling, WV: Wild Animal Propagation Trust.

Dunbar, R. I. M. 1992. Time: A hidden constraint on the behavioural ecology of baboons. *Behaviour Ecology and Sociobiology* 31: 35–49.

Eisenberg, J. F. 1978. Comparative ecology and reproduction of New World monkeys, In D. G. Kleiman (ed.), *The biology and conservation of the Callitrichidae,* pp. 23–37. Washington, DC: Smithsonian Institution Press.

———. 1980. The density and biomass of tropical mammals. In M. E. Soulé and B. A. Wilcox (eds.), *Conservation biology: An evolutionary-ecological perspective,* pp. 123–29. Sunderland, MA: Sinauer Associates.

Eisenberg, J. F. and Kleiman, D. G. 1972. Olfactory communication in mammals. *Annual Review of Ecology and Systematics* 3: 1–32.

Elliot, D. G. 1913. *A review of the primates, 1.* Monograph, New York: American Museum of Natural History.

Ellis, D. H., Dobrott, S. J., and Goodwin, J. G. 1977. Reintroduction techniques for masked bobwhites. In S. A. Temple (ed.), *Endangered birds: Management techniques for preserving threatened species,* pp. 345–54. Madison: University of Wisconsin Press.

Elowson, A. M., Snowdon, C. T., and Sweet, C. J. 1992. Ontogeny of trill and J-call vocalizations in the pygmy marmoset, *Cebuella pygmaea. Animal Behaviour* 43:703–15.

Epple, G. 1972. Social behavior of laboratory groups of *Saguinus fuscicollis*. In D. D. Bridgwater (ed.), *Saving the lion marmoset,* pp. 50–58. Wheeling, WV: Wild Animal Propagation Trust.

———. 1975a. Parental behavior in *Saguinus fuscicollis* ssp. (Callithricidae). *Folia Primatologica* 24: 221–38.

———. 1975b. The behavior of marmoset monkeys (Callithricidae). In L. A. Rosenblum (ed.), *Primate behavior,* vol. 4, pp. 195–239. New York: Academic Press.

———. 1978. Reproductive and social behavior of marmosets with special reference to captive breeding. In N. Gengozian and F. W. Deinhardt (eds.), *Marmosets in experimental medicine,* pp. 50–62. Basel: S. Karger.

Epple, G., Belcher, A. M., Küderling, I., Zeller, U., Scolnick, L., Greenfield, K. L., and Smith, A. B. 1993. Making sense out of scents: Species differences in scent glands, scent-marking behavior, and scent-mark composition in the Callitrichidae. In A. B. Rylands (ed.), *Marmosets and tamarins: Systematics, behaviour, and ecology,* pp. 123–51. Oxford: Oxford University Press.

Erwin, J. and Deni, R. 1979. Strangers in a strange land: Abnormal behaviors or abnormal environments? In J. Erwin, T. L. Maple, and G. Mitchell (eds.), *Captivity and behavior,* pp. 1–29. New York: Van Nostrand Reinhold.

Evans, D. T., Piekarczyk, M. S., Cadavid, L., Hinshaw, V. S., and Watkins, D. I. 1998. Two different primate species express an identical functional MHC class I allele. *Immunogenetics* 47: 206–11.

Evans, S. and Hodges, J. K. 1984. Reproductive status of adult daughters in family groups of common marmosets (*Callithrix jacchus*). *Folia Primatologica* 42: 127–33.

Faaborg, J. and Patterson, C. B. 1981. The characteristics and occurrence of cooperative polyandry. *Ibis* 123: 477–84.

Federsoni, P., Jr., Vitiello, N., and Calixto, S. 1997. Educação ambiental com animais peçonhentos: "Na natureza não existem vilões." In S. M. Padua and M. F. Tabanez (eds.), *Educação ambiental: Caminhos trilhados no Brasil,* pp. 211–20. Nazaré Paulista, SP: Instituto de Pesquisas Ecológicas (IPÊ).

Feio, J. L. de A. 1953. Contribuição ao conhecimento da história da zoogeografia do Brasil. *Papeis Avulsas do Museu Nacional, Rio de Janeiro* 12: 1–22.

Feistner, A. T. C. 1985. Food sharing in the cotton-top tamarin *Saguinus oedipus*. Master's thesis, University of Stirling, Scotland.

Feistner, A. T. C. and Chamove, A. S. 1986. High motivation toward food increases food-sharing in cotton-top tamarins. *Developmental Psychobiology* 19: 439–52.

Feistner, A. T. C. and McGrew, W. C. 1989. Food sharing in primates: A critical review. In P. K. Seth and S. Seth (eds.), *Perspectives in primate biology,* pp. 21–36. New Delhi: Today and Tomorrow Press.

Feistner, A. T. C. and Price, E. C. 1990. Food-sharing in cotton-top tamarins (*Saguinus oedipus*). *Folia Primatologica* 54: 34–45.

———. 1991. Food offering in New World primates: Two species added. *Folia Primatologica* 57: 165–68.

———. 1999. Cross-generic food-sharing in tamarins. *International Journal of Primatology* 20: 47–54.

———. 2000. Food sharing in black lion tamarins (*Leontopithecus chrysopygus*). *American Journal of Primatology* 52: 47–54.

Feldman, M. W. and Christiansen, F. B. 1984. Population genetic theory of the cost of inbreeding. *American Naturalist* 123: 642–53.

Felsenstein, J. 1985. Confidence limits on phylogenies: An approach using the bootstrap. *Evolution* 39: 783–91.

Fernandes, R. V., Rambaldi, D. M., Bento, M. I. da S., and Matsuo, P. M. 2000. A RPPN—Reserva Particular do Patrimônio Natural—como mecanismo de proteção legal para o habitat do mico-leão-dourado (*Leontopithecus rosalia*). In M. S. Milano and V. Theulen (eds.), *II Congresso Brasileiro de Unidades de Conservação. Anais. Vol. II. Trabalhos técnicos,* pp. 613–17. Rede Nacional Pró-Unidades de Conservação, Campo Grande, Fundacão O Boticário de Proteção à Natureza, São José dos Pinhais, Paraná.

Ferrari, S. F. 1988. The behaviour and ecology of the buffy-headed marmoset, *Callithrix flaviceps* (O. Thomas, 1903). Doctoral thesis, University College, London.

———. 1987. Food transfer in a wild marmoset group. *Folia Primatologica* 48: 203–6.

———. 1991. Preliminary report on a field study of *Callithrix flaviceps*. In A. B. Rylands and A. T. Bernardes (eds.), *A primatologia no Brasil—3,* pp. 159–71. Belo Horizonte: Sociedade Brasileira de Primatologia and Fundação Biodiversitas.

———. 1992. The care of infants in a wild marmoset, *Callithrix flaviceps,* group. *American Journal of Primatology* 26: 109–18.

Ferrari, S. F. and Diego, V. H. 1992. Long-term changes in a wild marmoset group. *Folia Primatologica* 58: 215–18.

Ferrari, S. F. and Lopes Ferrari, M. A. 1989. A re-evaluation of the social organisation of the Callitrichidae, with reference to the ecological differences between genera. *Folia Primatologica* 52: 132–47.

———. 1990. Predator avoidance behaviour in the buffy-headed marmoset, *Callithrix flaviceps. Primates* 31: 323–38.

Ferrari, S. F. and Rylands, A. B. 1994. Activity budgets and differential visibility in field studies of three marmosets (*Callithrix* spp.). *Folia Primatologica* 63: 78–83.

Ferrari, S. F. and Strier, K. B. 1992. Exploitation of *Mabea fistulifera* nectar by marmosets (*Callithrix flaviceps*) and muriquis (*Brachyteles arachnoides*) in south-east Brazil. *Journal of Tropical Ecology* 8: 225–39.

Ferreira, B. R., Pissinatti, A., Cruz, J. B., and Bechara, G. H. 1995. Benign prostatic hyperplasia in the nonhuman primate *Leontopithecus* (Lesson, 1840) Callitrichidae—Primates. *Folia Primatologica* 65: 48–53.

Ferreira, B. R., Bechara, G. H., Pissinatti, A., and Cruz, J. B. da. 1997. Benign prostatic hyperplasia in the nonhuman primate *Leontopithecus* (Lesson, 1840), Callitrichidae, Primates. In S. F. Ferrari and H. Schneider (eds.), *A primatologia no Brasil—5,* pp. 217–25. Belém, PA: Universidade Federal do Pará, Sociedade Brasileira de Primatologia.

Ferri, M. G. 1980. *A vegetação brasileira.* 1st ed. São Paulo, SP: Editora da Universidade de São Paulo.

Fonseca, G. A. B. da. 1985. The vanishing Brazilian Atlantic forest. *Biological Conservation* 34: 17–34.

Fonseca , G. A. B. da and Pinto, L. P. de S. 1996. O papel das ONGs. In Fundação Getúlio Vargas (ed.), *Gestão ambiental no Brasil: Experiência e sucesso,* pp. 294–346. Rio de Janeiro: Fundação Getúlio Vargas.

Fonseca, G. A. B. da, Rylands, A. B., Costa, C. M. R., Machado, R. B., and Leite Y. L. R. (eds.). 1994. *Livro dos mamíferos brasileiros ameaçados de extinção.* Belo Horizonte, MG: Fundação Biodiversitas.

Fonseca, G. A. B. da, Cavalcanti, R., Santos, I. B., and Braga, R. 1995. Priority areas for conservation in the Atlantic forest of north-east Brazil. *Neotropical Primates* 3: 54–55.

Fonseca G. A. B. da, Herrman, G., and Leite, Y. L. R. 1999. Macrogeography of Brazilian mammals. In J. F. Eisenberg and K. H. Redford (eds.), *Mammals of the neotropics. The central neotropics. Vol. 3. Ecuador, Peru, Bolivia, Brazil,* pp. 549–63. Chicago: University of Chicago Press.

Foose, T. J., Lande, R., Flesness, N. R., Rabb, G., and Read, B. 1986. Propagation plans. *Zoo Biology* 5: 139–46.

Foose , T. J., de Boer, L., Seal, U. S., and Lande, R. 1995. Conservation management strategies based on viable populations. In J. D. Ballou, M. Gilpin, and T. J. Foose (eds.), *Population management for survival and recovery,* pp. 273–94. New York: Columbia University Press.

Ford, S. M. 1986. Systematics of the New World monkeys. In D. R. Swindler and J. Erwin (eds.), *Systematics, evolution, and anatomy. Vol. I. Comparative primate biology,* pp. 73–135. New York: Alan R. Liss.

Forman, L., Kleiman, D. G., Bush, M., Dietz, J. M., Ballou, J. M., Phillips, L. G., Coimbra-Filho, A. F., and O'Brien, S. J. 1986. Genetic variation among lion tamarins. *American Journal of Physical Anthropology* 71: 1–11.

Foster, R. B. 1980. Heterogeneity and disturbance in tropical vegetation. In M. E. Soulé and B. A. Wilcox (eds.). *Conservation biology: An evolutionary perspective,* pp. 75–92. Sunderland, MA: Sinauer Associates.

Frankham, R., Hemmer, H., Ryder, O. A., Cothran, E. G., Soulé, M. E., Murray, N. D., and Snyder, M. 1986. Selection in captive populations. *Zoo Biology* 5: 127–38.

Franklin, S. and Dietz, J. M. 2001. Predation in Poço das Antas: What's eating the tamarins? *Tamarin Tales,* no. 5: 1–2.

French, J. A. 1984. Lactation and fertility: An examination of nursing and interbirth intervals in tamarins (*Saguinus oedipus*). *Folia Primatologica* 40: 276–82.

———. 1997. Proximate regulation of singular breeding in callitrichid primates. In N. G. Solomon and J. A. French (eds.), *Cooperative breeding in mammals,* pp. 34–75. Cambridge: Cambridge University Press.

French, J. A. and Cleveland, J. 1984. Scent-marking in the tamarin, *Saguinus oedipus:* Sex differences and ontogeny. *Animal Behaviour* 32:615–23.

French, J. A. and Inglett, B. J. 1989. Female-female aggression and male indifference in response to unfamiliar intruders in lion tamarins. *Animal Behaviour* 37: 487–97.

French, J. A. and Schaffner, C. M. 2000. Social modulation of reproductive behavior in monogamous primates. In K. Wallen and J. Schneider (eds.), *Reproduction in context,* pp. 325–53. Cambridge, MA: MIT Press.

French, J. A. and Stribley, J. A. 1985. Patterns of urinary oestrogen excretion in female golden lion tamarins (*Leontopithecus rosalia*). *Journal of Reproduction and Fertility* 75: 537–46.

———. 1987. Synchronization of ovarian cycles within and between social groups in the golden lion tamarin. *American Journal of Primatology* 12: 469–78.

French, J. A., Abbott, D. H., Scheffler, G., Robinson, J. A., and Goy, R. W. 1983. Cyclic excretion of urinary oestrogens in female tamarins (*Saguinus oedipus*). *Journal of Reproduction and Fertility* 68: 177–84.

French, J. A., Abbott, D. H., and Snowdon, C. T. 1984. The effect of social environment on estrogen excretion, scent marking, and sociosexual behavior in tamarins (*Saguinus oedipus*). *American Journal of Primatology* 6: 155–67.

French, J. A., Inglett, B. J., and Dethlefs, T. M. 1989. The reproductive status of nonbreeding group members in captive golden lion tamarin social groups. *American Journal of Primatology* 18: 73–86.

French, J. A., deGraw, W. A., Hendricks, S. E., Wegner, F., and Bridson, W. E. 1992. Urinary and plasma gonadotropin concentrations in golden lion tamarins (*Leontopithecus rosalia*). *American Journal of Primatology* 26: 53–59.

French, J. A., Brewer, K. J., Shaffner, C. M., Schalley, J., Hightower-Merritt, D. L., Smith, T. E., and Bell, S. M. 1996a. Urinary steroid and gonadotropin excretion across the reproductive cycle in female Wied's black tufted-ear marmosets (*Callithrix kuhli*). *American Journal of Primatology* 40: 231–45.

French, J. A., Pissinatti, A., and Coimbra-Filho, A. F. 1996b. Reproduction in captive lion tamarins (*Leontopithecus*): Seasonality, infant survival and sex ratios. *American Journal of Primatology* 39: 17–33.

French, J. A., Bales, K. Baker, A. J., and Dietz, J. M. Submitted. Reproductive status of daughters and subordinate females in groups of free-ranging lion tamarins (*Leontopithecus rosalia*).

Fundação SOS Mata Atlântica/INPE. 1998. *Atlas da evolução dos remanescentes florestais e ecossistemas associados no domínio da mata Atlântica no período 1990–1995.* Fundação SOS Mata Atlântica, São Paulo, Instituto Nacional de Pesquisas Espaciais (INPE), São José dos Campos.

Garber, P. A. 1993. Seasonal patterns of diet and ranging in two species of tamarin monkeys: Stability versus variability. *International Journal of Primatology* 14: 145–66.

———. 1997. One for all and breeding for one: Cooperation and competition as a tamarin reproductive strategy. *Evolutionary Anthropology* 5: 187–99.

Garber, P. A., Moya, L., and Malaga, C. 1984. A preliminary field study of the moustached tamarin monkey (*Saguinus mystax*) in northeastern Peru: Questions concerned with the evolution of a communal breeding system. *Folia Primatologica* 42: 17–32.

Garlick, D. S., Marcus, L. C., Pokras, M., and Schelling, S. H. 1996. *Baylisascaris* larva migrans in a spider monkey (*Ateles* sp.). *Journal of Medical Primatology* 25: 133–36.

Gengozian, N., Batson, J. S., and Smith, T. A. 1978. Breeding of tamarins (*Saguinus* spp.) in the laboratory. In D. G. Kleiman (ed.), *The biology and conservation of the Callitrichidae,* pp. 207–13. Washington, DC: Smithsonian Institution Press.

Geoffroy Saint-Hilaire, I. 1827. Ouistiti—Jacchus. *Dict. Class. Hist. Nat.* 12: 512–20.

Gibbs, C. J. and Gajdusek, D. C. 1976. Studies on the viruses of subacute spongiform encephalopathies using primates, their only available indicator. In *1st Inter-American Conference on Conservation and Utilization of American Nonhuman Primates in Biomedical Research, PAHO Scientific Publication No. 17,* pp. 83–109. Washington, DC: Pan American Health Organization.

Gilpin, M. E. 1987. Spatial structure and sub-population vulnerability. In M. E. Soulé (ed.), *Viable sub-populations for conservation*, pp. 125–39. Cambridge: Cambridge University Press.

Glenn, F. and McSherry, C. K. 1970. The baboon and experimental cholelithiasis. *Archives of Surgery* 100: 105–8.

Goff, M. L., Whittaker J. O., Jr., and Dietz, J. M. 1986. A new species of *Microtrombicula* (Acari: Trombiculidae) from the golden lion tamarin in Brazil. *International Journal of Acarology* 12: 171–73.

———. 1987. The genus *Speleocola* (Acari: Trombiculidae), with description of a new species from Brazil and a key to the species. *Journal of Medical Entomology* 24: 198–200.

Goldizen, A. W. 1987a. Tamarins and marmosets: Communal care of offspring. In B. B. Smuts, D. L. Cheney, R. M. Seyfarth, R. W. Wrangham, and T. T. Struhsaker (eds.), *Primate Societies*, pp. 34–43. Chicago: University of Chicago Press.

———. 1987b. Facultative polyandry and the role of infant-carrying in wild saddle-back tamarins (*Saguinus fuscicollis*). *Behavioral Ecology and Sociobiology* 20: 99–109.

———. 1989. Social relationships in a cooperatively polyandrous group of tamarins (*Saguinus fuscicollis*). *Behavioral Ecology and Sociobiology* 24: 79–89.

Goldizen, A. W. and Terborgh, J. 1989. Demography and dispersal patterns of a tamarin: Possible causes of delayed breeding. *American Naturalist* 134: 208–24.

Goldizen, A. W., Mendelson, J., van Vlaardingen, M., and Terborgh, J. 1996. Saddle-back tamarin (*Saguinus fuscicollis*) reproductive strategies: Evidence from a thirteen-year study of a marked population. *American Journal of Primatology* 38: 57–83.

Gonçalves, W. M., Magalhães, H., Costa, C. H. C., Ronconi, M. A., Cruz, J. B. da, and Pissinatti, A. 1997. Actinomicose cervicofacial de caráter disseminado em *Leontopithecus chryspopygus* (Mikan, 1823). Primates: Callitrichidae. In M. B. C. de Sousa and A. L. L. Menezes (eds.), *A primatologia no Brasil—6*, pp. 241–48. Natal, RN: Universidade Federal do Rio Grande do Norte and Sociedade Brasileira de Primatologia.

Grativol, A. D., Ballou, J. D., and Fleischer, R .C. 2001. Microsatellite variation within and among recently fragmented populations of the golden lion tamarin (*Leontopithecus rosalia*). *Conservation Genetics* 2: 1–9.

Green, K. M. 1979. Vocalizations, behavior, and ontogeny of the golden lion tamarin, *Leontopithecus rosalia rosalia*. D.Sc. dissertation, Johns Hopkins University, Baltimore.

———. 1980. An assessment of the Poço das Antas Reserve, Brazil, and prospects for the survival of the golden lion tamarin, *Leontopithecus r. rosalia*. Unpublished report. Washington, DC: World Wildlife Fund–U.S.

Griffith, B., Scott, J. M., Carpenter, J. W., and Reed, C. 1989. Translocation as a species conservation tool: Status and strategy. *Science* 245: 477–80.

Griner, L. A. 1983. *Pathology of zoo animals*. San Diego: Zoological Society of San Diego.

Groves, C. P. 1993. Order Primates. In D. E. Wilson and D. M. Reeder (eds.), *Mammal species of the world: A taxonomic and geographic reference*, 2nd ed., pp. 243–77. Washington, DC: Smithsonian Institution Press.

———. 2001. *Primate taxonomy*. Washington, DC: Smithsonian Institution Press.

Guillaumon, J. R., Negreiros, O. C., Faria, A. J., Dias, A. C., Brettas, D. E., Carvalho,

C. T. de, Domingues, E. N., Serio, F. C., Ogawa, H. Y., and Pfeifer, R. M. 1983. *Proposta para o manejo da Reserva Estadual do Morro do Diabo, São Paulo.* São Paulo, SP: Instituto Florestal.

Guimarães, E. F., Mautone, L., Mattos Filho, A. de, Dietz, J. M., Pinder, L., and Pessamílio, D. M. 1985. A vegetação da Reserva Biológica de Poço das Antas, habitat do mico-leão-dourado (*Leontopithecus rosalia* L.). In *Resumos. XXXVI Congresso Nacional de Botânica,* p. 122. Curitiba, PR: Universidade Federal do Paraná.

Haazen, W. 1988. Studie van het vokale repertoire van het goudkopleewaapje, *Leontopithecus chrysomelas.* Honors thesis, biology, University of Antwerp, Antwerp.

Hall, L. M., Jones, D. S., and Wood, B. A. 1998. Evolution of the gibbon subgenera inferred from cytochrome b DNA sequence data. *Molecular Phylogenetics and Evolution* 10: 281–86.

Halloy, M. and Kleiman, D. G. 1994. Acoustic structure of long-calls in free-ranging groups of golden lion tamarins *Leontopithecus rosalia. American Journal of Primatology* 32: 303–10.

Hamilton, W. D. 1964. The genetical evolution of social behaviour. *Journal of Theoretical Biology* 7: 1–52.

Hampton, J. K., Jr., Hampton, S. H., and Landwehr, B. T. 1966. Observations on a successful breeding colony of the marmoset *Oedipomidas oedipus. Folia Primatologica* 4: 265–87.

Hanski, I. and Gilpin, M. 1991. Metapopulation dynamics: Brief history and conceptual domain. *Biological Journal of the Linnaean Society* 42: 3–16.

Harada, M. L., Schneider, H., Schneider, M. P. C., Sampaio, I., Czelusniak, J., and Goodman, M. 1995. DNA evidence on the phylogenetic systematics of New World monkeys: Support for the sister-grouping of *Cebus* and *Saimiri* from two unlinked nuclear genes. *Molecular Phylogenetics and Evolution* 4: 331–49.

Harding, R. D., Hulme, M. J., Lunn, S. F., Henderson, C., Aitken, R. J. 1982. Plasma progesterone levels throughout the ovarian cycle of the common marmoset (*Callithrix jacchus*). *Journal of Medical Primatology* 11: 43–51.

Hardner, J. 1996. Tendência da alta no preço do cacau pode não garantir o cultivo no sul da Bahia. In *Alternativas econômicas para conservação e desenvolvimento da região de Una, Bahia. Resumo das pesquisas, 1994–1995,* pp. 18–20. Ilhéus, BA: Instituto de Estudos Sócio-Ambientais do Sul da Bahia (IESB).

Hauser, M. 1997. *The evolution of communication.* Boston: MIT press.

Hearn, J. P. 1983. The common marmoset (*Callithrix jacchus*). In J. P. Hearn (ed.), *Reproduction of New World Monkeys,* pp. 181–215. Lancaster, UK: MTP Press.

Hediger, H. 1964. *Wild animals in captivity: An outline of the biology of zoological gardens.* London: Butterworth's Scientific Publications, Ltd., 1950. Reprint, New York: Dover Publications.

Heistermann, M. and Hodges, J. K. 1995. Endocrine monitoring of the ovarian cycle and pregnancy in the saddle-back tamarin (*Saguinus fuscicollis*) by measurement of steroid conjugates in urine. *American Journal of Primatology* 35: 117–28.

Heistermann, M., Prove, E., Wolters, J. H., and Mika, G. 1987. Urinary oestrogen and progesterone excretion before and during pregnancy in pied bare-face tamarins (*Saguinus bicolor bicolor*). *Journal of Reproduction and Fertility* 80: 635–40.

Hersek, M. J. and Owings, D. H. 1993. Tail flagging by adult California ground squirrels: A tonic signal that serves different functions for males and females. *Animal Behaviour* 46: 129–38.

Hershkovitz, P. 1977. *Living New World monkeys (Platyrrhini) with an introduction to Primates, Vol. 1*. Chicago: University of Chicago Press.

———. 1987. A history of the recent mammalogy of the neotropical region from 1492 to 1850. In B. D. Patterson and R. M. Timm (eds.), *Studies in neotropical mammalogy: Essays in honor of Philip Hershkovitz, Fieldiana, Zoology, New Series* 39: 11–98.

Heymann, E. W. 2000. The number of adult males in callitrichine groups and its implications for callitrichine evolution. In P. M. Kappeler (ed.), *Primate males: Causes and consequences of variation in group composition*, pp. 64–71. Cambridge: Cambridge University Press.

Hiddleston, W. A. 1976. Large scale production of a small laboratory primate—*Callithrix jacchus*. In T. Antikatzides, S. Erichsen, and A. Spiegel (eds.), *The laboratory animal in the study of reproduction*, pp. 51–57. Stuttgart: Gustav Fischer Verlag.

Hill, C. A. 1970. The last of the golden marmosets. *Zoonooz* 43: 12–17.

Hill, W. C. O. 1957. *Primates. Comparative anatomy and taxonomy III. Pithecoidea, Platyrrhini (families Hapalidae and Callimiconidae)*. Edinburgh: Edinburgh University Press.

Hillis, D. M. and Dixon, M. 1991. Ribosomal DNA: Molecular evolution and phylogenetic inference. *Quarterly Review of Biology* 64: 411–53.

Hillis, D. M., Moritz, C., and Mable, B. K. (eds.). 1996. *Molecular systematics*. 2nd ed. Sunderland, MA: Sinauer Associates.

Hilton-Taylor, C. 2000. *2000 IUCN Red List of Threatened Species*. Gland, Switzerland: World Conservation Union (IUCN).

Hoage, R. J. 1977. Biosocial development in the golden lion tamarin, *Leontopithecus rosalia rosalia*. Primates: Callitrichidae. Ph.D. dissertation, University of Pittsburgh, Pittsburgh.

———. 1978. Parental care in *Leontopithecus rosalia rosalia:* Sex and age differences in carrying behavior and the role of prior experience. In D. G. Kleiman (ed.), *The biology and conservation of the Callitrichidae*, pp. 293–305. Washington, DC: Smithsonian Institution Press.

———. 1982. Social and physical maturation in captive lion tamarins *Leontopithecus rosalia rosalia*, Primates, Callitrichidae. *Smithsonian Contributions to Zoology* 354: 1–56.

Hodgen, G. 1996. GnRH agonists and antagonists in ovarian stimulation. *Human Reproduction* 11(suppl. 1): 123–32.

Hodges, J. K. and Eastman, S. A. K. 1984. Monitoring ovarian function in marmosets and tamarins by the measurement of urinary estrogen metabolites. *American Journal of Primatology* 6: 187–97.

Holzer, C., Bergmann, H. H., and McLean, I. A. 1996. Training captive-raised, naïve birds to recognize their predator. In U. Ganslosser, J. K. Hodges, and W. Kaumanns (eds.), *Research and captive propagation,* pp. 198–206. Fuerth: Findler Verlag.

Horovitz, I. and Meyer, A. 1995. Systematics of New World monkeys (Platyrrhini, Primates) based on 16S mitochondrial DNA sequences: A comparative analysis of different weighting methods in cladistic analysis. *Molecular Phylogenetics and Evolution* 4: 448–56.

Horwich, R. H. 1989. Use of surrogate parental models and age periods in a successful release of hand-reared sandhill cranes. *Zoo Biology* 8: 379–90.

Hunt, R. D. and Desroisiers, R. C. 1994. Study of spontaneous infectious diseases of primates: Contributions of the Regional Primate Research Centers' program to conservation and new scientific opportunities. *American Journal of Primatology* 34: 3–10.

Huntress, S. L. and Spraker, T. 1985. *Baylisascaris* infection in the marmoset. *Annual Proceedings of the American Association Zoo Veterinarians,* p. 78.

IESB. 1996. *Alternativas econômicas para conservação e desenvolvimento da região de Una, Bahia. Resumos das pesquisas, 1994–1995.* 2nd ed. Report. Ilhéus, BA: Instituto de Estudos Sócio-Ambientais do Sul da Bahia (IESB).

———. 2000. As pontes entre saberes: Pesquisadores desenvolvem educação ambiental junto a fazendeiros e pequenos produtores da região de Ilhéus. In I. Tamaio and D. Carreira (eds.), *Caminhos e aprendizagens: Educação ambiental, conservação e desenvolvimento,* pp. 67–70. Brasília, DF: WWF-Brasil.

Inglett, B. J. 1993. The role of social bonds and the female reproductive cycle on the regulation of social and sexual interactions in the golden lion tamarin (*Leontopithecus rosalia rosalia*). Ph.D. dissertation, University of Nebraska, Omaha.

Inglett, B. J., French, J. A., Simmons, L. G., and Vires, K. W. 1989. Dynamics of intrafamily aggression and social reintegration in lion tamarins. *Zoo Biology* 8: 67–78.

Inglett, B. J., French, J. A., and Dethlefs, T. M. 1990. Patterns of social preference across different contexts in golden lion tamarins (*Leontopithecus rosalia*). *Journal of Comparative Psychology* 104: 131–39.

IPÊ. 1995/1996. *Annual Report.* Nazaré Paulista, SP: Instituto de Pesquisas Ecológicas (IPÊ).

Irwin, D.M., Kocher, T.D., and Wilson, A.C. 1991. Evolution of the cytochrome *b* gene of mammals. *Journal of Molecular Evolution* 32(2): 128–44.

Isbell, L. A. and Young, T. P. 1993. Social and ecological influences on activity budgets of vervet monkeys, and their implications for group living. *Behavioral Ecology and Sociobiology* 32: 377–85.

IUCN. 1988. *Translocation of living organisms: Introductions, re-introductions, and re-stocking. IUCN Position Statement.* Gland, Switzerland: International Union for Conservation of Nature and Natural Resources (IUCN).

———. 1994. *IUCN Red List Categories.* Gland, Switzerland: International Union for Conservation of Nature and Natural Resources (IUCN), Species Survival Commission (SSC).

———. 1996. *1996 IUCN Red List of Threatened Animals.* Gland, Switzerland: World Conservation Union (IUCN), Species Survival Commission (SSC).

———. 1998. *IUCN Guidelines for Re-introductions*. Gland, Switzerland: World Conservation Union (IUCN), Species Survival Commission (SSC), Reintroduction Specialist Group.

Izawa, K. 1978. A field study of the ecology and behavior of the black-mantled tamarin (*Saguinus nigricollis*). *Primates* 19: 241–74.

Jacobson, S. K. 1991. Evaluation model for developing, implementing, and assessing conservation education programs: Examples from Belize and Costa Rica. *Environmental Management* 15(2): 143–50.

———. 1999. *Communication skills for conservation professionals*. Covelo, CA: Island Press.

Jacobson, S. K. and Padua, S. M. 1992. Pupils and parks: Environmental education using national parks in developing countries. *Childhood Education* 68: 290–93.

———. 1995. Systems model for conservation education in parks: Examples from Malaysia and Brazil. In Jacobson, S. K. (ed.), *Conserving wildlife: International education and communication approaches,* pp. 3–15. New York: Columbia University Press.

Janzen, D. H. 1973. Sweep samples of tropical foliage insects: Effects of seasons, vegetation types, elevation, time of day, and insularity. *Ecology* 54: 687–708.

Jones, M. 1973. *International studbook for the golden lion tamarin,* Leontopithecus r. rosalia. *1972*. Wheeling, WV: American Association of Zoological Parks and Aquariums (AAZPA).

Jurke, M. H., Pryce, C. R., Hug-Hodel, A., and Doebli, M. 1995. An investigation into the socioendocrinology of infant care and postpartum fertility in Goeldi's monkey (*Callimico goeldii*). *International Journal of Primatology* 16: 453–74.

Jurke, M. H., Czekala, N. M., and Fitch-Snyder, H. 1997. Non-invasive detection and monitoring of estrus, pregnancy and the postpartum period in pygmy loris (*Nycticebus pygmaeus*) using fecal estrogen metabolites. *American Journal of Primatology* 41: 103–15.

Karr, S. L. and Wong, M. M. 1975. A survey of *Sarcocystis* in nonhuman primates. *Laboratory Animal Science* 25: 641–45.

Kay, R. F. 1990. The phyletic relationships of extant and fossil Pitheciinae (Platyrrhini, Anthropoidea). *Journal of Human Evolution* 19: 175–208.

Keith-Lucas, T., White, F. J., Keith-Lucas, L., and Vick, L. G. 1999. Changes in behavior in free-ranging *Lemur catta* following release in a natural habitat. *American Journal of Primatology* 47: 15–28.

Keuroghlian, A. 1990. Observations on the behavioral ecology of the black lion tamarin (*Leontopithecus chrysopygus*) at Caetetus Reserve, São Paulo, Brazil. Master's thesis, West Virginia University, Morgantown.

Keuroghlian, A. and Passos, F. C. 2001. Prey foraging behavior, seasonality and time budgets in black lion tamarins, *Leontopithecus chrysopygus* (Mikan, 1823) (Mammalia, Callitrichidae). *Brazilian Journal of Biology* 61(3): 455–59.

Kierulff, M. C. M. 1993a. Avaliação das populações selvagens de mico-leão-dourado, *Leontopithecus rosalia,* e propostas de estratégia para sua conservação. Master's thesis, Universidade Federal de Minas Gerais, Belo Horizonte.

———. 1993b. Status and distribution of the golden lion tamarin in Rio de Janeiro. *Neotropical Primates* 1(4): 23–24.

———. 1999. New biological reserve created for golden lion tamarins. *Tamarin Tales,* no. 3:1.

———. 2000. Ecology and behaviour of translocated groups of golden lion tamarins (*Leontopithecus rosalia*). Ph.D. dissertation, University of Cambridge, UK.

Kierulff, M. C. M. and Procópio de Oliveira, P. 1994. Habitat preservation and the translocation of threatened groups of golden lion tamarins, *Leontopithecus rosalia. Neotropical Primates* 3(suppl.): 15–18.

———. 1996. Re-assessing the status and conservation of the golden lion tamarin *Leontopithecus rosalia* in the wild. *Dodo, Journal of the Wildlife Preservation Trusts* 32: 98–115.

Kierulff, M. C. M. and Rylands, A. B. Submitted. Census and distribution of the golden lion tamarin (*Leontopithecus rosalia*).

Kierulff, M. C. M. and Stallings, J. R. 1991. Levantamento das populações de mico-leão-dourado (*Leontopithecus rosalia*) no estado do Rio de Janeiro. In *Resumos. XVIII Congresso Brasileiro de Zoologia,* p. 393. Universidade Federal da Bahia, Salvador.

Kierulff, M. C. M., Kleiman, D. G., and Santos, E. M. dos. 1997. O uso de "play-back" para o levantamento de populações de mico-leão-dourado. In S. F. Ferrari and H. Schneider (eds.), *A primatologia no Brasil—5,* pp. 160–61. Universidade Federal do Pará, Sociedade Brasileira de Primatologia, Belém.

Kim, A. C. and Passos, F. C. 1994. A contribution to the study of arboreal vegetation of the Caetetus Ecological Station, São Paulo, Brazil. *Neotropical Primates* 2(suppl.): 42–44.

King, G. 1975. Feeding and nutrition of the Callitrichidae at the Jersey Zoological Park. *Jersey Wildlife Preservation Trust 12th Annual Report,* pp. 81–90.

Kinnaird, M. F. 1990. Behavioral and demographic responses to habitat change by the Tana River crested mangabey (*Cercocebus galeritus galeritus*). Ph.D. dissertation, University of Florida, Gainesville.

Kinnaird, M. F. and O'Brien, T. G. 2001. Who's scratching whom? Reply to Whitten et al. *Conservation Biology* 15(5): 1459–60.

Kinzey, W. G. 1981. The titi monkeys, genus *Callicebus.* In A. F. Coimbra-Filho and R. A. Mittermeier (eds.), *Ecology and behavior of neotropical primates,* vol. 1, pp. 241–76. Rio de Janeiro: Academia Brasiliera de Ciências.

———. 1982. Distribution of primates and forest refuges. In G. T. Prance (ed.), *Biological diversification in the tropics,* pp. 455–82. New York: Columbia University Press.

Kleiman, D. G. 1972. Recommendations on research priorities for the lion marmoset. In D. D. Bridgwater (ed.), *Saving the lion marmoset,* pp. 137–39. Wheeling, WV: Wild Animal Propagation Trust.

———. 1976. Will the pot of gold have a rainbow? Hope for Brazil's golden tamarins in North American Zoos. *Animal Kingdom* 79: 2–6.

———. 1977a. Progress and problems in lion tamarin *Leontopithecus rosalia rosalia* reproduction. *International Zoo Yearbook* 17: 92–97.

———. 1977b. 1975 world register of golden lion tamarins *Leontopithecus rosalia rosalia. International Zoo Yearbook* 17: 232.

———. 1977c. Monogamy in mammals. *Quarterly Review of Biology* 52: 39–69.
———. 1977d. Changes in the golden lion tamarin (*Leontopithecus rosalia rosalia*) population in 1976. *AAZPA Newsletter* 18(6): 20–21.
———. (ed.). 1978a. *The biology and conservation of the Callitrichidae*. Washington, DC: Smithsonian Institution Press.
———. 1978b. Characteristics of reproduction and sociosexual interactions in pairs of lion tamarins (*Leontopithecus rosalia*) during the reproduction cycle. In D. G. Kleiman (ed.), *The biology and conservation of the Callitrichidae,* pp. 181–90. Washington, DC: Smithsonian Institution Press.
———. 1978c. The development of pair preferences in the lion tamarin (*Leontopithecus rosalia*): Male competition or female choice? In H. Rothe, H.-J. Wolters, and J. P. Hearn (eds.), *Biology and behaviour of marmosets,* pp. 181–88. Göttingen: Eigenverlag H. Rothe.
———. 1979. Parent-offspring conflict and sibling competition in a monogamous primate. *American Naturalist* 114: 753–60.
———. 1980. The sociobiology of captive propagation in mammals. In M. Soulé and B. Wilcox (eds.), *Conservation biology,* pp. 243–61. Sunderland, MA: Sinauer Associates.
———. 1981. *Leontopithecus rosalia. Mammalian Species,* no. 148: 1–7.
———. 1982. Cooperative research and management agreement for the golden lion tamarins. *AAZPA Newsletter* 23: 16.
———. 1983. Social and parental behavior of a monogamous primate, the golden lion tamarin. *Journal of the American Veterinary Medicine Association* 183: 363.
———. 1984a. The National Zoo's role in an international primate conservation program. *IUCN/SSC Primate Specialist Group Newsletter,* no. 4: 45–46.
———. 1984b. The behavior and conservation of the golden lion tamarin, *Leontopithecus r. rosalia*. In M. T. de Mello (ed.), *A primatologia no Brasil,* pp. 35–53. Brasília, DF: Sociedade Brasileira de Primatologia.
———. 1985. Paternal care in New World primates. *American Zoologist* 25(3): 857–59.
———. 1989. Re-introduction of captive mammals for conservation: Guidelines for reintroducing endangered species into the wild. *Bioscience* 39: 152–61.
———. 1990a. Responses to long call playbacks: Differences between wild and reintroduced golden lion tamarins. Paper presented at the Animal Behavior Society Meeting, Binghamton, NY.
———. 1990b. Decision-making about a reintroduction: Do appropriate conditions exist? *Endangered Species Update* 8: 18–19.
———. 1994. Mammalian sociobiology and zoo breeding programs. *Zoo Biology* 13: 423–32.
———. 1996. Reintroduction programs. In D. G. Kleiman, M. E. Allen, K. V. Thompson, and S. Lumpkin (eds.), *Wild mammals in captivity: Principles and techniques,* pp. 297–305. Chicago: University of Chicago Press.
Kleiman, D. G. and Evans, R. 1980–1982. *International golden lion tamarin studbook.* Washington, DC: National Zoological Park.
Kleiman, D. G. and Jones, M. 1978. The current status of *Leontopithecus rosalia* in captiv-

ity with comments on breeding success at the National Zoological Park. In D. G. Kleiman (ed.), *The biology and conservation of the Callitrichidae*, pp. 215–18. Washington, DC: Smithsonian Institution Press.

Kleiman, D. G. and Mack, D. S. 1977. A peak in sexual activity during mid-pregnancy in the golden lion tamarin, *Leontopithecus rosalia,* Primates: Callitrichidae. *Journal of Mammalogy* 58: 657–60.

———. 1980. The effects of age, sex, and reproductive status on scent-marking frequencies in the golden lion tamarin *Leontopithecus rosalia. Folia Primatologica* 33: 1–14.

Kleiman, D. G. and Malcolm, J. 1981. The evolution of male parental investment in mammals. In D. J. Gubernick and P. H. Klopfer (eds.), *Parental care in mammals,* pp. 347–87. New York: Plenum Press.

Kleiman, D. G. and Mallinson, J. J. C. 1998. Recovery and Management Committees for lion tamarins: Partnerships in conservation planning and implementation. *Conservation Biology* 12: 1–13.

Kleiman, D. G., Gracey, D. W., and Hodgen, G. D. 1978. Urinary chorionic gonadotropin levels of pregnant golden lion tamarins. *Journal of Medical Primatology* 7: 333–38.

Kleiman, D. G., Ballou, J. D., and Evans, R. F. 1982. An analysis of recent reproductive trends in captive golden lion tamarins *Leontopithecus r. rosalia* with comments on their future demographic management. *International Zoo Yearbook* 22: 94–101.

Kleiman, D. G., Beck, B. B., Dietz, J. M., Dietz, L. A., Ballou, J. D., and Coimbra-Filho, A. F. 1985. A conservation program for golden lion tamarins. *Proceedings 1985 AAZPA Annual Conference,* pp. 60–62.

Kleiman, D. G., Beck, B. B., Dietz, J. M., Dietz, L. A., Ballou, J. D., and Coimbra-Filho, A. F. 1986. Conservation program for the golden lion tamarin: Captive research and management, ecological studies, educational strategies, and reintroduction. In K. Benirschke (ed.), *Primates: The road to self-sustaining populations,* pp. 959–79. New York: Springer Verlag.

Kleiman, D. G., Hoage, R. J., and Green, K. M. 1988. The lion tamarins, genus *Leontopithecus.* In R. A. Mittermeier, A. B. Rylands, A. F. Coimbra-Filho, and G. A. B. da Fonseca (eds.), *Ecology and behavior of neotropical primates,* vol. 2, pp. 299–347. Washington, DC: World Wildlife Fund–U.S.

Kleiman, D. G., Beck, B. B., Baker, A. J., Ballou, J. D., Dietz, L. A., and Dietz, J. M. 1990a. The conservation program for the golden lion tamarin *Leontopithecus rosalia. Endangered Species Update* 8(1): 82–85.

Kleiman, D. G., Dietz, J. M., Baker, A. J., French, J. A., Rambaldi, D., Dietz, L. A., and Montali, R. J. 1990b. Golden Lion Tamarin Working Group summary. In U. S. Seal, J. D. Ballou, and C. Valladares-Padua (eds.), Leontopithecus: *Population Viability Analysis Workshop report,* pp. 17–21. Apple Valley, MN: International Union for Conservation of Nature and Natural Resources/Species Survival Commission (IUCN/SSC) Captive Breeding Specialist Group (CBSG).

Kleiman, D. G., Beck, B. B., Dietz, J. M., and Dietz, L. A. 1991. Costs of a reintroduction and criteria for success: Accounting and accountability in the Golden Lion Tamarin Conservation Program. In J. H. W. Gipps (ed.), *Beyond captive breeding: Reintroducing endangered mammals to the wild,* pp. 125–42. Oxford: Clarendon Press.

Kleiman, D. G., Stanley-Price, M. R., and Beck, B. B. 1994. Criteria for reintroductions. In P. J. S. Olney, G. M. Mace, and A. T. C. Feistner (eds.), *Creative conservation: Interactive management of wild and captive animals,* pp. 287–303. London: Chapman and Hall.

Kleiman, D. G., Reading, R. P., Miller, B. J., Clark, T. W., Scott, J. M., Robinson, J., Wallace, R., Cabin, R., and Felleman, F. 2000. The importance of improving evaluation in conservation. *Conservation Biology* 14: 356–365.

Kocher, T. D., Thomas, W. K., Meyer, A., Edwards, S. V., Päbo, S., Villablanca, F. X., and Wilson, A. C. 1989. Dynamics of mitochondrial DNA evolution in animals: Amplification and sequencing with conserved primers. *Proceedings of the National Academy of Sciences* 86: 6196–200.

Koenig, A., Radespiel, U., Siess, M., Rothe, H., and Darms, K. 1990. Analysis of pairing-parturition and interbirth intervals in a colony of common marmosets (*Callithrix jacchus*). *Zeitschrift für Säugetierkunde* 55: 308–14.

Kolb, S. R. 1993. Islands of secondary vegetation in degraded pastures of Brazil: Their role in re-establishing Atlantic coastal forest. Doctoral thesis, University of Georgia, Athens.

Konstant, W. R. 1986. Illegal trade in golden-headed lion tamarins. *Primate Conservation,* no. 7: 29–30.

———. 1990. Project Mico-Leão Baiano. *On the Edge,* no. 41: 1–5.

———. 1996/1997. Funding for primate conservation: Where has it originated? *Primate Conservation,* no. 17: 30–36.

Konstant, W. R. and Mittermeier, R. A. 1982. Introduction, reintroduction and translocation of neotropical primates: Past experiences and future possibilities. *International Zoo Yearbook* 22: 69–77.

Koontz, F. W. and Roush, R. S. 1996. Effects of captivity on the behavior of wild mammals. In D. G. Kleiman, M. E. Allen, K. V. Thompson, and S. Lumpkin (eds.), *Wild mammals in captivity: Principles and techniques*, pp. 317–33. Chicago: University of Chicago Press.

Korte, W. E. and Lond, M. B. 1905. On the presence of *Sarcospiridium* in the thigh muscle of macacus rhesus. *Journal of Hygiene* 5: 451–52.

Küderling, I., Evans, C. S., and Abbott, D. H. 1995. Differential excretion of urinary oestrogens by breeding females and daughters in the red-bellied tamarin (*Saguinus labiatus*). *Folia Primatologica* 64: 140–45.

Kuehler, C., Kuhn, M., Kuhn, J., Lieberman, A., Harvey, N., and Rideout, B. 1996. Artificial incubation, hand-rearing, behavior and release of common amakihi (*Hemingnathus virens virens*): Surrogate research for restoration of endangered Hawaiian forest birds. *Zoo Biology* 15: 541–53.

Kuhl, H. 1820. *Beiträge zur Zoologie und vergleichenden Anatomie.* Frankfurt am Main: *Erste Abtheilung,* pp. 1–152.

Kummer, H. and Kurt, F. 1965. A comparison of social behavior in captive and wild hamadryas baboons. In H. Vagtborg (ed.), *The baboon in medical research,* vol. 1, pp. 65–80. Austin: University of Texas.

Lacy, R. C. 1989. Analysis of founder representation in pedigrees: Founder equivalents and founder genome equivalents. *Zoo Biology* 8: 111–23.

———. 1993. VORTEX: A computer simulation model for Population Viability Analysis. *Wildlife Research* 20: 45–65.

———. 1993/1994. What is Population (and Habitat) Viability Analysis? *Primate Conservation,* no. 14–15: 27–33.

Laemmert, H. W., Jr., Ferreira, L. de C., and Taylor, R. M. 1946. An epidemiological study of jungle yellow fever in an endemic area in Brazil. Part II—Investigations of vertebrate hosts and arthropod vectors. *American Journal Tropical Medicine* 26: 23–60.

Lee, E. T. 1992. *Statistical methods for survival data analysis.* New York: John Wiley and Sons.

Leite, J. F. 1979. As reservas ecológicas do sudeste Paulista. Paper presented at XXIII Congresso Estadual de Municípios, Praia Grande, São Paulo, May 1979. 38 pp.

Leus, K. 1993–1999. *International studbook for golden headed lion tamarins.* Antwerp: Royal Zoological Society of Antwerp.

———. 1999. *Leontopithecus* population control workshop at Antwerp. *EAZA News,* no. 25: 6–7.

Leus, K. and De Vleeschouwer, K. 2001. *1999 International studbook for golden headed lion tamarins.* No. 10. Antwerp: Royal Zoological Society of Antwerp.

Levins, R. 1969. Some genetic and demographic consequences of environmental heterogeneity for biological control. *Bulletin of the Entomological Society of America* 15: 237–40.

Lickliter, R. and Ness, J. W. 1990. Domestication and comparative psychology: Status and strategy. *Journal of Comparative Psychology* 104(3): 211–18.

Lima, H., Farias, D., and Farág, P. 1995. Lista das éspecies vegetais da dieta do mico-leão-dourado, *Leontopithecus rosalia.* Unpublished. Rio de Janeiro: Jardim Botânico do Rio de Janeiro.

Lima, M. G. de. 1990. Uma proposta para a conservação dos primatas da mata Atlântica do sul da Bahia. Master's thesis, Universidade Federal de Paraíba, João Pessoa.

Lindbergh, S. M. 1986. O mico-leão-de-cara-dourada na Reserva Biológica de Una, Bahia (Relatório de viagem de 22 de novembro a 02 de dezembro, 1986). Report, Departamento de Parques Nacionais e Reservas Equivalentes, Instituto Brasileiro de Desenvolvimento Florestal (IBDF), Brasília, DF.

Lindburg, D. G. 1977. Feeding behaviour and diet of rhesus monkey (*Macaca mulatta*) in a Siwalik forest in North India. In T. H. Clutton-Brock (ed.), *Primate ecology: Studies of feeding and ranging behaviour in lemurs, monkeys and apes,* pp. 223–49. London: Academic Press.

Linnaeus, Carolus. 1766. *Systema Naturae . . . Tomus I. Regnum Animalium.* 12th ed., reformed. Holm. 532 pp.

Lisboa, C. V., Dietz, J. M., Baker, A. J., Russel, N. E., and Jansen A. M. 2000. *Try-*

panosoma cruzi infection in *Leontopithecus rosalia* at the Reserva Biológica de Poço das Antas, Rio de Janeiro, Brazil. *Memórias do Instituto Oswaldo Cruz* 95(4): 445–52.

Lorini, M. L. and Persson, V. G. 1990. Uma nova espécie de *Leontopithecus* Lesson, 1840, do sul do Brasil (Primates, Callitrichidae). *Boletim do Museu Nacional, Rio de Janeiro, nova sér., Zoologia* 338: 1–14.

———. 1994a. Densidade populacional de *Leontopithecus caissara* Lorini & Persson, 1990, na Ilha de Superagüi/PR (Primates, Callitrichidae). In *Resumos, XX Congresso Brasileiro de Zoologia,* p.145. Universidade Federal do Rio de Janeiro, Rio de Janeiro, 24–29 July 1994.

———. 1994b. Status of field research on *Leontopithecus caissara:* The Black-Faced Lion Tamarin Project. *Neotropical Primates* 2(suppl.): 52–55.

Mace, G. M. and Lande, R. 1991. Assessing extinction threats: Toward a reevaluation of IUCN threatened species categories. *Conservation Biology* 5: 148–57.

Mace, G. M. and Mallinson, J. J. C. 1992. *International studbook golden-headed lion tamarin* Leontopithecus chrysomelas. Number 5. Trinity, Jersey, Channel Islands: Jersey Wildlife Preservation Trust.

Machado, A. B. M., Fonseca, G. A. B. da, Machado, R. B., Aguiar, L. M. de S., and Lins, L. V. 1998. *Livro vermelho das espécies ameaçadas de extinção da fauna de Minas Gerais.* Belo Horizonte, MG: Fundação Biodiversitas.

Mack, D. S. and Kleiman, D. G. 1978. Distribution of scent marks in different contexts in captive lion tamarins *Leontopithecus rosalia* (Primates). In H. Rothe, H.-J. Wolters, and J. P. Hearn (eds.), *Biology and behaviour of marmosets,* pp. 181–88. Göttingen: Eigenverlag H. Rothe.

Mager, W. B. and Griede, T. 1986. Using outside areas for tropical primates in the Northern Hemisphere: Callitrichidae, *Saimiri,* and *Gorilla.* In K. Benirschke (ed.), *Primates: The road to self-sustaining populations,* pp. 471–77. New York: Springer-Verlag.

Magnanini, A. 1975. Uma espécie ameaçada de extinção no Brasil (problemas e soluções no caso dos micos-leões, *L. rosalia*). *FBCN/Informativo,* Rio de Janeiro, no. 8: 21–23.

———. 1978. Progress in the development of Poço das Antas Biological Reserve for *Leontopithecus rosalia rosalia* in Brazil. In D. G. Kleiman (ed.), *The biology and conservation of the Callitrichidae,* pp. 131–36. Washington, DC: Smithsonian Institution Press.

Magnanini, A. and Coimbra-Filho, A. F. 1972. The establishment of a captive breeding program and a wildlife research center for the lion marmoset, *Leontopithecus,* in Brazil. In D. D. Bridgwater (ed.), *Saving the lion marmoset,* pp. 110–19. Wheeling, WV: Wildlife Propagation Trust.

Magnanini, A., Coimbra-Filho, A. F., Mittermeier, R. A., and Aldrighi, A. 1975. The Tijuca Bank of lion marmosets *L. rosalia:* A progress report. *International Zoo Yearbook* 15: 284–87.

Malaga, C. 1985. Nonstandard mating systems for *Saguinus mystax. American Journal of Physical Anthropology* 66: 201 (Abstract).

Mallinson, J. J. C. 1984. Golden-headed lion tamarin contraband a major conservation problem. *IUCN/SSC Primate Specialist Group Newsletter,* no. 4: 23–25.

———. 1986. The Wildlife Preservation Trusts' (J.W.P.T./W.P.T.I.) support for the conservation of the genus *Leontopithecus*. *Dodo, Journal of the Jersey Wildlife Preservation Trust* 23: 6–18.

———. 1987a. *A preliminary international studbook for golden-headed lion tamarin,* Leontopithecus chrysomelas. Trinity, Jersey, Channel Islands: Jersey Wildlife Preservation Trust.

———. 1987b. International efforts to secure a viable population of the golden-headed lion tamarin. *Primate Conservation,* no. 8: 121–25.

———. 1989. A summary of the work of the International Recovery and Management Committee for Golden-Headed Lion Tamarin *Leontopithecus chrysomelas,* 1985–1990. *Dodo, Journal of the Jersey Wildlife Preservation Trust* 26: 77–86.

———. 1994a. The Lion Tamarins of Brazil Fund: With reference to the International Management Committees for *Leontopithecus. Neotropical Primates* 2(suppl.): 4–5.

———. 1994b. Saving the world's richest rain forest. *Biologist* 41(2): 57–60.

———. 1995. Conservation breeding programmes: An important ingredient for species survival. *Biodiversity and Conservation* 4: 617–35.

———. 1996. The history of golden lion tamarin management and propagation outside of Brazil and current management practices. *Zoologische Garten N.F.* 66: 197–217.

———. 1997a. The Lion Tamarins of Brazil Fund. *Tamarin Tales,* no. 1: 15.

———. 1997b. A 'case study': Partnerships and conservation initiatives resulting from a Population Viability Assessment (PVA) Workshop for the genus *Leontopithecus. Zoologische Garten N. F.* 67: 355–63.

———. 2001a. Margot Marsh Biodiversity Foundation (MMBF) matching support for *in situ* conservation projects for lion tamarins in Brazil. *Tamarin Tales,* no. 5: 13.

———. 2001b. Saving Brazil's Atlantic rainforest: Using the golden-headed lion tamarin (*Leontopithecus chrysomelas*) as a flagship for a biodiversity hotspot. *Dodo* 37: 9–20.

Mamede-Costa, A. C. 1997. Ecologia de um grupo de micos-leões-pretos (*Leontopithecus chrysopygus* Mikan, 1823) na mata ciliar da Fazenda Rio Claro, Lenções Paulista, SP. Master's thesis, Universidade Estadual Paulista (UNESP), Rio Claro.

Mamede-Costa, A. C. and Gobbi, N. 1998. The black lion tamarin *Leontopithecus chrysopygus*—its conservation and management. *Oryx* 32(4): 295–300.

Mandour, A. M. 1969. *Sarcocystis nesbitti* n.sp., from the Rhesus monkey. *Journal of Protozoology* 16(2): 353–54.

Mansour, J. A. and Ballou, J. D. 1994. Capitalizing the ark: The economic benefit of adding founders to captive populations. *Neotropical Primates* 2(suppl.): 8–11.

Maple, T. L. and Finlay, T. W. 1989. Applied primatology in the modern zoo. *Zoo Biology Supplement* 1: 101–16.

Margarido, T. C. C., Pereira, L. C. M., and Nicola, P. A. 1997. Diagnóstico da mastofauna terrestre na APA de Guaraqueçaba, Paraná, Brasil. In M. S. Milano (ed.), *Anais do 1° Congresso Brasileiro de Unidades de Conservação*. Curitiba, pp. 757–68. Curitiba, PR: Universidade Livre do Meio Ambiente, Rede Nacional Pró-Unidades de Conservação, and Instituto Ambiental do Paraná (IAP).

Marler, P. and Mitani, J. 1988. Vocal communication in primates and birds: Parallels and contrasts. In D. Todt, P. Goedeking, and D. Symmes, *Primate vocal communication,* pp. 3–15. Berlin: Springer Verlag.

Marques, M. A., Vasconcellos, H. A. de, Queiroz, S., and Pissinatti, A. 1997. Morphological and morphometric study of biceps brachii, triceps brachii and dorsoepitrochlearis muscles, in three species of *Leontopithecus* (Lesson, 1840). *Brazilian Journal of Morphological Sciences* 14(2): 281–88.

Marsh, C. W. 1981. Ranging behaviour and its relation to diet selection in Tana River red colobus (*Colobus badius rufomitratus*). *Journal of Zoology, London* 195: 473–92.

Martin, C. R. 1985. *Endocrine physiology*. New York: Oxford University Press.

Martuscelli, P. and Rodrigues, M. G. 1992. Novas populações do mico-leão caiçara, *Leontopithecus caissara* (Lorini & Persson, 1990) no sudeste do Brasil (Primates-Callitrichidae). *Revista do Instituto Florestal, São Paulo* 4: 920–24.

Mason, W. A. 1985. Experiential influences on the development of expressive behaviors in rhesus monkeys. In G. Zivin (ed.), *The development of expressive behavior: Biology-environment interactions,* pp. 117–52. New York: Academic Press.

Matteri. R. L., Roser, J. F., Baldwin, D. M., Lipovetsky, V., and Papkoff, H. 1987. Characterization of a monoclinal antibody which detects luteinizing hormone from diverse mammalian species. *Domestic Animal Endocrinology* 4: 157–65.

Maxwell, J. M. and Jamieson, I. G. 1997. Survival and recruitment of captive-reared and wild-reared takahe in Fiordland, New Zealand. *Conservation Biology* 11: 683–91.

McCullough, D. R. 1996. *Metapopulations and wildlife conservation.* Washington, DC: Island Press.

McDonald, R. E. 1974. *Dentistry for the child and adolescent.* 2nd ed. St. Louis: C. V. Mosky Co.

McGrew, W. C. 1988. Parental division of infant caretaking varies with family composition in cotton-top tamarins. *Animal Behaviour* 36: 285–86

McLanahan, E. B. and Green, K. M. 1978. The vocal repertoire and an analysis of the contexts of vocalizations in *Leontopithecus rosalia.* In D. G. Kleiman (ed.), *The biology and conservation of the Callitrichidae,* pp. 251–69. Washington, DC: Smithsonian Institution Press.

McLean, I. G., Lundie-Jenkins, G., and Jarman, P. F. 1994. Training captive rufous hare-wallabies to recognize predators. In M. Serena (ed.), *Reintroduction biology of Australian and New Zealand fauna,* pp. 177–82. London: Surrey Beatty Sons.

———. 1996. Teaching an endangered mammal to recognise predators. *Biological Conservation* 56: 51–62.

McLean, I. G., Hölzer, C., and Studholme, B. J. S. 1999. Teaching predator-recognition to a naïve bird: Implications for management. *Biological Conservation* 87: 123–30.

Medici, P. 2001. Translocação como ferramenta para o manejo populacional de mico-leão-preto, *Leontopithecus chrysopygus* (Mikan, 1823). Master's thesis, Universidade Federal de Minas Gerais, Belo Horizonte.

Meireles, C. M. M., Sampaio, M. I. C., Schneider, H., and Schneider, M. P. C. 1992.

Protein variation, taxonomy, and differentiation in five species of marmosets. *Primates* 33: 227–38.

Melo, A. C. A., Sampaio, M. I. C., Schneider, M.P.C., and Schneider, H. 1992. Biochemical diversity and genetic distance in two species of the genus *Saguinus*. *Primates* 33: 217–36.

Ménard, N. and Vallet, D. 1997. Behavioral responses of barbary macaques (*Macaca sylvanus*) to variations in environmental conditions in Algeria. *American Journal of Primatology* 43: 285–304.

Mendes, S. L. 1997. Padrões biogeográficos e vocais em *Callithrix* do grupo *jacchus* (Primates: Callitrichidae). Doctoral thesis, Universidade Estadual de Campinas, Campinas.

Menzel, C. and Beck, B. B. 2000. Homing and detour behavior in golden lion tamarin social groups. In S. Boinski and P. A. Garber (eds.), *On the move: How and why animals travel in groups,* pp. 299–326. Chicago: University of Chicago Press.

Mesquita, C. A. B. 1996. Serrarias na região cacaueira são insustentáveis econômica e ambientalmente. In *Alternativas econômicas para conservação e desenvolvimento da região de Una, Bahia. Resumos das pesquisas, 1994–1995*, pp. 6–10. Ilhéus, BA: Instituto de Estudos Sócio-Ambientais do Sul da Bahia (IESB).

Mikan, J. C. 1823. *Delectus florae et faunae Brasiliensis*. Anthony Strauss, Vienna, 1820–1825. 24 pp.

Miller, B., Biggins, D., Wemmer, C., Powell, R., Calvo, L., and Wharton, T. 1991. Development of survival skills in captive-raised Siberian polecats (*Mustela eversmanni*) II: Predator avoidance. *Journal of Ethology* 8: 95–104.

Miller, B., Reading, R. P., and Forrest, S. 1995. *Prairie night: Black-footed ferrets and the recovery of endangered species*. Washington, DC: Smithsonian Institution Press.

Miller, B., Biggins, D., Vargas, A., Hutchins, M., Hanebury, L., Godbey, J., Anderson, S., Wemmer, C., and Oldemeier, J. 1998. The captive environment and reintroduction: The black-footed ferret as a case study with comments on other taxa. In D. J. Shepherdson, J. D. Mellen, and M. Hutchins (eds.), *Second nature: Environmental enrichment for captive animals,* pp. 97–112. Washington, DC: Smithsonian Institution Press.

Miller, K. E., Bush, A. G., and Dietz, J. M. In preparation. Influence of intrinsic and extrinsic constraints on feeding behavior in wild golden lion tamarins (*Leontopithecus rosalia*).

Milton, K. 1980. *The foraging strategy of howler monkeys: A study in primate economics*. New York: Columbia University Press.

———. 1984. Habitat, diet, and activity patterns of free-ranging woolly spider monkeys (*Brachyteles arachnoides* E. Geoffroy 1806). *International Journal of Primatology* 5: 491–514.

Mittermeier, R. A. 1978. A global strategy for primate conservation. Washington, DC: International Union for Conservation of Nature and Natural Resources/Species Survival Commission (IUCN/SSC) Primate Specialist Group.

———. 1982. Further survey work and development of a preliminary management plan for the endangered golden-rumped lion tamarin (*Leontopithecus chrysopygus*) in the Morro do Diabo State Reserve, São Paulo, Brazil. *Atlantic Forest Region of Eastern Brazil—Project 12. Report, IUCN/World Wildlife Fund–U.S.* 8 pp.

Mittermeier, R. A., Constable, I. D., and Coimbra-Filho, A. F. 1981. Conservation of eastern Brazilian primates. Unpublished report, Project 1614. Washington, DC: Primate Program of the World Wildlife Fund–U.S., period 1979/1980. 39 pp.

Mittermeier, R. A., Coimbra-Filho, A. F., Constable, I. D., Rylands, A. B., and Valle, C. 1982. Conservation of primates in the Atlantic forest region of eastern Brazil. *International Zoo Yearbook* 22: 2–17.

Mittermeier, R.A., Padua, C. V., Valle, C., and Coimbra-Filho, A. F. 1985. Major program underway to save the black lion tamarin in São Paulo, Brazil. *Primate Conservation,* no. 6: 19–21.

Mittermeier, R. A., Myers, N., Robles Gil, P., and Mittermeier, C. G. 1999a. *Hotspots: The earth's biologically richest and most endangered terrestrial ecoregions.* Mexico City: Cemex.

Mittermeier, R. A., Fonseca, G. A. B., Rylands, A. B., and Mittermeier, C. G. 1999b. Atlantic Forest. In R. A. Mittermeier, N. Myers, P. Robles Gil, and C. G. Mittermeier (eds.), *Hotspots: The earth's biologically richest and most endangered terrestrial ecoregions,* pp. 136–47. Mexico City: Cemex.

Monfort, S. L., Bush, M., and Wildt, D. E. 1996. Evaluation of natural and induced ovarian synchrony in golden lion tamarins (*Leontopithecus rosalia*). *Biology of Reproduction* 55: 875–82.

Montali, R. J. 1993a. *Pterygodermatites nycticebi* Tamarin, 1993. In T. C. Jones (ed.), *Nonhuman primates: Monographs on pathology of laboratory animals,* pp. 69–70. New York: International Life Sciences Institute and Springer Verlag.

———. 1993b. Congenital retrosternal diaphragmatic defects, golden lion tamarins. In T. C. Jones (ed.), *Nonhuman primates: Monographs on pathology of laboratory animals,* pp. 132–33. New York: International Life Sciences Institute and Springer Verlag.

Montali, R. J. and Bush, M. 1981. Rictulariasis in Callitrichidae at the National Zoological Park. Internationalen Symposium über die Erkrankungen der Zootiers. Halle/Saale.

———. 1999. Diseases of the Callitrichidae. In M. Fowler and E. Miller (eds.), *Zoo and wild animal medicine: Current therapy 4,* pp. 369–76. Philadelphia, PA: W. B. Saunders and Co.

Montali, R. J., Bush, M., Kleiman, D. G., and Evans, R. F. 1980. Familial diaphragmatic defects in golden lion tamarins. *Erkrankungen der Zootiere* 22: 173–79.

Montali, R. J., Gardiner, C. H., Evans, R. E., and Bush, M. 1983. *Pterygodermatites nycticebi* (Nematoda: Spirurida) in golden lion tamarins. *Laboratory Animal Science* 33: 194–97.

Montali, R. J., Scanga, C. A., Pernikoff, D., Wessner, D. R., Ward., D., and Holmes, K. V. 1993. A common-source outbreak of callitrichid hepatitis in captive tamarins and marmosets. *Journal of Infectious Diseases* 167: 946–50.

Montali, R. J., Connolly, B. M., Armstrong, D. L., Scanga, C. A., and Holmes, K. V. 1995a. Pathology and immunohistochemistry of callitrichid hepatitis, and emerging disease of captive New World primates caused by lymphocytic chroriomeningitis virus. *American Journal of Pathology* 148(5): 1441–49.

Montali, R. J., Bush, M., Hess, J., Ballou, J. D., Kleiman, D. G., and Beck, B. B. 1995b. Ex situ diseases and their control for the reintroduction of the endangered lion tamarin species (*Leontopithecus* spp.). *Erkrankungen der Zootiere* 34: 93–98.

Moore, M. T. 1997. Behaviour adaptation of captive-born golden-headed lion tamarins *Leontopithecus chrysomelas* to a free-ranging environment. *Dodo, Journal of the Wildlife Preservation Trusts* 33: 156–57.

Moreau, M., Silva, L. A. M., and Alger, K. 1996. Exploração da piaçava pode ser mais rentável nos municípios litorâneos da Região Cacuaeira. In *Alternativas econômicas para conservação e desenvolvimento da região de Una, Bahia. Resumos das pesquisas, 1994–1995,* pp. 30–33. Ilhéus, BA: Instituto de Estudos Sócio-Ambientais do Sul da Bahia (IESB).

Moreira, M. A. M. and Seuánez, H. N. 1999. Mitochondrial pseudogenes and phyletic relationships of *Cebuella* and *Callithrix* (Platyrrhini, Primates). *Primates* 40: 353–64.

Moreira, M. A. M., Almeida, C. A. S., Canavez, F., Olicio, R., and Seuánez, H. N. 1996. Heteroduplex mobility assays (HMAs) and analogous sequence analysis of a cytochrome B region indicate phylogenetic relationships of selected callitrichids. *Journal of Heredity* 87: 456–60.

Morland, H. S. 1993. Seasonal behavioral variation and its relationship to thermoregulation in ruffed lemurs (*Varecia variegata variegata*). In P. M. Kappeler and J. U. Ganzhorn (eds.), *Lemur social systems and their ecological basis,* pp. 193–203. New York: Plenum Press.

Morris, M. L., Jr. 1976. Prepared diets for zoo animals in the USA. *International Zoo Yearbook* 16: 13–17.

Morton, E. S. 1975. Ecological sources of selection on avian sounds. *American Naturalist* 109: 17–34.

———. 1977. On the occurrence and significance of motivation-structural rules in some bird and mammal sounds. *American Naturalist* 111: 855–69.

———. 1982. Grading, discreteness, redundancy, and motivation-structural rules. In D. E. Kroodsma, E. H. Miller, and H. Ouellet (eds.), *Acoustic communication in birds,* vol. 1, pp. 183–212. New York: Academic Press.

Moura, A. C. A. and Langguth, A. 1997. Estudo da partilha de alimento em *Leontopithecus chrysomelas* no cativeiro. In *Resumos. VIII Congresso Brasileiro de Primatologia e V Reunião Latino Americana de Primatologia,* p. 83. João Pessoa, Paraíba, Brazil, 10–15 August 1997.

Moura, A. C. A., Porfírio, S., and Alonso, C. 1997. Aggressive response toward intruders by captive male *Leontopithecus chrysomelas*. *Neotropical Primates* 5(4): 111–13.

Mourão, R. 1996. Ecoturismo é alternativa viável e lucrativa para a região, mas é necessário planejar. In *Alternativas econômicas para conservação e desenvolvimento da região de Una, Bahia. Resumos das pesquisas, 1994–1995,* pp. 26–28. Ilhéus, BA: Instituto de Estudos Sócio-Ambientais do Sul da Bahia (IESB).

Mundy, N. I. and Kelly, J. 2001. Phylogeny of lion tamarins (*Leontopithecus* spp.) based in interphotoreceptor retinal binding protein intron sequences. *American Journal of Primatology* 54: 33–40.

Murnane, R. D., Zdziarski, J. M., and Phillips, L. G. 1996. Melengestrol acetate–induced exuberant endometrial decidualization in Goeldi's marmosets (*Callimico goeldii*) and squirrel monkeys (*Saimiri sciureus*). *Journal of Zoo and Wildlife Medicine* 27: 315–24.

Murray, O. and Oldfield, M. 1984. The top ten endangered species of the world. *IUCN Bulletin* 15: 77.

Myers, N. and Knoll, A. H. 2001. The biotic crisis and the future of evolution. *Proceedings of the National Academy of Sciences* 98: 5389–92.

Myers, N., Mittermeier, R. A., Mittermeier, C. G., Fonseca, G. A. B. da, and Kent, J. 2000. Biodiversity hotspots for conservation priorities. *Nature, London* 403: 853–58.

Nagagata, E. Y. 1994a. Evaluation of community-based conservation education: A case study of the Golden-Headed Lion Tamarin Education Program in Bahia state, Brazil. Master's thesis, Michigan State University.

———. 1994b. Evaluation of community-based conservation education: A case study of the golden-headed lion tamarin education program in the state of Bahia, Brazil. *Neotropical Primates* 2(suppl.): 33–35.

Natori, M. 1989. An analysis of cladistic relationships of *Leontopithecus* based on dental and cranial characters. *Journal of Anthropological Society Nippon* 97: 157–67.

Nelson, B., Cosgrove, G. E., and Gengozian, N. 1966. Diseases of an imported primate *Tamarinus nigricollis*. *Laboratory Animal Care* 16: 255–75.

Neusser, M., Stanyon, R., Bigoni, F., Wienberg, J., and Muller, S. 2001. Molecular cytotaxonomy of New World monkeys (Platyrrhini–comparative analysis of five species by multi-color chromosome painting gives evidence for a classification of *Callimico goeldii* within the family of Callitrichidae. *Cytogenetics and Cell Genetics* 94(3–4): 206–15.

Newman, J. D. and Symmes, D. 1982. Inheritance and experience in the acquisition of primate acoustic behavior. In C. T. Snowdon, C. H. Brown, and M. R. Petersen (eds.), *Primate communication*, pp. 259–75. Cambridge: Cambridge University Press.

Neyman, P. F. 1980. Ecology and social organization of the cotton-top tamarin (*Saguinus oedipus*). Doctoral thesis, University of California, Berkeley.

Nielsen, L. 1988. Definitions, considerations, and guidelines for translocation of wild animals. In L. Nielsen and R. D. Brown (eds.), *Translocation of wild animals*, pp. 12–49. Milwaukee: Wisconsin Humane Society.

Nievergelt, C. M., Mundy, N. I., and Woodruff, D. S. 1998. Microsatellite primers for genotyping common marmosets (*Callithrix jacchus*) and other callitrichids. *Molecular Ecology* 7: 1431–39.

Nievergelt, C. M., Digby, L. J., Ramakrishnan, U., and Woodruff, D. S. 2000. Genetic analysis of group composition and breeding system in a wild common marmoset (*Callithrix jacchus*) population. *International Journal of Primatology* 21: 1–20.

Nisbet, R. M. and Gurney, W. S. 1982. *Modelling fluctuating sub-populations.* New York: John Wiley and Sons.

Norman, A. W. and Litwack, G. 1987. *Hormones.* New York: Academic Press.

Nunes, A. 1995. Foraging and ranging patterns in white-bellied spider monkeys. *Folia Primatologica* 65: 85–99.

Oates, J. F. 1987. Food distribution and foraging behavior. In B. B. Smuts, D. L. Cheney, R. M. Seyfarth, R. W. Wrangham, and T. T. Struhsaker (eds.), *Primate societies,* pp. 197–209. Chicago: University of Chicago Press.

———. 1999. *Myth and reality in the rain forest: How conservation strategies are failing in West Africa.* Berkeley: University of California Press.

O'Brien, S. J. and Mayr, E. 1991. Bureaucratic mischief: Recognizing endangered species and subspecies. *Science* 251: 1187–88.

O'Brien, S. J., Wildt, D. E., Goldman, D., Merril, C. R., and Bush, M. 1983. The cheetah is depauperate in genetic variation. *Science* 221: 459–62.

O'Brien, S. J., Roelke, M. E., Marker, L., Newman, A., Winkler, C. A., Meltzer, D., Colly, L., Evermann, J. F., Bush, M., and Wildt, D. E. 1985. Genetic basis for species vulnerability in the cheetah. *Science* 227: 1428–34.

O'Brien, S. J., Wildt, D. E., and Bush, M. 1986. The cheetah in genetic peril. *Scientific American* 5: 84–92.

O'Brien, T. G. and Kinnaird, M. F. 1997. Behavior, diet, and movements of the Sulawesi crested black macaque (*Macaca nigra*). *International Journal of Primatology* 18: 321–51.

Oliveira, K. and Pereira, L. C. M. 1990. Levantamento preliminar de primatas na Área de Proteção Ambiental (APA) de Guaraqueçaba—PR. In *Resumos. XVII Congresso Brasileiro de Zoologia,* p. 235. Universidade Estadual de Londrina, Londrina, Paraná.

Oliveira, M. S. de, Lopes, F. A., Alonso, C., and Yamamoto, M. E. 1999. The mother's participation in infant carrying in captive groups of *Leontopithecus chrysomelas* and *Callithrix jacchus. Folia Primatologica* 70: 146–53.

Oliver, W. L. R. and Santos, I. B. 1991. Threatened endemic mammals of the Atlantic forest region of south-east Brazil. *Wildlife Preservation Trust, Special Scientific Report,* no. 4: 1–125.

Omedes, A. 1979. Social communication of silvery marmosets (*Callithrix argentata*), and differences between two of the three subspecies (*C. a. argentata* and *C. a. melanura*). *Dodo, Journal of the Jersey Wildlife Preservation Trust* 16: 45–51.

Oring, L.W. 1986. Avian polyandry. In R. F. Johnston (ed.), *Current ornithology,* vol. 3, pp. 309–51. New York: Plenum Press.

Orzack, S. H. 1985. Population dynamics in variable environments V. The genetics of homeostasis revisited. *American Naturalist* 125: 550–72.

Ostro, L. E. T. 1998. The spatial ecology of translocated black howler monkeys (*Alouatta pigra*) in Belize. Ph.D. dissertation, Fordham University, New York.

Padua, S. M. 1991. Conservation awareness through an environmental education school program at the Morro do Diabo State Park, São Paulo state, Brazil. Master's thesis, University of Florida, Gainesville.

———. 1994a. Conservation awareness through an environmental education programme in the Atlantic forest of Brazil. *Environmental Conservation* 21: 145–51.

———. 1994b. Environmental education and the black lion tamarin, *Leontopithecus chrysopygus*. *Neotropical Primates* 2(suppl.): 45–49.

———. 1995. Environmental education programmes for natural areas in underdeveloped countries—a case study in the Brazilian Atlantic Forest. In *Planning education to care for the planet*, pp. 51–56. Gland, Switzerland: World Conservation Union (IUCN).

———. 1997. Uma pesquisa em educação ambiental: A conservação do mico-leão-preto (*Leontopithecus chrysopygus*). In C. Valladares-Padua, R. E. Bodmer, and L. Cullen, Jr. (eds.), *Manejo e conservação de vida silvestre no Brasil*, pp. 34–51. Brasília, DF: Conselho Nacional de Desenvolvimento Científico e Tecnológico (CNPq), Sociedade Civil Mamirauá.

Padua, S. M. and Jacobson, S. K. 1993. A comprehensive approach to an environmental education program in Brazil. *Journal of Environmental Education* 24(4): 29–36.

Padua, S. M. and Tabanez, M. (eds.). 1997a. *Educação ambiental: Caminhos trilhados no Brasil*. Fundo Nacional do Meio Ambiente, Brasília, and Instituto de Pesquisas Ecológicas (IPÊ), Nazaré Paulista, São Paulo.

———. 1997b. Uma abordagem participativa para a conservação de áreas naturais: Educação ambiental na mata Atlântica. In M. S. Milano (ed.), *Anais do Congresso de Unidades de Conservação*, pp. 371–79. Curitiba, PR: Fundação O Boticário de Proteção à Natureza, Secretaria do Meio Ambiente e Recursos Hídricos do Paraná e Instituto Ambiental do Paraná.

Padua, S. M. and Valladares-Padua, C. 1997. Um programa integrado para a conservação do mico-leão-preto (*Leontopithecus chrysopygus*)—pesquisa, educação e envolvimento comunitário. In S. M. Padua and M. Tabanez (eds.), *Educação ambiental: Caminhos trilhados no Brasil*, pp. 119–32. Fundo Nacional do Meio Ambiente, Brasília, Instituto de Pesquisas Ecológicas (IPÊ), Nazaré Paulista, São Paulo.

———. 1998. The conservation of black lion tamarins (*Leontopithecus chrysopygus*): From research and education to landscape ecology, agrarian reform movement and policy. In *Second Pan American Congress on the Conservation of Wildlife through Education*. Bronx, NY: Wildlife Conservation Society.

Padua, S. M., Dietz, L. A., Nagagata, B., and Alves C. 1990. Environmental education and the lion tamarins. In U. S. Seal, J. D. Ballou, and C. Valladares-Padua (eds.), Leontopithecus*: Population Viability Analysis Workshop report*, pp. 131–34. Apple Valley, MN: International Union for Conservation of Nature and Natural Resources/Species Survival Commission (IUCN/SSC) Captive Breeding Specialist Group (CBSG).

Padua, S. M., Mamede, C., Silva, M., and Martins, C. 1994. Do parents learn from their children? In *Proceedings of the 22nd Annual Conference of the North American Association of Environmental Education*, Montana, pp. 390–93. Troy, OH: North American Association of Environmental Education.

Padua, S. M., Tabanez, M., and Souza, M. G. 1997. How can nature trails be more effective? In *Environmental education for the next generation. Selected papers from the 25th Annual Conference of the North American Association of Environmental Education—NAAEE, San*

Francisco, CA, 1996, pp. 263–67. Troy, OH: North American Association of Environmental Education.

Padua, S. M., Tabanez, M., and Souza. M. G. 1998. More research on nature trails in Brazil. In *Weaving connections: Cultures and environments. Selected papers from the 26th Annual Conference of the North American Association of Environmental Education—NAAEE, Vancouver, 1997,* pp. 237–40. Troy, OH: North American Association of Environmental Education.

Padua, S. M., Mamede, C., Silva, M., and Martins, C. 2000. Os pais aprendem com os filhos? In I. Tamoio and S. Sinicco (eds.), *Educador ambiental—6 anos de experiências e debates,* pp. 40–43. São Paulo, SP: WWF-Brasil.

Passamani, M. 1998. Activity budget of Geoffroy's marmoset (*Callithrix geoffroyi*) in an Atlantic forest in southeastern Brazil. *American Journal of Primatology* 46: 333–40.

Passos, F. C. 1991. *Leontopithecus chrysopygus* (Primates) como dispersor de sementes. *Anais do Seminário Regional de Ecologia,* Universidade Federal de São Carlos 6: 505–14.

———. 1992. Habito alimentar do mico-leão-preto *Leontopithecus chrysopygus* (Mikan 1823) (Callitrichidae, Primates) na Estação Ecológica dos Caetetus, município de Gália, SP. Master's thesis, Universidade Estadual de Campinas, Campinas.

———. 1994. Behavior of the black lion tamarin, *Leontopithecus chrysopygus,* in different forest levels in the Caetetus Ecological Station, São Paulo, Brazil. *Neotropical Primates* 2(suppl.): 40–41.

———. 1997a. Padrão de atividades, dieta e uso do espaço em um grupo de mico-leão-preto (*Leontopithecus chrysopygus*) na Estação Ecológica dos Caetetus, SP. Doctoral thesis, Universidade Federal de São Carlos, São Carlos.

———. 1997b. Seed dispersal by the black lion tamarin, *Leontopithecus chrysopygus* (Primates, Callitrichidae), in southeastern Brazil. *Mammalia* 61: 109–11.

———. 1997c. A foraging association between the olivaceous woodcreeper *Sittasomus griseicapillus* and the black lion tamarin *Leontopithecus chrysopygus* in south-east Brazil. *Ciência e Cultura* 49: 144–45.

———. 1999. Dieta de um grupo de mico-leao-preto, *Leontopithicus chrysopygus* (Mikan) (Mammalia, Callitrichidae), na Estação Ecológica de Caetetus, São Paulo. *Revista Brasileira de Zoologia* 16(suppl. 1): 269–78.

Passos, F. C. and Alho, C. J. R. 2001. Importância de diferentes microhabitats no comportamento de forrageio por presas do mico-leão-preto, *Leontopithecus chrysopygus* (Mikan) (Mammalia, Callitrichidae). *Revista Brasileira de Zoologia* 18(suppl. 1): 335–42.

Passos, F. C. and Keuroghlian, A. 1999. Foraging behavior and microhabitats used by black lion tamarins, *Leontopithecus chrysopygus* (Mikan) (Primates, Callitrichidae). *Revista Brasileira de Zoologia* 16(suppl. 2): 219–22.

Passos, F. C. and Kim, A. C. 1999. Nectar feeding on *Mabea fistulifera* Mart. (Euphorbiaceae) by black lion tamarins, *Leontopithecus chrysopygus* Mikan, 1823 (Callitrichidae), during the dry season in southeastern Brazil. *Mammalia* 63: 519–21.

Pastorini, J., Forstner, M. R. J., Martin, R. D., and Melnick, D. J. 1998. A reexamination of the phylogenetic position of *Callimico* (Primates) incorporating new mitochondrial DNA sequence data. *Journal of Molecular Evolution* 47: 32–41.

Pelzeln, A. von. 1883. *Brasilische Säugethiere: Resultat von Johann Natterer's Reisen in den Jahren 1817 bis 1835*. Vienna. 140 pp.

Peres, C. A. 1986a. Costs and benefits of territorial defense in golden lion tamarins, *Leontopithecus rosalia*. Master's thesis, University of Florida, Gainesville.

———. 1986b. Golden Lion Tamarin Project. II. Ranging patterns and habitat selection in golden lion tamarins, *Leontopithecus rosalia* (Linnaeus, 1766) (Callitrichidae, Primates). In M. T. de Mello (ed.), *A primatologia no Brasil—2,* pp. 223–33. Brasília, DF: Sociedade Brasileira de Primatologia.

———. 1989a. Exudate eating by wild golden lion tamarins, *Leontopithecus rosalia*. *Biotropica* 21: 287–88.

———. 1989b. Costs and benefits of territorial defense in wild golden lion tamarins *Leontopithecus rosalia*. *Behavioral Ecology and Sociobiology* 25: 227–33.

———. 1994. Primate responses to phenological changes in an Amazonian terra firme forest. *Biotropica* 26: 98–112.

———. 2000a. Identifying keystone plant resources in tropical forests: The case of gums from *Parkia* pods. *Journal of Tropical Ecology* 16: 287–317.

———. 2000b. Territorial defense and the ecology of group movements in small-bodied neotropical primates. In S. Boinski and P. A. Garber (eds.), *On the move: How and why animals travel in groups*. Chicago: University of Chicago Press.

Peres, C. A., Patton, J. L., and Silva, M. N. F. da. 1997. Riverine barriers and gene flow in Amazonian saddle-back tamarins. *Folia Primatologica* 67: 113–24.

Perry, J. 1971. The golden lion marmoset. *Oryx* 11: 22–24.

Persson, V. G. and Lorini, M. L. 1991. Notas sobre o mico-leão-de-cara-preta, *Leontopithecus caissara* Lorini & Persson, 1990, no sul do Brasil (Primates, Callitrichidae). In *Resumos. XVII Congresso Brasileiro de Zoologia*, p. 385. Universidade Federal da Bahia, Salvador. 24 February–1 March 1991 (Abstract).

———. 1993. Notas sobre o mico-leão-de-cara-preta, *Leontopithecus caissara* Lorini & Persson, 1990, no sul do Brasil (Primates, Callitrichidae). In M. E. Yamamoto and M. B. C. de Sousa (eds.), *A primatologia no Brasil—4,* pp. 168–81. Natal, RN: Universidade Federal do Rio Grande do Norte, Sociedade Brasileira de Primatologia.

Pessamílio, D. M. 1994. Revegetation of deforested areas in the Poço das Antas Biological Reserve, Rio de Janeiro. *Neotropical Primates* 2(suppl.): 19–20.

Pessier, A. P., Stringfield, C., Tragle, J., Holshuh, H. J., Nichols, D. K., and Montali, R. J. 1997. Cerebrospinal nematodiasis due to *Baylisascaris* sp. in golden-headed lion tamarins, *Leontopithecus chrysomelas:* Implications for management. *Proceedings of the Annual Meeting of the American Association of Zoo Veterinarians,* pp. 245–47.

Pfister, R., Heider, B., Illgen, B., and Berlinger, R. 1990. *Trichospirura leptostoma:* A possible cause of wasting disease in the marmoset. *Zeitschrift für Versuchstierkunde*. 33: 157–61.

Phillips L. G., Jr. and Montali, R. J. 1996. Radiographic evaluation of diaphragmatic defects in golden lion tamarins (*Leontopithecus rosalia*): Implications for reintroduction. *Journal of Zoo and Wildlife Medicine* 27: 346–57.

Pinder, L. 1986a. Translocação como técnica de conservação em *Leontopithecus rosalia.* Master's thesis, Universidade Federal do Rio de Janeiro, Rio de Janeiro.

———. 1986b. Projeto Mico-Leão. III. Avaliação técnica de translocação em *Leontopithecus rosalia* (Linnaeus, 1766) (Callitrichidae, Primates). In M. T. de Mello (ed.), *A primatologia no Brasil—2,* pp. 235–41. Brasília, DF: Sociedade Brasileira de Primatologia.

Pinder, L. and Pissinatti, A. 1991. Malformações congênitas em *Leontopithecus rosalia* (Linnaeus, 1766) (Callitrichidae, Primates). In A. B. Rylands and A. T. Bernardes (eds.), *A primatologia no Brasil—3,* pp. 191–95. Belo Horizonte, MG: Fundação Biodiversitas and Sociedade Brasileira de Primatologia.

Pinto, L. P. de S. 1994. Distribuição geográfica, população e estado de conservação do mico-leão-de-cara-dourada, *Leontopithecus chrysomelas* (Callitrichidae, Primates). Master's thesis, Universidade Federal de Minas Gerais, Belo Horizonte.

Pinto, L. P. de S. and Rylands, A. B. 1997. Geographic distribution of the golden-headed lion tamarin, *Leontopithecus chrysomelas:* Implications for its management and conservation. *Folia Primatologica* 68: 161–80.

Pinto, L. P. de S. and Tavares, L. I.. 1994. Inventory and conservation status of wild populations of golden-headed lion tamarins, *Leontopithecus chrysomelas. Neotropical Primates* 2(suppl.): 24–27.

Pinto, L. P. de S., Sábato, M. A. L., Lamas, I. R., and Tavares, L. I. 1993. Inventário faunístico e conservação da mata Atlântica do sul da Bahia. Unpublished report, Fundação Biodiversitas, Belo Horizonte.

Pissinatti, A. 2001. Order Primates (Primates): Medicine, selected disorders. In M. E. Fowler and Z. S. Cubas (eds.), *Biology, medicine, and surgery of South American wild animals,* pp. 272–74. Ames: Iowa State University Press,

Pissinatti, A. and Tortelly, R. 1984. Alterações produzidas por *Porocephalus crotali* (Humboldt, 1811) em *Leontopithecus rosalia rosalia* (Linnaeus, 1766). In M. T. de Mello (ed.), *A primatologia no Brasil,* pp. 253–57. Brasília, DF: Sociedade Brasileira de Primatologia.

Pissinatti, A., Duarte, M. J. F., Rocha e Silva, R. da, and Coimbra-Filho, A. F. 1981. Miiase por *Cuterebra* sp. no sagui *Leontopithecus r. chrysopygus* (Mikan, 1823). *Revista Brasileira de Medicina Veterinária* 4: 17.

Pissinatti, A., Cruz, J. B., and Luz, V. L. F. 1984a. Agenesia cecal em *Leontopithecus rosalia* (Linnaeus, 1766). Callitrichidae—Primates. *Anais do Congresso Brasileiro de Medicina Veterinária,* p. 310. Belém.

Pissinatti, A., Silveira, A. K., Coimbra-Filho, A. F., and Rocha e Silva, R. da. 1984b. Acerca de cesáreas em símios do gênero *Leontopithecus* (Callitrichidae—Primates). *Revista da Faculdade Veterinária, Universidade Federal Fluminense* 1: 9–19.

Pissinatti, A., Tortelly, R., Coimbra-Filho, A. F., and Cruz, J. B. 1985. Alterações causadas por *Trichospirura leptostoma* em pâncreas de *Leontopithecus chrysopygus* (Mikan, 1823) Callitrichidae—Primates. *Revista Brasileira de Medicina Veterinária* 7: 184–85.

Pissinatti, A., Silva, E. C. da, Jr., Coimbra-Filho, A. F., Bertolazzo, W., and Cruz, J. B. 1992. Sexual dimorphism of the pelvis in *Leontopithecus* (Lesson, 1840). *Folia Primatologica* 58: 204–9.

Pissinatti, A., Cruz, J. B., Nascimento, M. D., Rocha e Silva, R. da, and Coimbra-Filho, A. F. 1993. Spontaneous gallstones in marmosets and tamarins (Callitrichidae—Primates). *Folia Primatologica* 59: 44–50.

Pissinatti, A., Burity, C. H. F., and Mandarim-de-Lacerda, C. A. 2000. Morphological and morphometric age-related changes of the upper thoracic aorta in *Leontopithecus* (Lesson, 1840) (Callitrichidae—Primates). *Journal of Medical Primatology* 29: 421–26.

Pola, Y. V. and Snowdon, C. T. 1975. The vocalizations of pygmy marmosets (*Cebuella pygmaea*). *Animal Behaviour* 23: 826–42.

Pope, T.R. 1996. Socioecology, population fragmentation, and patterns of genetic loss in endangered primates. In J. C. Avise and J. L. Hamrick (eds.), *Conservation genetics: Case studies from nature,* pp. 119–59. New York: Chapman and Hall.

Porter, C. A., Sampaio, I., Schneider, H., Schneider, M. P. C., Czelusniak, J., and Goodman, M. 1995. Evidence on primate phylogeny from e-globin gene sequences and flanking regions. *Journal of Molecular Evolution* 40: 30–55.

Porter, C. A., Czelusniak, J., Schneider, H., Schneider, M. P. C., Sampaio, I., and Goodman, M. 1997a. Sequences of the e-globin gene: Implications for the systematics of the marmosets and other New World primates. *Gene* 205: 59–71.

Porter, C. A., Page, S. L., Czelusniak, J., Schneider, H., Schneider, M. P. C., Sampaio, I., and Goodman, M. 1997b. Phylogeny and evolution of selected primates as determined by sequences of the e-globin locus and 5′ flanking regions. *International Journal of Primatology* 18: 261–95.

Portman, O. W., Alexander, M., and Tanaka, N. 1980. Relationships between cholesterol gallstones, biliary function and plasma lipoprotein in squirrel monkeys. *Journal of Laboratory and Clinical Medicine* 96: 90–101.

Potkay, S. 1992. Diseases of the Callitrichidae: A review. *Journal of Medical Primatology* 21: 189–236.

Prado, F. 1999. Ecologia, comportamento e conservação do mico-leão-da-cara-preta (*Leontopithecus caissara*) no Parque Nacional do Superagüi, Guaraqueçaba, Paraná. Master's thesis, Universidade Estadual Paulista, Botucatu.

Prado, F. and Valladares-Padua, C. 1997. Orçamento temporal de um grupo de mico-leão-de-cara-preta *Leontopithecus caissara* Lorini and Persson, 1990 (Platyrrhini, Primates, Callitrichidae), no Parque Nacional do Superagüi, Guaraqueçaba—PR. In *VIII Congresso Brasileiro de Primatologia e V Reunião Latino-Americano de Primatologia—Programa e resumos,* p. 49. Sociedade Brasileira de Primatologia, João Pessoa.

Prado, F., Valladares-Padua, C., Padua, S. M., and Navas, S. 2000. Conservação do mico-leão-da-cara-preta. In *Cadernos do Litoral,* no. 3: 23–25. Sociedade de Pesquisa em Vida Silvestre e Educação Ambiental, Curitiba, Paraná.

Price, E. C. 1990. Infant carrying as a courtship strategy of breeding male cotton-top tamarins. *Animal Behaviour* 40: 784–86.

———. 1991. Stability of wild callitrichid groups. *Folia Primatologica* 57: 111–14.

———. 1992a. The costs of infant carrying in captive cotton-top tamarins. *American Journal of Primatology* 26: 23–33.

———. 1992b. Adaptation of captive-bred cotton-top tamarins (*Saguinus oedipus*) to a natural environment. *Zoo Biology* 11: 107–20.

———. 1997. Group instability following cessation of breeding in marmosets and tamarins. *Dodo, Journal of the Wildlife Preservation Trusts* 33: 157–58.

———. 1998. Incest in captive marmosets and tamarins. *Dodo, Journal of the Wildlife Preservation Trusts* 34: 25–31.

Price, E. C. and Feistner, A. T. C. 1993. Food-sharing in lion tamarins: Test of three hypotheses. *American Journal of Primatology* 31: 211–21.

———. 2001. Food sharing in pied bare-faced tamarins (*Saguinus bicolor bicolor*): Development and individual differences. *International Journal of Primatology* 22: 231–41.

Price, E. C. and McGrew, W. C. 1991. Departures from monogamy in colonies of captive cotton-top tamarins. *Folia Primatologica* 57: 16–27.

Price, E. O. 1984. Behavioral aspects of animal domestication. *Quarterly Review of Biology* 59: 1–32.

Procópio de Oliveira, P. 1993. Sucessão vegetacional e estruturação de comunidades de pequenos mamíferos em áreas afetadas pelo fogo na Reserva Biológica Poço das Antas, RJ. Master's thesis, Universidade Federal de Minas Gerais, Belo Horizonte.

Provost, P. J., Villarejos, V. M., and Hilleman, M. R. 1977. Suitability of the rufiventer marmoset as a host animal for human hepatitis A virus (39790). *Proceedings of the Society for Experimental Biology and Medicine* 155: 283–86.

Pryce, C. R., Schwarzenberger, F., and Döbeli, M. 1994. Monitoring fecal samples for estrogen excretion across the ovarian cycle in Goeldi's monkey (*Callimico goeldii*). *Zoo Biology* 13: 219–30.

Pryce, C. R., Schwarzenberger, F., Dobeli, M., and Etter, K. 1995. Comparative study of oestrogen excretion in female New World monkeys: An overview of non-invasive ovarian monitoring and a new application in evolutionary biology. *Folia Primatologica* 64: 107–23.

Raboy, B. E., Bach, A., and Dietz, J. M. 2001. Birth seasonality and infant survival in wild GHLTs. *Tamarin Tales,* no. 5: 4–5.

Rambaldi, D. 2000. Além do mico-leão-dourado—a partir de trabalho de proteção da espécie, Associação muda mentalidades e o uso do solo em região do Rio de Janeiro. In I. Tamaio and D. Carreira (eds.), *Caminhos e aprendizagens: Educação ambiental, conservação e desenvolvimento,* pp. 77–82. Brasília, DF: WWF-Brasil.

Randolph, J., Bush, M., Abramowitz, M., Kleiman, D. G., and Montali, R. J. 1981. Surgical correction of familial diaphragmatic hernia of Morgagni in the golden lion tamarin. *Journal of Pediatric Surgery* 16: 396–401.

Rapaport, L. G. 1997. Food sharing in golden lion tamarins (*Leontopithecus rosalia*): Provisioning of young, maintenance of social bonds, and resource constraints. Doctoral dissertation, University of New Mexico, Albuquerque.

———. 1999. Provisioning of young in golden lion tamarins (Callitrichidae, *Leontopithecus rosalia*): A test of the information hypothesis. *Ethology* 105: 619–36.

Rapaport, L. G. and Ruiz-Miranda, C. R. In press. Tutoring in wild golden lion tamarins. *International Journal of Primatology*.

Ratcliffe, H. L. 1966. Diets for zoological gardens: Aids to conservation and disease control. *International Zoo Yearbook* 6: 4–23.

Rathbun, C. D. 1979. Description and analysis of the arch-display in the golden lion tamarin, *Leontopithecus rosalia rosalia*. *Folia Primatologica* 32: 125–48.

Reading, R.P., Vargas, A., Miller, B. J., Clark, T. W., Hanebury, L. R., and Biggins, D. 1996. Recent directions in black-footed ferret recovery. *Endangered Species Update* 13: 1–6.

Redshaw, M. E. and Mallinson, J. J. C. 1991a. Learning from the wild: Improving the psychological and physical well-being of captive primates. *Dodo, Journal of the Jersey Wildlife Preservation Trust* 27: 18–26.

———. 1991b. Stimulation of natural patterns of behaviour: Studies with golden lion tamarins and gorillas. In H. O. Box (ed.), *Primate responses to environmental change*, pp. 217–38. London: Chapman and Hall.

Rêgo, A. A. 1980. Pentastomídeos de mamíferos da coleção helmintológica do Instituto Oswaldo Cruz. *Revista Brasileira de Biologia* 40(4): 783–91.

Reid, J. and Blanes, J. 1996. A pecuária extensiva, da forma como é realizada na Região Cacuaeira, não é rentável. In *Alternativas econômicas para conservação e desenvolvimento da região de Una, Bahia. Resumos das pesquisas 1994–1995*, pp. 22–25. Ilhéus, BA: Instituto de Estudos Sócio-Ambientais do Sul da Bahia (IESB).

Renner, M. J. 1988. Learning during exploration: The role of behavioral topography during exploration in determining subsequent adaptive behavior. *International Journal of Comparative Psychology* 2: 43–56.

Renner, M. J. and Rosenzweig, M. R. 1987. *Enriched and impoverished environments: Effects on brain and behavior*. New York: Springer Verlag.

Rettberg Beck, B. 1990. Protocolo de Manejo Mico-Leão-Dourado (*Leontopithecus rosalia*). Golden Lion Tamarin Management Committee, National Zoological Park, Washington, DC.

Ribas Lange, M. B. 1997. Programa Guaraqueçaba—seis anos de atuação da Sociedade de Pesquisa em Vida Silvestre (SPVS) na Áreaq de Proteção Ambiental de Guaraqueçaba, Paraná. In M. S. Milano (ed.), *Anais do 1° Congresso Brasileiro de Unidades de Conservação*, pp. 607–16. Curitiba, PR: Universidade Livre do Meio Ambiente, Rede Nacional Pró-Unidades de Conservação, and Instituto Ambiental do Paraná (IAP).

Ribeiro, E. A. A. 1994. Uma análise da relaçào entre comportamento reprodutivo e níveis de progestinas fecais em um grupo silvestre do mico-leão-dourado, *Leontopithecus rosalia* (Callitrichidae, Primates). Doctoral dissertation, Universidade de Sao Paulo, São Paulo.

Richard, A. F. 1978. *Behavioral variation: Case study of a Malagasy lemur.* London: Bucknell University Press.

Rocha e Silva, R. da. 1984. Elaboração e distribuição de dietas para calitriquídeos em cativeiro. In M. T. de Mello (ed.), *A primatologia no Brasil,* pp. 137–42. Brasília, DF: Sociedade Brasileira de Primatologia.

———. 1986. Novo modelo de comedouro/bebedouro para primatas calitriquídeos em cativeiro. In M. T. de Mello (ed.), *A primatologia no Brasil—2,* pp. 419–22. Brasília, DF: Sociedade Brasileira de Primatologia.

———. 2001. Order Primates (Primates): Nutrition. In M. E. Fowler and Z. S. Cubas (eds.), *Biology, medicine, and surgery of South American wild animals,* pp. 261–63. Ames: Iowa State University Press.

Rocha e Silva, R. da, Coimbra-Filho, A. F., and Pissinatti, A. 1991. Registro de dados biológicos no Centro de Primatologia do Rio de Janeiro-FEEMA. In A. B. Rylands and A. T. Bernardes (eds.), *A primatologia no Brasil—3,* pp. 451–56. Belo Horizonte, MG: Fundação Biodiversitas and Sociedade Brasileira de Primatologia.

Roda, S. A. 1989. Ocorrência de duas fêmeas reprodutivas em grupos selvagens de *Callithrix jacchus* (Primates, Callitrichidae). In *Resumos do XVI Congresso Brasileiro de Zoologia,* p. 122. João Pessoa: Universidade Federal de Paraíba (Abstract).

Rodrigues, M. G. 1998. Análise do status de conservação das unidades da paisagem do complexo estuarino-lagunar de Iguape-Cananéia-Guaraqueçaba, Paraná. Master's thesis, Universidade de São Paulo, São Paulo.

Rodrigues, M. G., Katsuyama, S., and Rodrigues, C. A. G. 1992. Estratégias para conservação do mico-leão caiçara, *Leontopithecus caissara:* Análise da situação econômico-social da comunidade do Ariri—Parte I. *Anais—2° Congresso Nacional sobre Essências Nativas,* pp. 1118–25. São Paulo.

Rosenberger, A. L. 1981. Systematics: The higher taxa. In A. F. Coimbra-Filho and R. A. Mittermeier, *Ecology and behavior of neotropical primates,* vol. 1, pp. 9–27. Rio de Janiero: Academia Brasileira de Ciências.

Rosenberger, A. L. and Coimbra-Filho, A. F. 1984. Morphology, taxonomic status and affinities of the lion tamarins, *Leontopithecus* (Callitrichinae, Cebidae). *Folia Primatologica* 42: 149–79.

Rosenberger, A. L. and Stafford, B. J. 1994. Locomotion in captive *Leontopithecus* and *Callimico:* A multimedia study. *American Journal of Physical Anthropology* 94: 379–94.

Rothe, H. 1974. Further observations on the delivery behavior of the common marmoset (*Callithrix jacchus*). *Zeitschrift für Säugetierkunde* 39: 135–42.

———. 1975. Some aspects of sexuality and reproduction in groups of captive marmosets (*Callithrix jacchus*). *Zeitschrift für Tierpsychologie* 37: 255–73.

Rothe, H. and Koenig, A. 1991. Variability of social organisation in captive common marmosets (*Callithrix jacchus*). *Folia Primatologica* 57: 28–33.

Rothe, H., Wolters, H.-J., and Hearn, J. P. (eds.). 1978. *Biology and behaviour of marmosets.* Göttingen: Eigenverlag H. Rothe.

Rowell, T. E. 1967. A quantitative comparison of the behavior of a wild and a caged baboon group. *Animal Behaviour* 15: 499–509.

Ruch, T. C. 1959. *Diseases of laboratory primates.* Philadelphia, PA: Saunders.

Rudran, R. 1978. Socioecology of the blue monkeys (*Cercopithecus mitis stuhlmanni*) of the Kibale Forest, Uganda. *Smithsonian Contributions to Zoology* 249: 1–88.

Ruiz, J. C. 1990. Comparison of affiliative behaviors between old and recently established pairs of golden lion tamarin, *Leontopithecus rosalia. Primates* 31: 197–204.

Ruiz-Miranda, C. R., Kleiman, D. G., Dietz, J. M., Moraes, E., Grativol, A. D., Baker, A. J., and Beck, B. B. 1999. Food transfers in wild and reintroduced golden lion tamarins, *Leontopithecus rosalia. American Journal of Primatology* 48: 305–20.

Ruiz-Miranda, C. R., Affonso, A. G., Martins, A., and Beck., B. B. 2000. Distribuição do sagüi (*Callithrix jacchus*) nas áreas de ocorrência do mico-leão-dourado (*Leontopithecus rosalia*) no estado do Rio de Janeiro. *Neotropical Primates* 8(3): 98–101.

Ruiz-Miranda, C. R., Kleiman, D. G., Moraes, E., and Grativol, A. D. In preparation. Differences between wild and reintroduced golden lion tamarins in the structure and usage of long calls.

Ruschi, A. 1964. Macacos do Espírito Santo. *Boletim do Museu de Biologia Mello Leitão, Zoologia* 23A: 1–23.

Rylands, A. B. 1980. The behavioural ecology of the golden-headed lion tamarin, *Leontopithecus rosalia chrysomelas* (Callitrichidae, Primates) in Brazil. Interim report, World Wildlife Fund–U.S., the Fauna and Flora Preservation Society, and the Conder Conservation Trust. September 1980. 9 pp.

———. 1982. Behaviour and ecology of three species of marmosets and tamarins (Callitrichidae, Primates) in Brazil. Doctoral thesis, University of Cambridge, Cambridge.

———. 1983. The behavioural ecology of the golden-headed lion tamarin, *Leontopithecus rosalia chrysomelas,* and the marmoset, *Callithrix kuhli* (Callitrichidae, Primates). Final report, World Wildlife Fund–U.S., the Fauna and Flora Preservation Society, and the Conder Conservation Trust. December 1983. 47 pp.

———. 1986a. Infant carrying in a wild marmoset group, *Callithrix humeralifer:* Evidence for a polyandrous mating system. In M. T. de Mello (ed.), *A primatologia no Brasil—2,* pp. 131–44. Brasília, DF: Sociedade Brasileira de Primatologia.

———. 1986b. Ranging behaviour and habitat preference of a wild marmoset group, *Callithrix humeralifer* (Callitrichidae, Primates). *Journal of Zoology, London* 210: 489–514.

———. 1987. Primate communities in Amazonian forests: Their habitats and food resources. *Experientia* 43:265–79.

———. 1989a. Evolução do sistema de acasalamento em Callitrichidae. In C. Ades (ed.), *Etologia de animais e de homens,* pp. 87–108. São Paulo, SP: Edicon, University of São Paulo Press.

———. 1989b. Sympatric Brazilian callitrichids: The black tufted-ear marmoset, *Callithrix kuhli,* and the golden-headed lion tamarin, *Leontopithecus chrysomelas. Journal of Human Evolution* 18: 679–95.

———. 1990. Scent-marking behaviour of wild marmosets, *Callithrix humeralifer* (Callitrichidae, Primates). In D. W. Macdonald, D. Müller-Schwarze, and S. E. Natynczuk (eds.), *Chemical signals in vertebrates 5,* pp. 415–29. Oxford: Oxford University Press.

———. 1993. The ecology of the lion tamarins, *Leontopithecus:* Some intrageneric differences and comparisons with other callitrichids. In A. B. Rylands (ed.), *Marmosets and tamarins: Systematics, behaviour, and ecology,* pp. 296–313. Oxford: Oxford University Press.

———. 1993/1994. Population viability analyses and the conservation of the lion tamarins, *Leontopithecus,* of south-east Brazil. *Primate Conservation,* nos.14–15: 34–42.

———. 1996. Habitat and the evolution of social and reproductive behavior in Callitrichidae. *American Journal of Primatology* 38: 5–18.

Rylands, A. B. and Faria, D. S. de. 1993. Habitats, feeding ecology, and home range size in the genus *Callithrix.* In A. B. Rylands (ed.), *Marmosets and tamarins: Systematics, behaviour, and ecology,* pp. 262–72. Oxford: Oxford University Press.

Rylands, A. B. and Rodríguez-Luna, E. (eds.). 1994. Proceedings of the 2nd Symposium on *Leontopithecus* held during the Annual Meeting of the International Committees for the Preservation and Management of the Four Lion Tamarin Species, May 1994. *Neotropical Primates* 2(suppl.). 59 pp.

———. 2000. Threatened primates of Mesoamerica and South America—The Red List 2000. *Neotropical Primates* 8(3): 115–19.

Rylands, A. B., Spironelo, W. R., Tornisielo, V. L., Lemos de Sá, R. M., Kierulff, M. C., and Santos, I. B. 1988. Primates of the Rio Jequitinhonha valley, Minas Gerais, Brazil. *Primate Conservation,* no. 9: 100–109.

Rylands, A. B., Santos, I. B., and Mittermeier, R. A. 1991/1992. Distribution and status of the golden-headed lion tamarin, *Leontopithecus chrysomelas,* in the Atlantic forest of southern Bahia, Brazil. *Primate Conservation,* nos. 12–13: 15–23.

Rylands, A. B., Coimbra-Filho, A. F., and Mittermeier, R. A. 1993. Systematics, distributions and some notes on the conservation status of the Callitrichidae. In A. B. Rylands (ed.), *Marmosets and tamarins: Systematics, behaviour, and ecology,* pp. 11–77. Oxford: Oxford University Press.

Rylands, A. B., Fonseca, G. A. B. da, Leite, Y. L. R., and Mittermeier, R. A. 1996. Primates of the Atlantic forest: Origin, endemism, distributions and communities. In M. A. Norconk, A. L. Rosenberger, and P. A. Garber (eds.), *Adaptive radiations of neotropical primates,* pp. 21–51. New York: Plenum Press.

Rylands, A. B., Mittermeier, R. A., and Rodríguez-Luna, E. 1997. Conservation of neotropical primates: Threatened species and an analysis of primate diversity by country and region. *Folia Primatologica* 68(3–5): 134–60.

Rylands, A. B., Schneider, H., Langguth, A., Mittermeier, R. A., Groves, C. P., and Rodríguez-Luna, E. 2000. An assessment of the diversity of New World primates. *Neotropical Primates* 8(2): 61–93.

Saatchi, S., Agosti, D., Alger, K., Delabie, J., and Musinsky, J. 2001. Examining fragmentation and loss of primary forest in the southern Bahian Atlantic rain forest of Brazil with radar imagery. *Conservation Biology* 15: 867–75.

Saltz, D. and Rubenstein, D. 1995. Population dynamics of a reintroduced Asiatic wild ass (*Equus hemionus*) herd. *Ecological Applications* 5: 327–35.

Saltzman, W., Schultz-Darken, N. J., Scheffler, G., Wegner, F. H., and Abbott, D. H. 1994. Social and reproductive influences on plasma cortisol in female marmoset monkeys. *Physiology and Behavior* 56: 801–10.

Santos, C. V., French, J. A., and Otta, E. 1997. A comparative study of infant carrying behavior in callitrichid primates: *Callithrix* and *Leontopithecus. International Journal of Primatology* 18: 889–907.

Santos, G. R. dos. 1995. Sistematização de dados para planejamento e monitoramento de programa de educação ambiental direcionado aos fazendeiros do município de Una, Bahia, Brazil. Unpublished report, World Wildlife Fund, Brasília, DF.

Santos, G. R. dos and Blanes, J. 1997. Community environmental education in Brazil. *Dodo, Journal of the Wildlife Preservation Trusts* 33: 118–26.

———. 1999. Environmental education as a strategy for conservation of the remnants of Atlantic forest surrounding Una Biological Reserve, Brazil. *Dodo* 35: 151–57.

Santos, I. B. 1983. Sul da Bahia, Relatório No. 1. 04/08/83–23/08/83. Unpublished report, Departamento de Parques Nacionais e Reservas Equivalentes, Instituto Brasileiro de Desenvolvimento Florestal (IBDF), Brasília, and World Wildlife Fund–U.S. Primate Program, Washington, DC. 35 pp.

———. 1984. Sul da Bahia, Relatório No. 2. Unpublished report, Departamento de Parques Nacionais e Reservas Equivalentes, Instituto Brasileiro de Desenvolvimento Florestal (IBDF), Brasília, and World Wildlife Fund–U.S. Primate Program, Washington, DC. 29 pp.

Santos, I. B., Mittermeier, R. A., Rylands, A. B., and Valle, C. M. C. 1987. The distribution and conservation status of primates in southern Bahia, Brazil. *Primate Conservation,* no. 8: 126–42.

Saracura, V. F. 1997. Plano de Manejo. Reserva Biológica de Una—BA. Versão preliminar. Brasília, DF: Programa Nacional do Meio Ambiente (PNMA), Instituto Brasileiro do Meio Ambiente e dos Recursos Naturais Renováveis (IBAMA).

Sarich, V. M. and Cronin, J. E. 1980. South American mammal molecular systems, evolutionary clocks, and continental drift. In R. L. Ciochon and A. B. Chiarelli (eds.), *Evolutionary biology of the New World monkeys and continental drift,* pp. 399–421. New York: Plenum Press.

Savage, A., Snowdon, C. T., Giraldo, L. H., and Soto, L. H. 1996a. Parental care patterns and vigilance in wild cotton-top tamarins (*Saguinus oedipus*). In M. A. Norconk, A. L. Rosenberger, and P.A. Garber (eds.), *Adaptive radiations of Neotropical primates,* pp. 187–99. New York: Plenum Press.

Savage, A., Giraldo, L. H., Soto, L. H., and Snowdon, C. T. 1996b. Demography, group composition, and dispersal in wild cotton-top tamarin (*Saguinus oedipus*) groups. *American Journal of Primatology* 38: 85–100.

Scanga, C. A., Holmes, K. V., and Montali, R. J. 1993. Serologic evidence of infection with lymphocyticchoriomeningitis virus, the agent of callitrichid hepatitis, in primates in zoos, primate research centers, and a natural reserve. *Journal of Zoo and Wildlife Medicine* 24: 469–74.

Scanlon, C. E., Chalmers, N. K., and Monteiro da Cruz, M. A. O. 1988. Changes in the size, composition and reproductive condition of wild marmoset groups (*Callithrix jacchus*) in northeast Brazil. *Primates* 29: 295–305.

Schiller, C. A., Wolff, M. J., Munson, L., and Montali, R. J. 1989. *Streptococcus zooepidemicus* infections of possible horsemeat source in red-bellied tamarins and Goeldi's monkeys. *Journal of Zoo and Wildlife Medicine* 20(3): 322–27.

Schleidt, W. M. 1973. Tonic communication: Continual effects of discrete signs in animal communication systems. *Journal of Theoretical Biology* 42: 359–86.

Schneider, H., Schneider, M. P. C., Sampaio, M. I. C., Harada, M. L., Stanhope, M., Czelusniak, J., and Goodman, M. 1993. Molecular phylogeny of the New World monkeys (Platyrrhini, Primates). *Molecular Phylogenetics and Evolution* 2: 225–42.

Schneider, H., Sampaio, I., Harada, M. L., Barroso, C. M. L., Schneider, M. P. C., Czelusniak, J., and Goodman, M. 1996. Molecular phylogeny of the New World monkeys (Platyrrhini, Primates) based on two unlinked nuclear genes: IRBP intron 1 and e-globin sequences. *American Journal of Physical Anthropology* 100: 153–79.

Schneider, H., Canavez, F. C., Sampaio, I., Moreira, M. A. M., Tagliaro, C. H., and Seuánez, H. N. 2001. Can molecular data place each neotropical monkey in its own branch? *Chromosoma* 109: 515–23.

Schott, D. 1975. Quantitative analysis of the vocal repertoire of squirrel monkeys (*Saimiri sciureus*). *Journal of Comparative Ethology* 38: 225–50.

Schröpel, M. 1998. Multiple simultaneous breeding females in a pygmy marmoset group (*Cebuella pygmaea*). *Neotropical Primates* 6(1): 1–7.

Schultz, A. H. 1960. Age changes and variability in the skulls and teeth of the Central American monkeys. *Alouatta, Cebus* and *Ateles. Proceedings of the Zoological Society of London* 133: 337–90.

———. 1972. Developmental abnormalities. In R. N. Fiennes (ed.), *Pathology of simian primates. Part I. General pathology,* pp. 158–89. Basel: S. Karger.

Schuster, D., Burstein, S., and Cooke, B. A. 1976. *Molecular endocrinology of the steroid hormones*. London: John Wiley and Sons.

Scobie, P. 1994. SPARKS—Single Population Animal Records Keeping System. Apple Valley, MN: ISIS.

Seal, U. S., Ballou, J. D., and Valladares-Padua, C. (eds.). 1990. Leontopithecus: *Population Viability Analysis Workshop report.* Apple Valley, MN: International Union for Conservation of Nature and Natural Resources/Species Survival Commission (IUCN/SSC) Captive Breeding Specialist Group (CBSG).

Sério, F. C. 1986. Conservação da natureza na Reserva Estadual do Morro do Diabo. In M. T. de Mello (ed.), *A primatologia no Brasil—2,* pp. 261–68. Brasília, DF: Sociedade Brasileira de Primatologia.

Serra-Filho, R., Cavalli, A. C., Guillaumon, J. R., Chiarini, J. V., Nogueira, F. P., Montfort, I. C. M. de, Barbier, J. L., Donzeli, P. L. S. C., and Bittencourt, I. 1975. Levantamento da cobertura natural e de reflorestamento no estado de São Paulo. *Boletim Técnico, Instituto Florestal, São Paulo* 11: 1–53.

Seuánez, H. N., Forman, L., and Alves, G. 1988. Comparative chromosome morphology in three callitrichid genera: *Cebuella, Callithrix* and *Leontopithecus. Journal of Heredity* 79: 418–24.

Seuánez, H. N., Forman, L., Matayoshi, T., and Fanning, T.G. 1989. The *Callimico goeldii* (Primates, Platyrrhini) genome: Karyology and middle repetitive (LINE-1) DNA sequences. *Chromosoma* 98: 389–95.

Sheil, D. 2001. Conservation and biodiversity monitoring in the tropics: Realities, priorities and distractions. *Conservation Biology* 15(4): 1179–82.

Shepherdson, D. J. 1994. The role of environmental enrichment in the captive breeding and re-introduction of endangered species. Unpublished manuscript.

Silva, L. F. B. M. 1993. Ecologia do rato-de-bambú, *Kannabateomys amblyonyx* (Wagner, 1845) na Reserva Biológica de Poço das Antas, Rio de Janeiro. Master's thesis, Universidade Federal de Minas Gerais, Belo Horizonte.

Simberloff, D., Farr, J., Cox, J., and Mehlman, D. W. 1992. Movement corridors: Conservation bargains or poor investments? *Conservation Biology* 6: 493–504.

Simek, M. A. 1988. Food provisioning of infants in captive social groups of common marmosets (*Callithrix jacchus jacchus*) and cotton-top tamarins (*Saguinus oedipus oedipus*). Master's thesis, University of Tennessee, Knoxville.

Simon, F. 1988. Mico-leão Preto. Cadastro Provisório de Linhagens. Comitê Internacional de Preservação e Manejo, Fundação Parque Zoológico de São Paulo, São Paulo.

———. 1989. *Leontopithecus chrysopygus* studbook 1989. Comitê Internacional de Preservação e Manejo do Mico-Leão Preto, Fundação Parque Zoológico de São Paulo, São Paulo.

Smith, T. E. and French, J. A. 1997a. Social and reproduction conditions modulate urinary cortisol excretion in black tufted-ear marmosets (*Callithrix kuhli*). *American Journal of Primatology* 42: 253–67.

———. 1997b. Psychosocial stress and urinary cortisol excretion in marmoset monkeys (*Callithrix kuhli*). *Physiology and Behavior* 62: 225–32.

Smith, T. E., Schaffner, C. M., and French, J. A. 1997. Social modulation of reproductive function in female black tufted-ear marmosets (*Callithrix kuhli*). *Hormones and Behavior* 31: 159–68.

Snowdon, C. T. 1988. Communication as social interactions: Its importance in ontogeny and adult behavior. In D. Todt, P. Goedeking, and D. Symmes (eds.), *Primate vocal communication*, pp. 108–22. Berlin: Springer Verlag.

———. 1989a. The criteria for successful captive propagation of endangered primates. *Zoo Biology* suppl.1: 149–61.

———. 1989b. Vocal communication in New World monkeys. *Journal of Human Evolution* 18: 611–33.

———. 1993. A vocal taxonomy of the callitrichids. In A. B. Rylands (ed.), *Marmosets and tamarins: Systematics, behaviour, and ecology*, pp. 78–94. Oxford: Oxford University Press.

Snowdon, C. T., Hodun, A., Rosenberger, A. L., and Coimbra-Filho, A. F. 1986. Long-call structure and its relation to taxonomy in lion tamarins. *American Journal of Primatology* 11: 253–61.

Snyder, N. R. F., Koenig, S. E., Koschmann, J., Snyder, H. A., and Johnson, T. B. 1994. Thick-billed parrot releases in Arizona. *Condor* 96: 845–62.

Snyder. P. A. 1974. Behavior of *Leontopithecus rosalia* (golden lion marmoset) and related species: A review. *Journal of Human Evolution* 3: 109–22.

Soulé, M., Gilpin, M., Conway, W., and Foose, T. 1986. The Millenium Ark: How long a voyage, how many staterooms, how many passengers? *Zoo Biology* 5: 101–13.

Sousa, S. N. F. de, Dietz, J. M., and Dietz, L. A. 1988. Apoio para assegurar o domínio e proteção da Reserva Biológica Federal de Una, habitat do mico-leão-de-cara-dourada (*Leontopithecus chrysomelas*). Funding proposal, Departamento de Parques Nacionais e Reservas Equivalentes, Instituto Brasileiro de Desenvolvimento Florestal (IBDF), Brasília, DF. 14 pp.

Stafford, B. J. and Ferreira, F. M. 1995. Predation attempts on callitrichids in the Atlantic coastal rain forest of Brazil. *Folia Primatologica* 65: 229–33.

Stafford, B. J., Rosenberger, A. L., and Beck, B. B. 1994. Locomotion of free-ranging golden lion tamarins (*Leontopithecus rosalia*) at the National Zoological Park. *Zoo Biology* 13: 333–44.

Stanley-Price, M. R. 1991. A review of mammal re-introductions, and the role of the Re-introduction Specialist Group of IUCN/SSC. In J. H. W. Gipps (ed.), *Beyond captive breeding: Re-introducing endangered mammals to the wild,* pp. 9–25. Oxford: Clarendon Press.

Stevenson, M. F., Foose, T. J., and Baker, A. (compilers). 1991. *Global Captive Action Plan for Primates. Discussion edition, 15 September 1991.* International Union for Conservation of Nature and Natural Resources/Species Survival Commission (IUCN/SSC) Captive Breeding Specialist Group (CBSG), Apple Valley, MN, IUCN/SSC Primate Specialist Group (PSG), and Conservation International (CI), Washington, DC. 138 pp.

———(compilers). 1992. *Primate Conservation Assessment and Management Plan.* Apple Valley, MN: International Union for Conservation of Nature and Natural Resources/Species Survival Commission (IUCN/SSC) Captive Breeding Specialist Group (CBSG).

Stevenson, P. R., Quiñones, M. J., and Ahumada, J. A. 1994. Ecological strategies of woolly monkeys (*Lagothrix lagotricha*) at Tinigua National Park, Colombia. *American Journal of Primatology* 32: 123–40.

Stith, B. M. 1990. Satellites, landscapes, and GIS: A case study in the Atlantic forest of Brazil. Master's thesis, University of Florida, Gainesville.

Stoinski, T. S. 2000. Behavioral differences between captive-born, reintroduced golden lion tamarins (*Leontopithecus rosalia rosalia*) and their wild-born offspring. Doctoral thesis, Georgia Institute of Technology, Atlanta.

Stoinski, T. S. and Beck, B. B. 2001. Behavioral differences between captive-born, reintroduced golden lion tamarins and their wild-born offspring. Paper presented at the XVIIIth Congress of the International Society of Primatologists, Adelaide, Australia, 7–12 January.

Stoinski, T. S. and Beck, B. B. In preparation. Behavioral change of captive-born reintroduced golden lion tamarins (*Leontopithecus rosalia*).

Stoinski, T. S., Beck, B. B., Bowman, M., and Lehnhardt, J. 1997. The Gateway Zoo Program: A recent initiative in golden lion tamarin zoo introductions. In J. Wallis (ed.), *Primate conservation: The role of zoological parks,* pp. 113–29. Norman, OK: American Society of Primatologists.

Stoner, K. E. 1996. Habitat selection and seasonal patterns of activity and foraging of mantled howling monkeys (*Alouatta palliata*) in northeastern Costa Rica. *International Journal of Primatology* 17: 1–31.

Stribley, J. A., French, J. A., and Inglett, B. J. 1987. Mating patterns in the golden lion tamarin (*Leontopithecus rosalia*): Continuous receptivity and concealed estrus. *Folia Primatologica* 49: 137–50.

Strier, K. B. 1987. Activity budgets of woolly spider monkeys, or muriquis (*Brachyteles arachnoides*). *American Journal of Primatology* 13: 385–95.

———. 1992. *Faces in the forest: The endangered muriqui monkeys of Brazil.* Oxford: Oxford University Press.

Strum, S. C. and Southwick, C. H. 1986. Translocation of primates. In K. Benirschke (ed.), *Primates: The road to self-sustaining populations*, pp. 949–57. New York: Springer Verlag.

Summers, P. M., Wennink, C. J., and Hodges, J. K. 1985. Cloprostenol-induced luteolysis in the marmoset monkey (*Callithrix jacchus*). *Journal of Reproduction and Fertility* 73: 133–38.

Sussman, R. W. and Garber, P. A. 1987. A new interpretation of the social organization and mating system of the Callitrichidae. *International Journal of Primatology* 8: 73–92.

Sussman, R. W. and Kinzey, W. G. 1984. The ecological role of the Callitrichidae: A review. *American Journal of Physical Anthropology* 64: 419–49.

Sutcliffe, A. G. and Poole, T. B. 1984. An experimental analysis of social interaction in the common marmoset (*Callithrix jacchus jacchus*). *International Journal of Primatology* 5: 591–607.

Sutherland, W. J. 1998. The importance of behavioral studies in conservation biology. *Animal Behaviour* 56: 801–9.

Swofford, D. L. 2000. PAUP*. Phylogenetic Analysis Using Parsimony (*and Other Methods). Version 4.0. Sunderland, MA: Sinauer Associates.

Tabanez, M., Padua, S. M., Souza, M. G., Cardoso, M. M., and Gurgel Garrido, L. 1997. Avaliação de trilhas interpretativas para educação ambiental. In S. M. Padua and M. Tabanez (eds.), *Educação ambiental: Caminhos trilhados no Brasil,* pp. 89–102. Fundo Nacional do Meio Ambiente, Brasília, Instituto de Pesquisas Ecológicas (IPÊ), Nazaré Paulista, São Paulo.

Tardif, S. D. 1996. The bioenergetics of parental behavior and the evolution of alloparental care in marmosets and tamarins. In N. Solomon and J. A. French (eds.), *Cooperative breeding in mammals,* pp. 11–33. Cambridge: Cambridge University Press.

Tardif, S. D. and Bales, K. 1997. Is infant carrying a courtship strategy in callitrichid primates? *Animal Behaviour* 53: 1001–7.

Tardif, S. D., Richter, C. B., and Carson, R. L. 1984. Reproductive performance of three species of Callitrichidae. *Laboratory Animal Science* 34: 272–75.

Tardif, S. D., Carson, R. L., and Gangaware, B. L. 1986. Comparison of infant-care in family groups of the common marmoset (*Callithrix jacchus*) and the cotton-top tamarin (*Saguinus oedipus*). *American Journal of Primatology* 11: 103–10.

———. 1990. Infant-care behavior of mothers and fathers in a communal-care primate, the cotton-top tamarin (*Saguinus oedipus*). *American Journal of Primatology* 22: 73–85.

Tardif, S. D., Harrison, M. L., and Simek, M. A. 1993. Communal infant care in marmosets and tamarins: Relation to energetics, ecology, and social organization. In A. B. Rylands (ed.), *Marmosets and tamarins: Systematics, behaviour, and ecology*, pp. 220–34. Oxford: Oxford University Press.

Templeton, J. J. 1998. Learning from others' mistakes: A paradox revisited. *Animal Behaviour* 55: 79–85.

Terborgh, J. 1983. *Five New World primates: A study in comparative ecology*. Princeton, NJ: Princeton University Press.

Terborgh, J. and Goldizen, A. W. 1985. On the mating system of the cooperatively breeding saddle-backed tamarin (*Saguinus fuscicollis*). *Behavioral Ecology and Sociobiology* 16: 293–99.

Terborgh, J. and Stern, M. 1987. The surreptitious life of the saddle-backed tamarin. *American Scientist* 75: 260–69.

Thomas, C. 1995. A comparative study of foraging behaviour and morphological adaptations in a group of free-living tamarins (*Leontopithecus chrysomelas*) and marmosets (*Callithrix argentata argentata*). BSc. thesis, University of Glasgow, UK.

Thomas, O. 1922. On the systematic arrangement of the marmosets. *Annals of the Magazine of Natural History, Series* 9: 196–99.

Thompson, S. D., Power, M. L., and Kalkstein, T. 1990. Thermoregulation, activity, and energy requirements of golden lion tamarins: Do big monkeys use more energy? *Proceedings 1990 AAZPA Annual Conference,* Wheeling, WV.

Thompson, S. D., Power, M. L., Rutledge, C. E., and Kleiman, D. G. 1994. Energy metabolism and thermoregulation in the golden lion tamarin (*Leontopithecus rosalia*). *Folia Primatologica* 63: 131–43.

Todt, D. 1988. Serial calling as a mediator of interaction processes: Crying in primates. In D. Todt, P. Goedeking, and D. Symmes (eds.), *Primate vocal communication,* pp. 88–107. Berlin: Springer Verlag.

Todt, D., Hammersschmidt, K., Ansorge, V., and Fischer, J. 1995. The vocal behavior of Barbary macaques (*Macaca sylvanus*): Call features and their performance in infants and adults. In E. Zimmermann (ed.), *Current topics in primate vocal communication,* pp. 141–60. New York: Plenum Press.

Trivers, R. L. 1971. The evolution of reciprocal altruism. *Quarterly Review of Biology* 46: 35–57.

Tucker, M. J. A. 1984. A survey of pathology of marmosets (*Callithrix jacchus*) under experiment. *Laboratory Animals* 18: 351–58.

Turleau, C., Cochet, C., and de Grouchy, J. 1978. Identité des caryotypes de *Papio papio* et *Macaca mulatta* en bandes R, G, C et Ag-NOR. *Annales de Génétique* 21: 149–51.

Valerio, D. A., Miller, R. L., Innes, J. R. M., Courtney, K. D., Pallota, A. J., and Guttmacher, R. M. 1969. Macaca mulatta: *Management of a laboratory breeding colony.* London: Academic Press.

Valladares-Padua, C. 1986. Clinical trials and the breeding of endangered species in captivity. In M. T. de Mello (ed.), *A primatologia no Brasil—2,* pp. 393–410. Brasília, DF: Sociedade Brasileira de Primatologia.

———. 1987. Black lion tamarin *Leontopithecus chrysopygus:* Status and conservation. Master's thesis, University of Florida, Gainesville.

———. 1993. The ecology, behavior and conservation of the black lion tamarin (*Leontopithecus chrysopygus,* Mikan, 1823). Doctoral dissertation, University of Florida, Gainesville.

———. 1997. Habitat analysis for the metapopulation conservation of black lion tamarins (*Leontopithecus chrysopygus,* Mikan, 1823). In M. B. C. Sousa and A. A. L. Menezes (eds.), *A primatologia no Brasil—6,* pp. 13–26. Natal, RN: Universidade Federal do Rio Grande do Sul and Sociedade Brasileira de Primatologia.

———. 2000. *Black lion tamarin studbook for 1999.* Nazaré Paulista, SP: Instituto de Pesquisas Ecológicas (IPÊ).

Valladares-Padua, C. and Ballou, J. D. 1998. *Leontopithecus chrysopygus* metapopulation management action plan. In J. D. Ballou, R. C. Lacy, D. G. Kleiman, A. B. Rylands, and S. Ellis (eds.), Leontopithecus *II: The second Population and Habitat Viability Assessment for lion tamarins* (Leontopithecus), Appendix B, 10 pp. Apple Valley, MN: World Conservation Union/Species Survival Commission (IUCN/SSC) Conservation Breeding Specialist Group (CBSG).

Valladares-Padua, C. and Cullen, L., Jr. 1994. Distribution, abundance and minimum viable metapopulation of the black lion tamarin (*Leontopithecus chrysopygus*). *Dodo, Journal of the Wildlife Preservation Trusts* 30: 80–88.

Valladares-Padua, C. and Mamede, R. 1996a. *1996 International studbook black lion tamarin* Leontopithecus chrysopygus. São Paulo, SP: International Committee for the Preservation and Management of the Black Lion Tamarin.

———. 1996b. Ecology and behavior of the black lion tamarins: A report to Duratex. Nazaré Paulista, SP: Instituto de Pesquisas Ecológicas (IPÊ).

Valladares-Padua, C. and Martins, C. S. 1996. Proposal for conservation and metapopulation management of the black lion tamarin (*Leontopithecus chrysopygus*). Unpublished. Instituto de Pesquisas Ecológicas (IPÊ), Nazaré Paulista, São Paulo.

Valladares-Padua, C. and Prado, F. 1996. Notes on the natural history of the black-faced lion tamarin. *Dodo, Journal of the Jersey Wildlife Preservation Trusts* 32: 123–25.

Valladares-Padua, C. and Simon, F. 1990. *1990 International studbook black lion tamarin* Leontopithecus chrysopygus. São Paulo, SP: International Management Committee for the Black Lion Tamarin.

Valladares-Padua, C., Padua, S. M., and Cullen, L., Jr. 1994. The conservation biology of the black-lion tamarin, *Leontopithecus chrysopygus:* First ten years' report. *Neotropical Primates* 2(suppl.): 36–39.

Valladares-Padua, C., Cullen, L., Jr., Padua, S. M., Ditt, E., Medici, P., Betini, G., and Luca, A. de. 1997. Resgatando a Grande Reserva do Pontal do Paranapanema: Reforma Agrária e Conservação da Biodiversidade. In M. S. Milano (ed.), *Anais do 1° Congresso Brasileiro de Unidades de Conservação,* pp. 783–92. Curitiba, PR: Universidade Livre do Meio Ambiente, Rede Nacional Pró Unidades de Conservação, and Instituto Ambiental do Paraná (IAP).

Valladares-Padua, C., Ditt, E., and Bassi, C. 1999. Rapid ecological assessment of forest fragments of the Atlantic Forest of the interior. Report, The John D. and Catherine T. MacArthur Foundation, Chicago.

Valladares-Padua, C., Prado, F., and Maia, R. G. 2000a. Survey of new populations of black-faced lion tamarin (*Leontopithecus caissara*) in São Paulo and Paraná states. Instituto de Pesquisas Ecológicas (IPÊ), Nazaré Paulista, São Paulo. Unpublished report, Margot Marsh Biodiversity Foundation, Virginia. 19 pp.

Valladares-Padua, C., Weffort, D. D., and Cullen, L., Jr. 2000b. Corredor Morro do Diabo (SP)—Ilha Grande (PR) proposta de conservação de uma ecorregião para a Mata Atlântica do interior e varjões do rio Paraná. In M. S. Milano and V. Theulen (eds.), *II Congresso Brasileiro de Unidades de Conservação. Anais. Vol. II. Trabalhos técnicos,* pp. 700–705. Rede Nacional Pró-Unidades de Conservação, Campo Grande, Fundacão O Boticário de Proteção à Natureza, São José dos Pinhais, Paraná.

Valladares-Padua, C., Martins, C. S., Wormell, D., and Setz, E. Z. F. 2001. Preliminary evaluation of the reintroduction of a mixed wild-captive group of black lion tamarins *Leontopithecus chrysopygus. Dodo* 36: 30–38.

Valle, C. and Rylands, A. B. 1986. Lion tamarins rescued. *Oryx* 20: 71–72.

Van Elsacker, L., de Meurichy, W., Verheyen, R. F., and Walraven, V. 1992. Maternal differences in infant carriage in golden-headed lion tamarins (*Leontopithecus chrysomelas*). *Folia Primatologica* 59: 121–26.

Van Elsacker, L., Heistermann, M., Hodges, J. K., De Laet, A., and Verheyen, R. F. 1994. Preliminary results on the evaluation of contraceptive implants in golden-headed lion tamarins, *Leontopithecus chrysomelas. Neotropical Primates* 2(suppl.): 30–32.

Van Heezik, Y. and Maloney, R. 1997. Update on the houbara bustard re-introduction programme in Saudi Arabia. *Re-introduction News,* no. 13: 3–4.

Vargas, A. 1994. Ontogeny of the endangered black-footed ferret (*Musteles nigripes*) and effects of captive upbringing on predatory behavior and post-release survival. Doctoral thesis, University of Wyoming.

Vargas, A. and Anderson, S. H. 1999. Effects of experience and cage enrichment on predatory skills of black-footed ferrets (*Mustela nigripes*). *Journal of Mammalogy* 80: 263–69.

———. In press. Ontogeny of black-footed ferret predatory behavior towards prairie dogs. *Canadian Journal of Zoology.*

Vargas, A., Lockhart, M., Marinari, P., Gober, P. 1996. The reintroduction process: Black-footed ferrets as a case study. *Proceedings of the American Zoo and Aquarium Association Western Regional Conference,* pp. 829–34.

Victor, M. A. M. 1975. *A devastação florestal em São Paulo.* São Paulo: Sociedade Brasileira de Silvicultura.

Vieira, C. da C. 1955. Lista remissiva dos mamíferos do Brasil. *Arquivos de Zoologia,* São Paulo 8: 341–474.

Vivekananda, G. 1994. The Superagüi National Park, problems concerning the protection of the black-faced lion tamarin, *Leontopithecus caissara. Neotropical Primates* 2(suppl.): 56–57.

———. 2001. Parque Nacional do Superagui: A presença humana e os objetivos de conservação. Master's thesis, Universidade Federal do Paraná, Curitiba.

von Dornum, M. 1997. DNA sequence data from mitochondrial COII and nuclear G6PD loci and a molecular phylogeny of the New World monkeys (Primates, Platyrrhini). Doctoral dissertation, Harvard University, Cambridge.

von Dornum, M. and Ruvolo, M. 1999. Phylogenetic relationships of the New World monkeys (Primates, Platyrrhini) based on nuclear G6PD DNA sequences. *Molecular Phylogenetics and Evolution* 11: 459–76.

Von Ihering, R. 1940. *Dicionário dos animais do Brasil.* São Paulo: Diretoria da Publicidade Agrícola.

Waddell, M. D. and Taylor, R. M. 1945. Studies on cyclic passage of yellow fever virus in South American mammals and mosquitoes. *American Journal of Tropical Medicine* 25: 225–30.

Wallace, M. 1997. Carcasses, people, and power lines. *Zoo View* 31: 12–17.

Walraven, V. and Van Elsacker, L. 1992. Scent-marking in New World primates: A functional review. *Acta Zoologica et Pathologica Antverpiensia* 82: 51–59.

Wamboldt, M. Z., Gelhardt, R. E., and Insel, T. R. 1988. Gender differences in caring for infant *Cebuella pygmaea:* The role of infant age and relatedness. *Developmental Psychobiology* 21: 187–202.

Waser, P. M. and. Brown, C. H. 1986. Habitat acoustics and primate communication. *American Journal of Primatology* 10: 135–54.

Waser, P. M. and Wiley, R. H. 1979. Mechanisms and evolution of spacing in animals. In P. Marler and J. G. Vanderbergh (eds.), *Handbook of behavioral neurobiology, 3. Social behavior and communication,* pp. 159–223. New York: Plenum Press.

Wasser, S. K., Risler, L., and Steiner, R. A. 1988. Excreted steroids in primate feces over the menstrual cycle and pregnancy. *Biology of Reproduction* 39: 862–72.

Watts, D. P. 1988. Environmental influences on mountain gorilla time budgets. *American Journal of Primatology* 15: 195– 211.

———. 1991. Strategies of habitat use by mountain gorillas. *Folia Primatologica* 56: 1–16.

Wayre, P. 1975. Conservation of eagle owls and other raptors through captive breeding and return to the wild. In R. D. Martin (ed.), *Breeding endangered species in captivity,*, pp. 125–31. New York: Academic Press.

West-Eberhard, M. J. 1975. The evolution of social behavior by kin-selection. *Quarterly Review of Biology* 50: 1–33.

Whitten, T., Holmes, D., and MacKinnnon, K. 2001. Conservation biology: A displacement behavior for academia? *Conservation Biology* 15(1): 1–3.

Widowski, T. M., Ziegler, T. E., Elowson, A. M., and Snowdon, C. T. 1990. The role of males in stimulating reproductive function in female cotton-top tamarins *Saguinus o. oedipus*. *Animal Behaviour* 40: 731–41.

Widowski, T. M., Porter, T. A., Ziegler, T. E., and Snowdon, C. T. 1992. The stimulatory effect of males on the initiation but not the maintenance of ovarian cycling in cotton-top tamarins (*Saguinus oedipus*). *American Journal of Primatology* 26: 97–108.

Wied-Neuwied, Prinz Maximilian zu. 1826. *Beiträge zur Naturgeschichte von Brasilien*, vol. 2. 620 pp.

———. 1940. *Viagem ao Brasil*. Translated by E. S. Mendonça and F. P. de Figueiredo, annotated by O. M. O. Pinto. São Paulo: Companhia Editora Nacional. 551 pp.

Wilcox, B. A. and Murphy, D. D. 1985. Conservation strategy: The effects of fragmentation on extinction. *American Naturalist* 125: 879–87.

Wiley, J. W., Snyder, N. F. R., and Gnam, R. S. 1992. Reintroduction as a conservation strategy for parrots. In S. R. Beissinger and N. F. R. Snyder (eds.), *New World parrots in crisis: Solutions from conservation biology*, pp. 165–200. Washington, DC: Smithsonian Institution Press.

Wiley, R. H. and Richards, D. G. 1982. Adaptations for acoustic communication in birds: Sound transmission and signal detection. In D. E. Kroodsma, E. H. Miller, and H. Ouellet (eds.), *Acoustic communication in birds*, pp. 132–81. New York: Academic Press.

Wilson, N., Dietz, J. M., and Whitaker, J. O. 1989. Ectoparasitic acari found on golden lion tamarins (*Leontopithecus rosalia rosalia*) from Brazil. *Journal of Wildlife Diseases* 25: 433–35.

Witte, S. M. and Rogers, J. 1999. Microsatellite polymorphisms in Bolivian squirrel monkeys (*Saimiri boliviensis*). *American Journal of Primatology* 487: 75–84.

Wolfe, L. G., Ogden, J. D., Deinhardt, J. B., Fisher, L., and Deinhardt, F. 1972. Breeding and hand-rearing marmosets for viral oncogenesis studies. In W. I. B. Beveridge (ed.), *Breeding primates*, pp. 145–57. Basel: S. Karger.

Wood, C., Ballou, J. D., and Houle, C. Submitted. Reversibility of melengestrol acetate contraceptive implants in golden lion tamarins (*Leontopithecus rosalia*).

Wormell, D. In press. The management and reproduction of the black lion tamarin *Leontopithecus chrysopygus* at Jersey Zoo. *Dodo* 37.

Xu, X. and Arnason, U. 1996a. The mitochondrial DNA molecule of the Sumatran orangutan and a molecular proposal for two (Bornean and Sumatran) species of orangutan. *Journal of Molecular Evolution* 43: 431–37.

———. 1996b. A complete sequence of the mitochondrial genome of the Western lowland gorilla. *Molecular Biology and Evolution* 3: 691–98.

Yamamoto, M. E. 1993. From dependence to sexual maturity: The behavioural ontogeny of Callitrichidae. In A. B. Rylands (ed.), *Marmosets and tamarins: Systematics, behaviour and ecology*, pp. 235–56. Oxford: Oxford University Press.

Yamamoto, M. E., Box, H. O., Albuquerque, F. S., and Arruda, M. de F. 1996. Carrying behaviour in captive and wild marmosets (*Callithrix jacchus*): A comparison between two colonies and a field site. *Primates* 37: 297–304.

Yoder, A. D., Cartmill, M., Ruvolo, M., Smith, K., and Vilgalys, R. 1996. Ancient single origin for Malagasy primates. *Proceedings of the National Academy of Sciences* 93: 5122–126.

Zhang, S. 1995. Activity and ranging patterns in relation to fruit utilization by brown capuchins (*Cebus apella*) in French Guiana. *International Journal of Primatology* 16: 489–507.

Ziegler, T. E., Savage, A., Scheffler, G., and Snowdon, C. T. 1987a. The endocrinology of puberty and reproductive functioning in female cotton-top tamarins (*Saguinus oedipus*) under varying social conditions. *Biology of Reproduction* 37: 618–27.

Ziegler, T. E., Bridson, W. E., Snowdon, C. T., and Eman, S. 1987b. Urinary gonadotropin and estrogen excretion during the postpartum estrus, conception and pregnancy in the cotton-top tamarin (*Saguinus oedipus*). *American Journal of Primatology* 12: 127–40.

Ziegler, T. E., Snowdon, C. T., and Bridson, W. E. 1990. Reproductive performance and excretion of urinary estrogens and gonadotropins in the female pygmy marmoset (*Cebuella pygmaea*). *American Journal of Primatology* 22: 191–204.

Ziegler, T. E., Scheffler, G., Wittwer, D. J., Schultz-Darken, N. J., Snowdon, C. T., and Abbott, D. H. 1996. Metabolism of reproductive steroids during the ovarian cycle in two species of callitrichids, *Saguinus oedipus* and *Callithrix jacchus,* and estimation of the ovulatory period from fecal steroids. *Biology of Reproduction* 54: 91–99.

Zuckerman, S. M. 1931. The menstrual cycle of the primates, part III. The alleged breeding season of primates with special reference to the chacma baboon (*Papio porcarius*). *Proceedings of the Zoological Society of London* (1931): 338–39.

INDEX

Page numbers followed by a t indicate pages with tables; numbers in italics refer to pages with figures.

C

G

H

I

J

S